Access to Knowledge

New Information Technologies and the Emergence of the Virtual University

ISSUES IN HIGHER EDUCATION

Series Editor: GUY NEAVE, International Association of Universities, Paris, France

Other titles in the series include

GOEDEGEBUURE *et al.*
Higher Education Policy: An International Comparative Perspective

NEAVE & VAN VUGHT
Government and Higher Education Relationships Across Three Continents: The Winds of Change

SALMI & VERSPOOR
Revitalizing Higher Education

YEE
East Asian Higher Education: Traditions and Transformations

DILL & SPORN
Emerging Patterns of Social Demand and University Reform: Through a Glass Darkly

MEEK et al.
The Mockers and the Mocked: Comparative Perspectives on Differentiation, Convergence and Diversity in Higher Education

BENNICH-BJORKMAN
Organizing Innovative Research: The Inner Life of University Departments

HUISMAN & MAASSEN
Higher Education and the Nation State

CLARK
Creating Entrepreneurial Universities: Organizational Pathways of Transformation

GURI-ROSENBLIT
Distance and Campus Universities: Tensions and Interactions

HIRSCH & WEBER
Challenges Facing Higher Education at the Millennium

Access to Knowledge

New Information Technologies and The Emergence of the Virtual University

Edited by

F. Ted Tschang
United Nations University, Institute of Advanced Studies, Tokyo, Japan

and

Tarcisio Della Senta
United Nations University, Institute of Advanced Studies, Tokyo, Japan

UNU/IAS
Institute of Advanced Studies

An imprint of Elsevier Science

Amsterdam – London – New York – Oxford – Paris – Shannon – Tokyo

ELSEVIER SCIENCE Ltd
The Boulevard, Langford Lane
Kidlington, Oxford OX5 1GB, UK.

First edition 2001

Library of Congress Cataloging in Publication Data
A catalog record from the Library of Congress has been applied for.

British Library Cataloguing in Publication Data
A catalogue record has been applied for.

ISBN: 0 08 043670 6

∞The paper used in this publication meets the requirements of ANSI/NISO Z39.48-1992 (Permanence of Paper).
Printed in the Netherlands.

The IAU

The International Association of Universities (IAU), founded in 1950, is a worldwide organization with member institutions in over 120 countries, which cooperates with a vast network of international, regional and national bodies. Its permanent Secretariat, the International Universities Bureau, is located at UNESCO, Paris, and provides a wide variety of services to Member Institutions and to the international higher education community at large.

Activities and Services

* IAU–UNESCO Information
* International Information Networks
* Meetings and Seminars
* Research and studies
* Promotor of academic mobility and cooperation
* Credential evaluation
* Consultancy
* Exchange of publications and materials

Publications

* International Handbook of Universities
* World List of Universities
* Issues in Higher Education (monographs)
* Papers and Reports
* Higher Education Policy (quarterly)
* IAU Bulletin (bimonthly)

Contents

Introduction to Issues in Higher Education

For the past quarter century, higher education has been high on the agenda of governments and central to the fortune of nations. Similarly, this same period has seen quite massive changes in direction, in the complexity of systems, in the sheer size of the enterprise in terms of students, staff and budgets, not to mention social and economic purpose. It is not surprising then that the study of higher education itself has broadened and now encompasses some 20 different disciplines, ranging from Anthropology through to Women's Studies, each with its own particular paradigms, methodologies and perspectives.

Against this background, the comparative analysis of higher education policy, which has always occupied a crucial place in understanding the contextual setting of reform in individual countries, has acquired a new significance as the pace of "internationalization" itself quickens. There are many reasons why this should be so: the creation of new economic blocs and, in the case of Europe, the gradual emergence of a trans-national policy for higher education and in how other nations are going about tackling often similar issues through different ways. This series has the purpose of examining issues and testing theories in the field of higher education policy which are of current and practical concern to its main constituencies—national and institutional leadership, administrators, teachers, those researching in this domain and students. As a series, it will focus on both advanced industrial and also on developing systems of higher education.

Issues in Higher Education will be resolutely comparative in its approach and will actively encourage original studies which are firmly based around an international perspective. Individual volumes will be based on a minimum of two different countries so as to bring out the variations occurring in a given problèmatique. Every encouragement will be given to the drawing of clear and explicit comparisons between higher education systems covered.

As the editor, I wish to thank the members of the Educational Advisory Board for their part in developing this series. They are:

GUY NEAVE
International Association of Universities
Paris, France

Preface

This volume arose out of the United Nations University/Institute of Advanced Studies' Virtual University Initiative. One of the aims of this initiative was to study and analyze collected thoughts on the potential of virtual education to solve educational problems around the world. Another objective of the VU Initiative is the development of software systems for virtual education. These two activities were synergistic, in that we could confirm our ideas for the book through these other substantial research and implementation activities; and in how the book project itself informed the VU Initiative's other activities.

This book was initially planned as an input to the UNESCO World Conference on Higher Education, held in Paris in October 1998. But it became much larger than that. In such a conference, much of the time is often taken up by political and institutional issues, and general principals and speeches. Our effort then became one of intellectually contributing a coherent and structured "map" to the trends and issues in virtual education.

To accomplish this, we chose to undertake the book's production in a "virtual" fashion, namely, by using the Internet and electronic mail to contact and attract many talented and capable scholars around the world who have been confronting difficult problems of systems development and implementation, or who have been reflecting on these topics. It was our good fortune to attract many of these scholars, and our ability to coordinate this activity over the Internet was partly due to the trust and reliance that people active in virtual communities are able to forge over electronic means – itself a sign of the times.

We believe that we have managed to collect together and structure a substantial amount of thought on virtual universities. However, in the era of the Internet, there are inherent problems with the printed medium. For one, the long time it takes a book to come to press, coupled with the rapidity of change, required us to update the chapters at least once, and in some cases a few times. Eventually, one has to recognize that with the Internet infrastructure and its applications growing at an explosive exponential rate, it is impossible to stay 100 percent current through the print medium. The ideal solution would have been to provide updates on chapters and topics on a Web site, but this requires much greater effort on the part of authors to stay in touch and update things themselves. Our approach to address this issue and to add greater lasting value to the volume was to develop within the book a general conceptual framework for understanding the trends and nature of virtual education. Thus, the outline of the chapters form one such framework, and the discussions in many chapters also sought to provide general frameworks of reference for their specific topics.

Two other issues that needed to be dealt with were: the need to get consistency across many different fields of study without writing too generally; and the natural limitation on the number of topics that could be covered. (In fact, this is apparently the largest volume ever in the *Issues in Higher Education* series!) The first occurs because this topic intertwines technologies with many other educational, social and economic areas – and this requires a multidisciplinary perspective. We also attempted to synthesize a new "knowledge-based" approach to help unify the fragmented thinking across disciplines. In doing so, we hope to have provided some fresh thinking on the topics, while still maintaining as much of each disciplinary perspective, be it education, cognition, or computer science, in each chapter.

As far as our limit on topics, there were many important topics that we would have liked to have included, such as the credentialling issue in virtual education, the impacts of information technology on university administration, and so on. Fortunately, there are many such fine volumes or works being published elsewhere on some of these topics, and we are sure the coverage will increase in time. In addition, this volume was originally envisaged to be the first in a United Nations University/Institute of Advanced Studies series on virtual education or on the Internet. We plan to do more research and publish more on these topics in the future, continuing our emphasis on the developing world.

As a last comment, the title *Access to Knowledge* was meant to open up the discussion of technologically-enhanced open or distance learning as means of broadening educational opportunities. However, the issues of equity and access also need to be addressed in terms of financial ability, physical access or other capabilities, which we did not do enough justice to. The problem of the "Digital Divide" is of increasing concern. Nevertheless, we hope that the experiences in this volume illustrate many of the general topics that need to be addressed in any system, whether it caters to students in particularly impoverished areas, or to students in developed regions trying a new educational modality. In fact, in this day and age, with the appropriate vision and the backing of some financial resources, even less well off communities in developing countries can and do take part in the global Internet and knowledge revolution, as shown by cases such as the secondary school-level vocational institutes in Indonesia.

In closing, we would like to thank our publisher's staff for their kind and efficient efforts, and the *Issues in Higher Education* series editor, Guy Neave, for his great effort and patience in shepherding the manuscript through.

Ted Tschang
Tarcisio Della Senta

Tokyo, Japan

Authors

Ricardo Miranda Barcia [rbarcia@uol.com.br]
Prof. Barcia is a professor at Engineering Production and Systems at Federal University of Santa Catarina. He is also a senior researcher at Cnpq—National Council of Scientific and Technological Development. He holds a Ph.D. from University of Waterloo (1984), and leads the Applied Intelligence and Media and Knowledge research group, his principal area of interest being "Engineering and Computer Science". Main lines of research are in: theory and techniques of Applied Intelligence, Innovation and Technological Management, Applied Intelligence in Services Management, Applied Intelligence in Production Systems Management and Environment, and Virtual Organizations and Teleworking.

Christine Borgman [cborgman@ucla.edu]
Christine L. Borgman, Professor, holds the Presidential Chair in Information Studies at UCLA and is Visiting Professor of Information Science at Loughborough University, England. Her teaching and research interests include digital libraries, telelearning, human-computer interaction, electronic publishing, information seeking behavior, scholarly communication, bibliometrics, and information technology policy. Prof. Borgman is the author of more than 130 publications in information studies, computer science, and communication and serves on 7 editorial boards in these fields. Her most recent book is *From Gutenberg to the Global Information Infrastructure: Access to Information in a Networked World* (MIT Press, 2000). She is a Fellow of the American Association for the Advancement of Science and serves on advisory boards to foundations and public interest groups concerned with libraries, computer and information science research, and information policy.

Peter Brusilovsky [plb@cs.cmu.edu]
Peter Brusilovsky, Assistant Professor, School of Information Sciences, University of Pittsburgh and Adjunct Faculty, Human-Computer Interaction Institute, Carnegie Mellon University (CMU). Ph.D. (1987) Moscow State University. Formerly Director of Computer Managed Instruction at Carnegie Technology Education, a nonprofit subsidiary of CMU. Main publications: intelligent tutoring

systems, adaptive hypermedia, adaptive Web-based systems, Web-based education. Recently edited: Adaptive Hypertext and Hypermedia (Kluwer Academic Publishers, 1998) and special issues of "New Review of Hypermedia and Multimedia" and "User Modeling and User-Adapted Interaction" on adaptive hypermedia.

Dênia Falcão de Bittencourt [denia@eps.ufsc.br]
Doctoral student in the Graduate Program in Production Engineering [PPGEP], at the Federal University of Santa Catarina [UFSC], works as a researcher and pedagogical co-ordinator in distance education certificate courses using the Internet. Graduated in psychology; Masters' Degree in Production Engineering, with a dissertation on the development and management of Internet courses.

Marialice de Moraes [mariali@led.ufsc.br]
Doctoral student in the Graduate Program in Production Engineering, at the Federal University of Santa Catarina, and working as a researcher and instructional designer in its Distance Education Laboratory. Bachelors in Social Sciences; Master's from the Roskilde University Centre, Denmark in European Studies of Society, Science and Technology. Writes on the use of Internet and videoconferencing, student support services, and evaluation in distance education.

Tarcisio Della Senta [dellasenta@ias.unu.edu]
Director, United Nations University's Institute of Advanced Studies, a newly established think tank. Ed.D. in Educational Planning and Management, Harvard University. He has taught in universities, and was director of the Department of Higher Education in the Ministry of Education and Head of the Planning and Budget Division of the National Research Council, Government of Brazil. At UNU he was Director of the Planning and Development Division, Vice-Rector, a.i. of the Academic Division. He has published in the areas of education, science and technology, and in issues of development.

Michael Dillinger [mdillinger@logos-usa.com]
Director of Linguistic Development, Logos Corporation, Rockaway, NJ, USA; former Associate Dean of Graduate Studies and Research, Faculty of Letters and Professor of Linguistics, Federal University of Minas Gerais, Brazil; Researcher in applied cognitive science; editorial board, Applied Psycholinguistics; Consultant to the United Nation's Universal Networking Language Project and Brazilian national and state research agencies; Main publications in text-based reasoning and text-based learning, linguistics and research methods; Current projects focus on developing a global machine translation infrastructure for Internet-based learning.

Magdallen N. Juma [avuku@ku.ac.ke]
Senior Lecturer, Department of Educational Administration and Curriculum Development, Director, African Virtual University and Institute of Distance Education, Kenyatta University—Kenya. Member of the Educational Research Network in

Kenya, (ERNIKE), founder member of the African Association of Distance Learning. Publishes widely on gender and education, and higher education in Africa. Consultant to major international organizations: World Bank, Department for International Development (DFID), The British Council, UNESCO, Commonwealth Secretariat, and the Commonwealth of Learning (COL). Currently involved in a Carnegie Corporation project: "A Study on the Use of Information Communication Technology (ICT's): Application in Higher Education in Africa".

Gloria Mark [gmark@ics.uci.edu]
Gloria Mark is an assistant professor at the University of California, Irvine, in the Department of Information and Computer Science. Formerly of GMD-FIT (the German National Institute for Applied Information Technology). She is very active in the field of computer supported cooperative work (CSCW), and her main research interests are in the design methodology and usability evaluation of collaborative technologies. She is on the editorial board of the Journal of Collaborative Computing and e-Service *Qu@rterly*, and has served on program committees for the CSCW, ECSCW, and DIS conferences.

Marisa Martín Pérez [mmartin@itesm.mx]
Director of Educational Research and Development at the Monterrey Institute of Technology, Mexico (ITESM). Ph.D. in Educative Sciences, Salamanca University, Spain. She has taught at the Virtual University of the ITESM, and now collaborates on ITESM's educative change process. She has published in instructional design teaching and learning technologies and evaluation. Previously director at the Montessori Elementary School of Monclova, Mexico.

Robin Mason [r.d.mason@open.ac.uk]
Professor of Educational Technology in the Institute of Educational Technology, The UK Open University; Director of the Masters Programme in Open and Distance Education. Author of Globalising Education: Trends and Applications (Routledge, 1998), amongst others; evaluator of several major Virtual University projects and other web-based teaching initiatives; researcher on the cultural issues arising from online courses; keynote speaker on topics such as globalisation of education, online teaching and learning, evaluation of technology-based learning; leader of the TeleLearning Research Group.

Philip L. Miller [plm@carnegietech.org]
Executive Vice President, Carnegie Technology Education, A nonprofit subsidiary of Carnegie Mellon University (CMU). Ph.D. (1975) Ohio State University. Formerly Chief Technical Officer and Chief Operating Officer, Virtual University International, Inc. Atlanta Georgia; director for introductory programming, computer science department, CMU. Developed or co-developed several academic software systems, including various programming language environments like Object Pascal Genie, Pascal Genie and Karel Genie, and CMU Online—Carnegie Mellon's distance learning delivery system.

Ng S. T. Chong [chong@ias.unu.edu]
Directs the VU Media and Technology Laboratory and the Network Operations Center at the Institute of Advanced Studies, United Nations University. His research interests include applied computing in education, Internet computing and operations. Currently heads the development of the *Classroom Anywhere* (CA) platform and its modules, an extensible and collaboration-based environment for distributed virtual learning. Recently published articles at the ACM Symposium on Applied Computing, Italy (2000), the Third IASTED International Conference on Internet and Multimedia Systems, the International Conference on Computers in Education, Japan (1999), and at INET (2000).

Om Vikas
Senior Director, Information Technology Group, Department of Electronics, Government of India. Ph.D. in Electrical Engineering, Indian Institute of Technology. Involved in studies of Indian IT and technology transfer, and projects on machine language translation. Sat on the national IT curricula revisions board. Has published on IT education in journals and conferences in India.

Rosângela Schwarz Rodrigues [rosangela@led.ufsc.br]
Doctoral student at Federal University of Santa Catarina, at the Production Engineering Graduate Program, in the Media and Knowledge area. Holds a Master's degree in the same program and a bachelors in Mass Communication. Works at the Laboratory of Distance Education/UFSC as a institutional researcher, internal consultant in planning and evaluation of courses in the research and development group

Fernando Spanhol [spanhol@led.ufsc.br]
Doctoral student in the Graduate Program in Production Engineering, at the Federal University of Santa Catarina. Works a researcher and technical coordinator of the videoconference systems at the Laboratory of Distance Education/ UFSC. Graduate in Pedagogy, with a master's degree dissertation in Technical and Environment Structures for Videoconference Systems for Distance Education.

Janusz Szczypula [js1m@andrew.cmu.edu]
Lecturer in and Director of the Management Information Systems Program at Carnegie Mellon University. Ph.D. in Information and Decision Systems, the Heinz School of Public Policy and Management, Carnegie Mellon. His research interests include decision support systems, social networks, forecasting methods, Bayesian forecasting, rule based forecasting, and residential use of the Internet. Written for journals like Management Science, Information Systems Research, International Journal of Forecasting, International Journal of Public Administration and Journal of Mathematical Sociology.

F. Ted Tschang [ttschang@adbi.org; tschang@ias.unu.edu]
Visiting scholar at the Asian Development Bank Institute (Tokyo). Ph.D. in Public Policy and Management, Carnegie Mellon University; Ph.D. fellow, Harvard

University. He was a research associate at the United Nations University's Institute of Advanced Studies on the institute's Virtual University Initiative. His current research and recent publications are in knowledge management, the socio-economic implications of information technology, and innovation and technology policy. Currently editing a manuscript on the digital economy in Asia.

João Vianney Valle dos Santos [vianney@led.ufsc.br]
Executive Coordinator of the Brazilian Virtual University. Ex-coordinator of the Laboratory of Distance Education at Federal University of Santa Catarina (UFSC) (1995–1999). His main areas of interest are in the development of strategies and methodologies for the implementation of Distance Education. The characteristics of this approach are the intensive use of information and communication technologies in the concept of media convergence. A psychologist and journalist by training; holds a certificate course in psychology of Communication, and a Masters degree in Political Sociology. Doctoral candidate in the Graduate Program in Production Engineering at the USFC (thesis in Distance Education).

Aya Yoshida [aya@ocean.nime.ac.jp]
Associate Professor of Sociology of Education at the National Institute of Multimedia Education (NIME), Japan. Leads NIME's *Campus IT Survey in Japan* research project, and a research project on the *Historical Functional Transformation of Secondary Education in Modern Japan,* sponsored by the Ministry of Education. Previously completed a survey on distance learning institutions in Asia for UNESCO. Recent publications include "Women's Secondary School and Students in Modern Japan" (2000) in *Modernity and Culture in Japan 8*, Iwanami Syoten: Tokyo (in Japanese); and "The New Movement of Graduate Program of Study in Distance Education in Japan" (1999) in the *Chinese Journal of Distance Education*.

1

Introduction

F. TED TSCHANG and Tarcisio DELLA SENTA

It has long been known that many education systems around the world face accumulating economic and other pressures, and are unable to meet the needs of increasingly knowledge-intensive economies. With the emergence of the Internet, there is the potential to completely alter the landscape for commerce, education, and social-cultural interaction. Whereas IT merely enhanced productivity or the control over information, the Internet is a fundamentally different technology that reshapes our social and economic relationships. The uncontrolled growth of information is flooding universities and learners alike with knowledge from all kinds of sources.

At the same time, increasing competitiveness and globalization is creating needs for learning that extend far beyond a textbook's or a degree's scope, such as the need for lifelong and open learning. The change in the conception of knowledge and needs for learning has been characterized as shifts in learning towards a more action-oriented, distributed, mass-customized and multi-mode learning, complex and adaptive paradigm[1].

Given these changes, universities have to fundamentally rethink their roles. Universities and their elements need to be re-conceptualized within a broader setting that shows how they can be shaped to the needs of knowledge-based societies. The notions that need to be revised foremost are the ideas that they are static, slowly changing entities, that they are less responsive to societal needs than to their own, and that students are essentially containers to which contents must be "delivered".

Many of these trends and needs for change were recognized in the recent World Declaration on Higher Education (UNESCO, 1998) which, amongst others, stated that higher education should: be equally accessible to all; be linked in a seamless educational system starting from childhood; provide for lifelong learning; be relevant to society; diversify educational models; provide for vigorous staff development, ensure quality in its many dimensions; place students and their needs at the center of its attention; ensure women's participation; and embrace the potential for IT and networking. This plethora of issues suggests that it is critical to view the changes wrought by IT and Internet-based education with a multi-dimensional lens.

Taking a slice of this huge and complex agenda, this book examines how the higher education environment is being reshaped by new information technologies (ITs), with a particular focus on technologies based on the Internet. This trend can be encapsulated in the term virtual university (VU), which describes the new forms of universities that enhance learning through technology, in particular, to make knowledge or more knowledge accessible to students that could be located, not just on-campus, but literally anywhere. While the term VU connotates the existence of an institution, we are using it more loosely to suggest the emergence of a multiplicity of actors, institutions, learning channels and styles, and organizational mechanisms, which underlie the different kinds of learning systems.

This book involves reflection on current virtual universities' and other institutions' experiences with the new technologies and educational systems, their potential and limitations, and the implications for the future. Universities are being buffeted by, or having to take advantage of, the same forces that are shaking up the real world. They are responding in many ways, e.g. adopting changes towards network learning, reformulating their knowledge systems to stretch beyond the confines of campuses, and changing the underlying philosophy towards the understanding of knowledge and its valuation. However, while technology is at the heart of many of these transformations, social, economic, political and other factors will certainly have an impact, either in limiting success, in helping systems become successful, or in mitigating the effects of dehumanizing technology.

Ultimately, our intent is that the book leads to greater inquiry, and better informs efforts to design VUs and virtual education systems. In doing this, we hope to provide a map to the possible transition paths between systems that currently exist, and systems that could be realized. The authors have been selected from a variety of fields, institutions and geographic regions, to give a comprehensive as possible a perspective on the emerging trends. Many of the chapters have quite an extensive scope, in part because the authors recognized the important interfaces between their discussions of a particular area and the changes in the broader environment. At the same time, because of the nascent nature of the field, an effort has been made to provide diverse views on many topics.

Virtual Universities and the Internet

Educational institutions and systems have gradually been transformed by technologies, leading to a range of systems. We can characterize three broad types of educational institutions: the first is the traditional campus-based university, which is increasingly enhancing classrooms with information technology. A second type is the open learning environment that serves off-campus or part-time students. This is often manifested as open universities or distance education (sometimes called distance learning or telelearning) programs. These commonly use technologies to facilitate either synchronous or asynchronous distance education for students' independent study. This book will focus on how new information technologies are not only reshaping the first two environments, but even more so, are bringing about a third type of educational system—the virtual university. In the most advanced sense, we can define a VU as a campus-less university that uses Internet technology for its main delivery mode. However, very few universities currently exist in this purest sense, given that VUs need to operate within the existing institutional boundaries set by campus-based universities or other institutions, and to base their operations on the existing human resources in those institutions. In this chapter, we will use the term virtual university in its broader connotation, and will use the term true virtual university to describe campus-less universities that use the Internet as their main mode of course delivery. (We will also use various terms to describe the actual activities, such as virtual education, distance education and so on). This latter is still largely an ideal, since most self-described VUs are still comprised of traditional campus-based universities or open universities at their heart.[2]

This raises the question of how exactly a university can "virtualize" its operations. The most commonly thought of way is by becoming campus-less, servicing students' learning needs from a distance, and by bringing knowledge to the student wherever he or she is. The Internet is bringing about an important new variant of this, by allowing schools to focus their educational and other activities around the Internet or by "putting" those activities on it. This makes much more information available, and consequently, can make the teaching/learning pedagogy more learner-centered. The Internet and its associated media also has the potential to increase the interactivity of students amongst themselves, or with information, which overcomes the limitations of traditional distance education media. All of this makes it easier to "virtualize" the traditional teaching and non-teaching activities of a university, such as the lectures, research opportunities, library collections, and communal and collaborative forms of learning.

Another related trend in many organizations in the business world, and which some universities are following, is to connect independent organizations into a virtual "super" organization in which they share their resources with each other, e.g. through partnerships or flexible institutional networks. Typically, this is one way of increasing the services offered by one or all of the organizations in the network. In its more advanced form, these networks of organizations or individuals could employ the Internet for coordinating their work more closely, as is done by many researchers in globally distributed teams.

In addition to the virtual universities, there are a variety of newer institutions on the horizon, and these may become competitors to the traditional educational institutions. These include virtual organizations, consisting of brokers who source lectures and other material from independent course providers (i.e. independent teachers that are contracted), corporate universities that provide on the job training for their employees and other students, and publishers and libraries that are starting to offer better-packaged content and even courses.

Each of these types of institutions could be based on different pedagogies and technologies, including Internet technologies. In fact, many on-campus universities themselves are also using new Internet technologies and pedagogies to varying extents. As a consequence, there is increasing convergence in practices across these different types of institutions.

Each type of institution also has different comparative advantages. For instance, in order to adapt to the demands of the rapidly-changing world, many campus-based universities are relying on their already significant research and teaching resources to develop new tools and course content oriented around the Internet and its impacts on society. Many open universities are also using new technologies to complement their existing distance education programs. The penultimate could be the true virtual university, which completely relies on the Internet to offer its courses. In doing so, it fundamentally alters the structure of its organization and resources, the contents offered, and the pedagogy used. Since these changes may get reflected in other institutions that partially use the Internet, the difference between the true VU and these other institutions may be a matter of degree. At the same time, a total reliance on the Internet could open up whole new vistas as to how universities socially create and diffuse knowledge.

The Major Issues Confronting Universities with Technology-induced Virtualization

This book focuses on technology as the main factor in a university's virtualization. Thus, we have to examine in what ways a university could interact with technological changes. That is, the virtualization of a university has to be understood in the context of a university, including its roles, the modes by which it fulfills those roles, and the transformations it is likely to undergo (including its eventual demise, as some suggest). (Brown and Duguid, 1996).

As far as the university's roles are concerned, there are many historical and current perspectives on social roles, including ones like duties to preserve and advance the societal knowledge base, and recent ones like the contribution to national innovation and economic growth. Its duties to individuals include enabling them to lead productive, fulfilling and reflective lives. We will not address these in detail here, but instead, since our focus is on how technologies enable learning, we will focus on how a university accomplishes its tasks, and the impacts of technology on them. From the perspective of the learner as a consumer of knowledge, this consists of the work of mediating learning resources for students' learning needs (including the diffusion of knowledge, and facilitating access to scholars, and ideally, the formation of communities of scholars); enabling learning processes; selecting and tracking students; and providing the necessary oversight and credentialling for learners to be respectively qualified and accepted by society.[3]

The task of describing the virtualization of universities then becomes one of characterizing the effects of technology on these modes:

The mediation of resources. In general, technology improves the ability of institutions to mediate learning resources, i.e. it helps to get the right types of content and learning to the individual. However, as with the impacts of the Internet on business, there are serious ramifications on the intermediaries, that is, the teachers and the institutions. While the mediation of resources is usually done through both teaching and administrative staff acting as intermediaries, the Internet could diffuse and allocate knowledge via quite different means, such as intermediate brokers, or technologies like delivery systems and software agents.

Learning processes. In the case of learning processes, the main issue concerns how technology can be used to improve learning. This may be at the individual level, or at the group level, e.g. how traditional communities of learning may be displaced, or how new ones are enabled. The issue of enhancing learning is complex since it depends on how appropriately technology can be implemented in the learning context, and not simply added on as another feature or replacement for learning. In this book, we discuss how pedagogy can be improved by technology, as well as more generally, how the "work" of the university can be accomplished.

Credentialling. The issue of credentialling is yet to be resolved. There is some reason to believe that that increasing virtualization (and the lower costs of virtual education) will cause campus-based institutions to lose their stranglehold on credentialling, as employers (or any rational buyer for that matter) will increasingly recognize the innate credentials of the knowledge gained, and not the institution itself. The changes in how credentialling, oversight and learning are done will depend a lot on how institutions evolve or transform themselves, and the relative success of virtual forms of education vis a vis traditional ones. As is discussed in the chapter by Barcia et al. as well as elsewhere, credentialling by professional organizations can be restrictive, and not necessarily useful to universities.

Costs and other operations. Two other issues that concern universities as institutions are: the bottom line (e.g. lower costs and higher productivity, and the continued existence of the university), and general operation, including administration, student selection and general welfare, and so on. We will focus on the cost issue, which dominates many decisions. The important point to recognize is that while the cost issue often occupies the modern university's administrators in a competitive environment, the issue is easily made misleading by different perceptions of what constitutes costs, productivity and learning. By this, we mean that IT has transformed the nature of work and learning, and the Internet has further transformed them. This circumstance is best seen through its analogy with how the impacts of information technology (IT), and now the Internet, were felt and measured in the business domain. In the case of IT, the economic puzzle of the productivity paradox questioned why increasing inputs of information technology had not led to concomitant increases in organisational productivity. The real answer partly lay not in the issue of measuring productivity, but in seeing how IT transformed the nature of work. In the case of the Internet, the transformation of relationships and the nature of information also add immeasurable value. Universities are no less vulnerable to these questions than business was.

Access to education. The final issue with which this book is concerned is the accessibility of education to learners. While it is certain that many learners will learn more, and more learners will be reached by new forms of virtual education, other less privileged learners, such as poorer people, could also continue to be shut out because they cannot connect to the new technological infrastructure for financial or other reasons. Rural populations in developing countries are one such group. The irony is that many distance education programs that were set up on the ideal of increasing access to the working class—such as the British Open University—actually only increased opportunities for the middle class, whereas the lower classes have not shared in this.

In the end, the degree to which the campus-based university will eventually metamorphosize into a virtual institution, given the pressure from the increasing number of virtual universities, remains to be seen. While it appears that increasing numbers of institutions are adding distance education components or programs of study, these are not yet replacing their traditional educational arms.

In terms of learning, the application of technology can be seen to have many effects, such as the following:

- Technology as enabler—technology can enable new learning processes, such as making learning more interactive and expanding access to information.
- Technology as multiplier—technology can multiply good outcomes, as when IT allows increased number of students to be taught at low incremental costs. But technology can be a multiplier of bad outcomes if it is used to replicate bad teaching practice for many students. These can be the result of institutional and other forces that act on or react to the technological system.
- Technology-dependent learning paths—there is the likelihood that technology could dictate what and how knowledge is learnt in these programs, as opposed to traditional modes of teaching that are only enhanced with technology. For instance, online learning may take a different form from classroom teaching, depending on more increasingly sophisticated tools and instruction.

In addition, there are other considerations associated with the application of technology:

- Mode of inserting technology—technology can either be added onto an existing learning system, as when a technology is used to supplement teaching, or it can be integrated holistically into the system. Examples of technology integration strategies are ones where a stronger emphasis is placed on pedagogical outcomes or other learning goals, and the technology is simply viewed as one tool to help enable those goals.

- Effects of the institutional environment and other factors—various factors affect how well technology can achieve institutional objectives, so the appropriate institutional environment needs to be in place (usually a technology integration strategy). These factors can act to embrace or resist technological emplacement, e.g. when institutions and faculty fail to change to suit the new systems.

We will not be able to address all of these in the book, but readers should be aware of these characteristics as they arise in various chapters. In the next section, we show how the book is organized to address the more salient of these issues.

The Organization of the Book: Five Clusters

This book is organized to cover five clusters of issues, based on the issues discussed in the last section: institutional, learning, functional, economic/market, and technology. Some of these can be seen to focus on different levels of aggregation, such as the learner and "classroom" level, institutional level, and inter-institutional (e.g. market) level. The institutional cluster refers to the university as an institution, including changes in its roles and its form of organization (in some instances becoming less institutions than virtual entities). The learning cluster covers new learning pedagogy and curricula at the learner and classroom level, which the Internet is greatly influencing. The functional cluster refers to the core educational functions (or learning activities) that universities employ to fulfill their missions, including the teaching (e.g. lecture), library, and research (e.g. laboratory) functions. The economic cluster refers to issues of costs as well as to how to organize these new institutional structures, through market, cooperative or other mechanisms. The technology cluster includes systems used to support educational functions, such as delivery systems, as well as the technological infrastructure.

This book is divided into three parts. The first two parts cover the clusters in a general way without focusing on particular institutions, while the third part examines specific cases of institutions within different geographic regions. Figure 1 shows how these clusters are covered by chapters in the first two parts of the book, and how some chapters connect different clusters together.

For reasons of space, we will only focus on how institutions can provide certain educational functions. Other critical roles of universities, such as research and reflection on social issues, will only be highlighted.

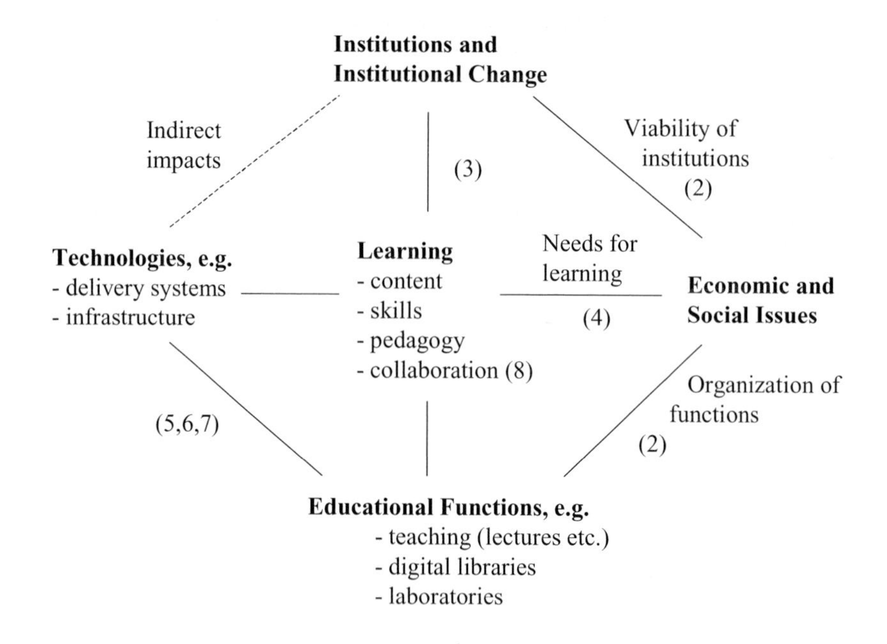

FIGURE 1.1
Organization of the Book.

Part I. Knowledge-based perspectives on learning and institutional change

The first part looks at the universities' interfaces with the economy and the world as influenced by technology and the emergence of the knowledge-based society/economy. The changing roles of universities are discussed in the chapters using different perspectives.

Chapter 2 illustrates the forms that virtual universities are currently taking, the factors causing their growth or constraining their evolution, and some issues involved in designing larger systems. The chapter begins with the context of the emerging knowledge-based economy, and an examination of how virtual universities can address its learning needs. At the same time, the structures of universities may change under technology, such as in how technology can alter the boundaries of educational systems, allowing institutions to provide different educational functions from what they traditionally provided. Next, the experiences of selected types of institutional players and their advantages are compared, based on some emerging ideas about knowledge and learning. Finally, the chapter addresses the design of the system by illustrating how various forms of socio-economic organization can be used to organize resources to provide educational functions. Various kinds of system issues are raised, including the effects of technologies on organizational and functional boundaries, and the costs and benefits of virtual education systems.

Chapter 3 (Dillinger) provides a rationale for dismantling the traditional university, and takes initial steps towards a framework describing a new kind of learning environment. It discusses issues of learning at three distinct levels: the economy, the institution, and the individual, and provides a framework for organizing our thoughts about why each level is important to another, and why and how the traditional university must change to suit the needs of the information economy. In the first approach, the requirements of the information economy for learning are laid out.[4] The university sits within a "sea" of information, and can only succeed by assisting the learner in his or her management of it. This view diminishes the role and operating mode of the traditional campus-based university, putting it in crisis. Next, Dillinger addresses learning from the perspective of the individual learner, in particular, the link between the learner and the learning environment. Cognitive and psychological perspectives suggest learning processes that are quite different from those of the traditional socio-environmental perspectives on education. This indicates that the needs for learning and skills development are quite different from what is actually provided in traditional classrooms, requiring skills for independent inquiry and collaborative learning. The chapter shows how this forms the beginnings of an ideal conceptualization of a learning environment, that abstracts from and even "eliminates" the traditional institution.

If Chapter 3 described a complete change in how learning should be done, Chapter 4 looks at what is being done as campus-based universities reform their curricula (defined broadly), for better or for status quo, in order to meet the new demands for skills and education. In light of the tight coupling between "how" to learn and "what" to learn, a framework is developed that considers curricula as broadly consisting of content, thinking abilities, and technological tools (the latter facilitates new pedagogy as well as skill formation). (A more detailed discussion of tools is given in Chapter 6, as well as in some of the other technology chapters). The chapter illustrates how the Internet is influencing several fields of study, and how institutions are making the corresponding changes in curricula content. The Internet also varies in its influence in different regions, according to their unique cultural and socio-economic contexts. With this in mind, the chapter examines the case of India as illustrative of a forward-looking emerging economy that coexists with a traditional society.

Part II. Enabling a virtual university's educational functions with new technologies

Part II discusses the effects and potential of technology for shaping virtual universities. The chapters focus on the technologies themselves, the educational functions, and the application of the technologies to improving those functions. A VU can be thought of as having to fulfill several educational functions. The two main functions we focus on are the "delivery" of content such as lectures and learning exercises (which may be interactive or non-interactive), and digital libraries. Note that in Chapter 9, Mason also discusses some of the current technologies (aside from the Internet) that are used in distance education.

Chapter 5 (Chong) discusses the evolution of the Internet and its implications for learning and other socio-economic activities. The chapter identifies key current and emerging component technologies that form the infrastructure as well as the backbone of Internet-based "classrooms" and learning networks. This also forms a link between the technology chapters and the applications in other chapters. The discussion includes Internet technologies that are only just hitting the market, or still under research, that could affect the Internet infrastructure and its evolution. The chapter examines the Internet's past, its present situation (with its corresponding limits and constraints), and its future, including the hardware and software technologies that could expand its potential, such as technologies for digital library and collaborative learning systems.

While some of the technologies described here can duplicate the functions of the technologies that Mason describes (e.g. videoconferencing can be done over the Internet rather than through dedicated lines), they may never totally replace those technologies. Nevertheless, Internet technologies do provide new channels for learning.

Chapter 6 (Brusilovsky and Miller) provides a comprehensive examination of the range of technologies and software that can deliver the "lecture" and other contents, via different modes, such as interactive or non-interactive, and asynchronously (as opposed to synchronous modes, such as real-time videoconferencing). The systems examined are categorized according to whether they meet certain characteristics of base, state-of-the-art, or research level systems, as defined by the authors. There are two basic types of systems: electronic textbooks and presentations, including electronic lectures and guided tours. While issues such as interactivity or learning-by-doing are often emphasized in discussions of specific systems, this chapter covers a more systematic and broader set of issues that impinge on the need to deliver contents. This includes issues such as authoring (i.e. the ability to incorporate the content into the system), the adaptivity of systems to students' needs, and the structure of the media, such as hypertext. Other issues of concern to large-scale production and quality assessment are also covered, such as student assessment and use of databases.

Chapter 7 (Borgman) provides an overview of digital libraries and their implications for VUs. Digital libraries are typically treated as technologies by computer science researchers, or as collections by institutions interested in making information available to their students. This chapter seeks to provide a broader conceptualization by examining a range of social, institutional, information and other issues. This includes ones such as the range of activities enabled, e.g. electronic publishing; organizational issues like seamless operations across different libraries; processes that allow greater user control, e.g. search and navigation; means of organizing information, such as metadata; and important features of learning, such as group learning. Through its analysis of traditional libraries and enabling technologies and the functions they enable, the chapter provides a map for migrating from what we know of libraries and our current expectations of them, to libraries that could serve not only virtual universities, but a variety of purposes.

To close this part of the book, the issue of collaborative learning is examined. Many experts increasingly view collaboration as a key aspect of learning in campus-based universities, and increasingly, the social environment could be the key determinant of whether virtual universities can successfully substitute for on-campus learning environments. Since collaborative learning cuts across a number of systems and educational functions, many of the chapters in this book also make reference to it.

Chapter 8 looks at the broader context for collaboration, being that of the social learning that takes place in both traditional campus-based universities and informal learning environments. Collaboration can take place between student peers and teachers in the broader campus, colleagues in the workplace, or others in the broader society. The concept of a "Social Web" is advanced as a type of meta social learning environment that allows various spheres of activity to overlap and promote wider learning opportunities. A Social Web can promote various types of "interpersonal" learning, such as learning to solve problems collectively, communication and other tacit interaction, the learning of social norms and mores, and the development of people skills through interaction. The chapter examines various types of social information needed to realize the Social Web as a social environment, so as to establish presence, build a sense of community, allow the social processing of information, and build a cohesive environment.

Part III. Moving towards the virtual university: institutional experiences from different regions

While the first two parts of the book provide theoretical and conceptual perspectives on virtual education, including scenarios of what is technologically possible, Part 3 examines examples of actual virtual education programs in place around the world. These range from virtual universities to distance education programs and networks of universities, all of which are using Internet technology or mixtures of different technologies to change the prevailing educational conditions in their respective regions. The chapters examine the successes as well as the difficulties in implementation faced by these campuses. Many chapters in Part II provided a somewhat technology-centered or technology-push view of virtual education, in part to illustrate the possible future routes to virtualization. However, the cases in Part III brings the focus back to the learner and the institution, to illustrate a more humane and human-based understanding of how current advancements in distance education have occurred, with new technologies seen more as an enabling force.

Chapter 9 (Mason) provides a framework for understanding the advantages and disadvantages of various organizational forms that VUs may take, with a particular focus on VUs in developed countries. These includes brokers that act as intermediaries between individual course providers and students; umbrella organizations, networks of universities and partnerships (which are variants similar to one another); dual mode universities (i.e. campuses that offer programs in both distance and on-campus modes); and greenfield universities (i.e. completely new sites). The large variety of institutional forms suggests that no one mode is the optimal solution, but rather, the choice of form depends on the unique historical or initial (starting) conditions of each individual actor or set of actors wishing to enter the virtual education field.

Chapter 10 discusses the African Virtual University and the context into which it was borne. A brief history of the birth of distance higher education in Africa is followed by a discussion of the current problems facing African higher education and distance education. The chapter then provides a detailed look at how the World Bank-initiated African Virtual University (AVU) operates, with a particular focus on the Kenyatta University site—one of the success stories. The AVU networks institutions in non-African countries with campuses in Africa, enhancing the latters' access to knowledge and learning. While the price for connection is not cheap, the provision of financial resources can offer poor countries the chance to leapfrog knowledge gaps by providing access to knowledge from other areas. Furthermore, the case illustrates how, despite many societies being deprived of knowledge, there do exist a pool of people who are willing to pay for it, leading to the possibility of financial self-sustainability for universities.

Chapter 11 examines the Monterrey Institute of Technology's (ITESM's) experience in constructing a true VU in Mexico. Its key features were networked learning and student- and community-centered learning. The university's experience in the international arena has been one of building knowledge-sharing networks with other programs. However, it was also a laborious learning experience, starting with satellite technologies that were not quite satisfactory to students, and progressing to the current, largely Internet-based, environment. At the same time, technology appears to appeal to students because of their perceptions of its advanced nature, and its ability to connect students with information where a physical campus is not available. However, many issues have to be dealt with, some of which are more substantial, due to the large community being dealt with. This includes issues of pedagogy, including assisting teachers with their new non-traditional roles, assisting students with the new environments, heavy investments in technology, and standardization of technology.

Chapter 12 discusses distance education in Brazil, and in particular, one institution's experience. The first section sets the larger context within which distance education is viewed, including how it was constrained and even held back by political forces and vested institutional interests. The second section discusses the experience of the Federal University of Santa Caterina—a dual-mode university in the terminology of Chapter 9—and how it managed to create a new "culture" and modality of graduate-level distance education, involving substantial institutional, technological, and pedagogical change. The program's learning experience involved creating a system for producing local content and knowledge for distance education. It also shows how a nucleus of self-regenerating expertise can be built up through contact with other institutions from around the world.

Chapter 13 describes the range of distance education programs and VUs starting to emerge in Asia. The survey shows that while fully Internet-based VUs are not yet realized in many Asian countries, several schools are starting to offer educational opportunities over the Internet. Some of these are on-campus universities that also offer Internet-based courses (dual-mode enterprises in the terminology of Chapter 9). The chapter indicates that the future and prospects of virtual education in Asia can still be very much dependent on, or as in the case of Japan, constrained by, the government.

In the concluding chapter, we provide a synthesis of the various chapters' in order to develop a set of considerations for virtual education. A special emphasis is placed on what has not been said in the book, namely, the factors and policies necessary for effecting the transformation of educational systems in the more underdeveloped countries.

Notes

1 A Book in Progress, T. Morrison. http://adbi.org/pub_morr.htm.
2 Some chapters may deviate from this definition slightly, such as Chapter 14 (by Yoshida), which classifies *open universities* as being different from *virtual universities*, which the chapter considers as universities that primarily employ advanced Internet-based technologies.
3 This is partly based on conversations with T. Morrison of the Asian Development Bank Institute, and on Brown and Duguid (1996).
4 On this, Dillinger's view of the information *economy* is actually more like an information *society's* functions. In this view, the economy/society, the universities within it, and the faculty, essentially have three *information functions*: producing, packaging and disseminating information.

References

Brown, J. S. and Duguid P. (1996) The University in a Digital Age. *Change* July/August 1996: 12–19.
UNESCO (1998) *Higher Education in the Twenty-first Century: Vision and Action.* Final Report, World Conference on Higher Education, UNESCO, Paris 5–9 October 1998.

Part I

Knowledge-based Perspectives on Learning and Institutional Change

2

Virtual Universities and Learning Environments: Characterizing their Emergence and Design

F. TED TSCHANG

Introduction

In many developing and developed countries, educational systems are not up to the task of preparing students for modern societal demands. The needs for information professionals are increasing in information-intensive economies like the U.S., and educational curricula on information technology (IT) and Internet topics are barely keeping up with changes. Society now holds that education should be provided over the course of an individual's whole life (i.e. lifelong), and that learning should be open, that is, opportunities should be provided for learners to learn at their convenience (at any time), in any setting, be it at home or work, and for almost any purpose.

In general, the Internet and IT have facilitated a wholesale transformation in learning by making increasingly larger volumes and varieties of knowledge accessible, such as online books, journals and other information; changing how knowledge is represented and created, as through the use of multimedia and interactive simulation; and changing social interaction through the formation of virtual communities and other means for keeping in touch.

These changes can be clearly seen in the emerging forms of virtual education. New technologies can provide virtual education systems with comparative economic advantages over traditional education, as well as flexibility and access to more sources of knowledge. Nevertheless, traditional education systems (i.e. campus-based universities) do continue to hold advantages over virtual learning, in particular, the immersive aspects of on-campus learning support social ways of learning.

New *virtual universities* are emerging with the potential for student-centered "unstructured" learning (i.e. guiding students through materials but not teaching them), access to potentially unlimited knowledge, and interaction with broader virtual communities in a globally-distributed space. Virtual universities could be started by literally any organization. Both traditional campus-based and distance education institutions are experimenting with new forms of virtual education, and other providers are also entering the market, at least in the U.S.

This chapter makes a first attempt to conceptualize how virtual educational institutions are evolving to meet modern learning requirements. We do this by introducing concepts of knowledge and learning and by developing a broader socio-economic basis, so as to evaluate and design virtual universities and learning environments. This will involve tracing the interconnections between Internet technology and its impacts on knowledge as a whole, educational activities such as teaching and research, and the institutions, markets and other means of organization. Social organizations like communities and governmental support can also provide access to knowledge.

It is useful to keep in mind that there are at least three levels of analysis in a virtual educational system: the classroom or learner level (some might call this the learner community level as well), the institutional level and the system level. In this chapter, we provide a mainly theoretical examination of the factors that shape the institutional and system levels.

Outline of the chapter

Figure 2.1 shows the topics covered by each section. The second and third sections develop the conceptual knowledge-based perspective on what is happening in economies and educational systems. The second section examines the background for changes in educational systems, particularly trends in globalization, and the emergence of knowledge-based economies and their new needs for human resources and learning. The educational (supply) systems are also in flux as new technologies and opportunities provide means of satisfying these needs. The third section discusses some basic concepts of knowledge, learning, and selected implications of the Internet for institutions.

The fourth section focuses on the issue of *emergence*, looking at the various actors that are just now developing virtual universities or their elements. The section also develops scenarios of institutions and their competitive advantages, based on their use of technologies and types of learning they can support.

The fifth and last section examines the *design* of whole educational systems. This conceptual treatment is premised on the Internet's capability to "blur" institutional boundaries. The main idea in this approach is that of a *virtual learning environment*, which is a system consisting of the virtual institutions with their educational activities, organizational structure, boundaries, purposes, and so on. Since there is no standard approach for how to design these environments, we will

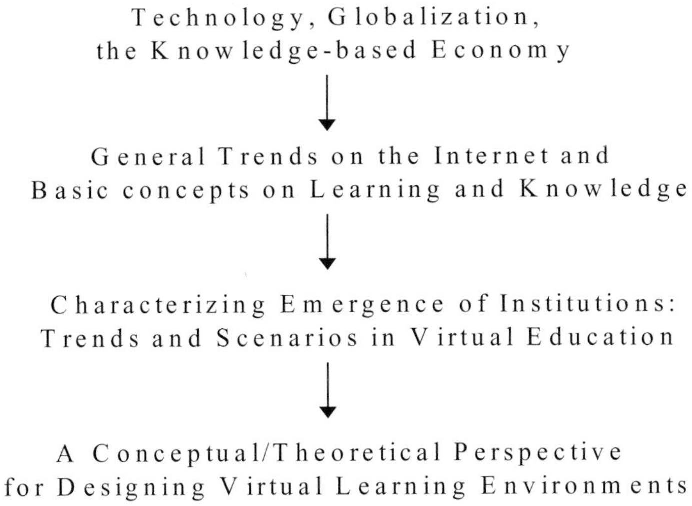

Technology, Globalization,
the Knowledge-based Economy

↓

General Trends on the Internet and
Basic concepts on Learning and Knowledge

↓

Characterizing Emergence of Institutions:
Trends and Scenarios in Virtual Education

↓

A Conceptual/Theoretical Perspective
for Designing Virtual Learning Environments

FIGURE 2.1
Main Sections of the Chapter

provide some basic concepts for doing so. The first is the idea that three forms of *organizational structures* can be used to organize virtual learning environments: markets, communities and organizational networks. These are applied to the organization of virtual teaching, library and research institutions. A second is the idea of system-level measures—such as benefit–cost analysis—of the effectiveness of virtual educational systems.

Trends in Globalization, Technology, and the Knowledge-based Economy: Implications for Education

The emergence of a knowledge-based economy is caused by various forces, two of the most central being technological change, and the increasing globalization and competitiveness of economies. These drive the evolution of educational systems in two ways: by their changing needs for knowledge and various types of capital, and by opening up new opportunities for providing education.

Technology, in particular IT, originally constituted a separate industrial sector, and was used to augment our information processing capacities, not unlike the way in which machines were used to augment human muscle power. However, a fundamental shift is now occurring with the Internet. The Internet facilitates a new virtual world for economic and social activities. In business, new Internet enterprises that deal solely with information are springing up, even as many traditional sectors, and their activities such as logistics, distribution, and design, are "informationalized". An example of this is the digitalization of library knowledge. IT now allows us to store vast quantities of "book" knowledge and other media for unlimited periods of time, and the Internet greatly increases accessibility to this store of knowledge. IT has altered economics, such as by reducing the barriers to entry of new competitors in many sectors. In the area of virtual education, new entrants such as virtual universities and educational providers from outside the traditional education industry threaten to overturn the status quo. New forms of socialization are also possible on the Internet, ranging from simple mailing lists to virtual communities.

The second factor is globalization. This has many different aspects, including the increasing influence of multinationals (particularly through their worldwide production and sales), and global trade, including the flow of human resources and knowledge between nations. In the global economy, the race to enhance the value added in production and to create new knowledge has led to a hierarchy of countries where some countries are leaders, some are followers, and many are falling further behind. Countries as well as subgroups within societies can become marginalized or isolated, and their specific knowledge, such as local cultural traditions, may become useless or valueless in the global economy.

These two factors—technology and globalization—are linked in various ways. Technology has altered the boundaries of production, for instance, by allowing firms to coordinate production across extensive distances, in some cases turning firms into completely virtual organizations, and the way in which economies are linked, e.g. allowing electronic funds to flow instantaneously across borders.

The knowledge-based economy idea is also a popular, if not unclear concept, in many academic fields. However, because it lacks a rigorous definition, and because of its all encompassing nature, researchers are only now scratching the surface with its many sides and dimensions. There are many issues to resolve, including the controversy about whether business productivity has been increased by IT (Brynjolfsson and Hitt, 1998; Dewan and Kraemer, 1998); problems on how information and knowledge are to be measured in economic terms or their value within the economy (Carter, 1996); lack of empirical evidence on the knowledge industries (Foray and Lundvall, 1996); as well as long-standing debates on what constitutes knowledge (Nonaka and Takeuchi, 1995; Tsoukas, 1996). Much of this is due to the intangible nature of knowledge, as well as the ill-preparedness of traditional fields like economics for dealing with less easily measured items like knowledge.

Educational needs of the knowledge-based economy

Knowledge-based economies have various needs for human resources and education, particularly needs for:

- Highly skilled, flexible labor forces—the needs for "knowledge workers" such as specialized information professionals are rapidly increasing in knowledge-intensive economies like that of the U.S. This requires changes in educational curricula that reflect more IT- and Internet-related subjects such as electronic commerce, as well as needs for technological and thinking skills.

- Open learning—The concept of open learning can be defined as learning that can take place anywhere, at any time (i.e. asynchronously), and from many sources. In addition, many societies now desire lifelong education. All this requires more flexible educational systems. Learning can also take place in various virtual and real-world settings, including schools, the workplace, virtual communities and the Web.
- Learning for many purposes—Learning also has to take place for various purposes, including the changing needs of workplaces, communities and individuals, and the broader society. Open learning can thus fulfill many purposes, such as degreed professional education done part-time (e.g. distance MBAs), courses for employees to update their skills, or distance education for self-improvement at home.

New trends in education systems

Many traditional educational systems are not meeting many of these new needs, let alone the existing needs of large sections of the population. This may be in part because they are too tradition-bound, are not flexible enough to adapt to new situations and requirements, or require high levels of investment and maintenance. However, rapid technological advances are changing the means for providing education, as well as its economics. Some of the main trends are:

- Increasingly virtual education—technology provides virtual education with a competitive edge for meeting many of the new needs for open learning and flexible education systems.
- Transformation of traditional universities by technology and the demands of knowledge-based economies—changes are being made in many universities and their classrooms, both at the curriculum level and in the use of technology. It is likely that traditional institutions will either have to adapt to the new uses of technologies, as many are, or risk losing ground (Massy and Zemsky, 1995).
- Changing market structure and new entrants—IT and the Internet are becoming major forces of change in educational markets. Increasing numbers of new players are entering the educational market to meet the new demands, making the market more competitive. There is strong evidence that the large numbers of traditional institutions are adding fuel to this fire by jumping in, and some estimates suggest that at least in the U.S., numbers of students that will enroll in distance education will jump, and numbers of universities offering distance courses are skyrocketing.

Access to knowledge on a global scale

As economies become increasingly global in scope and knowledge-intensive, and as digital forms of knowledge become more prevalent, the risk that developing countries could fall behind becomes greater. At the same time, the opportunities for them to leapfrog intermediate stages of development (e.g. by directly going to advanced technologies) increase. This leads to the basic question: which of the new technologies are most useful for fostering development in developing countries? At the very least, the Internet's tendency to support open access can lead to the greater global sharing of knowledge. However, any trends toward the privatization of knowledge or restrictions on flows of knowledge, such as when "techno-nationalist" feelings arise in advanced countries, could counteract these positive trends. Furthermore, it is unclear what situations could result from unfettered market forces. It is possible that non-market approaches could be equally effective, such as public subsidies for accessing knowledge, and cooperative networks between universities in the north and south.

One way to "leapfrog" technological stages is for universities to adopt commercial products such as IBM's Global Campus (see Marisa Martin's chapter for the experience of Mexico's ITESM with such products). There is at least some anecdotal evidence that at the classroom level, the broadcasting of guest lectures from overseas can also help local instructors improve their teaching practices (see the chapter by Magdallen Juma). All this requires thinking about broader issues, such as how to embed technology in a learning culture and institutional setting, and what broader technology strategies to use (and risks and rewards).

Many of the VU experiences in developed countries tend to be framed in a competitive market perspective (discussed later). In contrast, many developing countries still have government or publicly funded mechanisms for education. This presumes that the private sector is unable to provide public goods such as education, either because of the nascent state of the private sector, prohibitive costs for students, or other problems. For instance, many of the new virtual universities in Asia are being developed with government funding or assistance (see chapter by Aya Yoshida), which reflects much of the same thinking used by Asian countries in their paths to economic development. All this raises a number of interesting questions, such as whether to create market, government or mixed schemes, and whether these are more efficient, diverse and innovative, and provide more people with access to more knowledge. While we might generally expect more diversity (in practices and knowledge) in markets, markets that promote the success of economies of scale production model-type universities could also be less innovative. In light of this, there may be advantages to having government programs or other means of protecting local culture and practices in educational curricula.

Another way of addressing the issue of improving access and educational quality is to use a classification that distinguishes markets from cooperative social approaches. This will be examined in the following sections as part of our examination of virtual learning environments.

Internet trends and virtual learning: towards a conceptual framework

The increasing knowledge-intensity of societies and economies appears widespread, but this is a manifestation of complex economic and technological processes.[1] As we head into the "Digital Age", we have to be aware that business will *not* be as usual. Technology is helping induce the fragmentation of knowledge, causing knowledge to be broken up into ever smaller pieces, and dramatically changes the nature of knowledge. Learning through short "bite-sized" and "navigatable" video clips on the Internet with links to detailed sources, instead of one hour videotaped lectures with reference text books, may become the norm in distance education.

One way to examine how virtual learning environments could develop is to study the characteristics of virtual institutions in relation to the teaching and other educational activities that need to be provided. Different organizational structures can be used to organize learning environments. We use the term *organizational structure* to mean the mechanisms that allocate and connect resources, people and institutions within the environment. These could be competitive structures, like markets, or cooperative ones, like communities. The various writings on these structures are widespread and disconnected. In particular, few connections are made between discussions of market-based virtual education, community-based learning, and even institutional networks. Our approach is to distinguish between these three forms so as to understand the advantages of different organizational structures, the ways in which they fulfill needs for knowledge, and ultimately, the evolution of the system.

At the same time, to fully understand the learning environment and the evolution of virtual institutions, our framework will have to deal with various other dimensions. We examine the distinctions in the different educational activities, the institutions that supply those functions, the knowledge lifecycle stages, and, to a lesser extent, the different types of knowledge and learning.

To make this process concrete, we will seek to address the following questions in the next few sections:

- How is the Internet and IT affecting the economics, organization and character of knowledge? (Third section)
- What types of institutions are emerging to provide educational functions, and what are their advantages and possible scenarios? (Fourth section)
- How can we design learning environments for broader access, i.e. what types of organizational structures will organize institutions to meet learning goals, and what measures of effectiveness can we use? (Fifth section)

A Conceptual Framework for Knowledge Creation and Learning

The advent of the Internet has spawned a whole new environment in which huge amounts of information are available for a variety of educational purposes, including learning that ranges from casual inquiry to rigorous self-study, and serious research. The Internet is transforming knowledge in at least three ways:

- Reorganizing and changing the character of knowledge
- Creating new economic characteristics for knowledge
- Leading to convergence, or blurring of institutional and other boundaries.

The Internet's reorganization of knowledge

The Internet, the knowledge within it and even its "intelligence" all have a highly distributed structure. By this, we mean that geographic space is no longer a constraint, so knowledge and information are being scattered across different websites and embedded in new forms through hyperlinks and other new data structures. The number of websites around the world is increasing exponentially. The Lycos search engine indexed 8.5 million URLs in 1995, and 55 million in 1998.[2] However, most of these sites are not coordinated in any centralized fashion, causing some redundancy or difficulty in searching for and piecing together information. All this suggests that the evolution of the system may reflect the tendency of the individual "parts" to organize themselves into a structure, i.e. to be *self-organized*, rather than be organized by a centralized controlling force.

The structure of representing knowledge is changing as well, as witnessed, for example, by hyperlinked documents. No longer do we simply tack brief sheets of paper onto bulletin boards. We can now post links to software and huge amounts of data and material to virtual bulletin boards on the Web. These changes affect the way in which knowledge is created, and ultimately, how learning takes place.

The self-organized (and disorganized) structure of the Web, and the low economic costs of connecting to it and distributing knowledge, contributes to the diversity of knowledge and accessibility to it. It also creates problems, including: (1) fragmentation and redundancy of knowledge, as stored in different locations across the Internet; (2) decreasing reliability of knowledge, because of the lack of standards or certification procedures; and (3) increasing difficulty in searching for reliable or appropriate knowledge. One major problem is the inability of search engines to identify appropriate websites (i.e. sources of information). One search done on the Web using a fuzzy piece of information will yield thousands of Websites, mostly containing irrelevant material. This is largely due to the difficulty of identifying context, as well as the imprecise categorization of the websites. This has led some websites to offer "quality-controlled" information.

What is knowledge and learning?

There have been many discussions of knowledge and learning in various academic fields ranging from psychology and education to philosophy, and more recently, to management and economics. Each of these provides a valuable perspective that deserves recognition in any discussion of education. (Chapter 3 discusses both learning issues from the cognition-psychology perspective, while Chapter 4 discusses learning skills). In this section, we will examine two basic forms of knowledge and their economic character.

While knowledge has been classified into many types and forms, a popular recent distinction in the management literature has been that of tacit and explicit knowledge (Nonaka and Takeuchi, 1996). Explicit knowledge is knowledge that is easily codified into and learnt from objects like books, such as facts, formulae, and theories. Tacit knowledge is often associated with skills such as motor skills or skills learnt by doing (Polanyi, 1948). However, here, we also consider tacit knowledge to be any knowledge that is "locked-up" in our heads and therefore not easily articulated or codified, e.g. pattern recognition that requires experiences, certain thought processes, wisdom etc. This distinction has some relationship to how knowledge is stored in humans, acquired (or learnt) and transmitted.

A second important distinction relates to the knowledge lifecycle, such as the stages of creating, disseminating and using knowledge. Later, we will use the tacit-explicit distinction as a means for understanding the effectiveness of creating and disseminating knowledge—two key activities in education—along with their economic characteristics. Using a different approach, Dillinger (in Chapter 3) focuses on learning and societal information functions more from the point of view of the *usability* of the information.

Despite widespread recognition that knowledge is advanced in scientific and other means, the knowledge taught in a traditional classroom is still taken to be a fixed stock from which teachers can draw from. However, in the modern context, knowledge "flows" are as important as stocks, and learning how to tap into those flows may be as important as learning about those stocks. That is, in knowledge-intensive societies, rapid responses are needed to widely varying situations, so knowledge may be required to be supplied "on demand". Thus, knowledge has to be at the fingertips of organizations or people, and people have to learn how to access this knowledge rapidly and efficiently.

For our purposes, learning can be simply defined as the absorption of knowledge. The type of learning can be related to the type of knowledge. While there are many types of learning, we focus on two main types that some find to be a provocative way of understanding the effectiveness of learning (Marchese, 1997): deep learning and surface learning. Deep learning can be thought of as the understanding of the "why's" and "how's" of education, while surface learning can be thought of as understanding only the "what's" (i.e. facts), e.g. learning by rote.

The economics of knowledge creation

The Internet and IT causes the economic characteristics of knowledge to be different from those of traditional manufactured goods, and more similar to those of software and products of information-based industries. These include lower costs of entry, economies of scale in production, and possibilities for the "mass customization" of services and environments for users.[3]

While the Internet lowers the costs of entry for new virtual institutions, it also increases competitive pressures on firms. Literally anyone with an Internet connection can now create a website, and theoretically, offer learning opportunities. With this ability, competition to traditional educational institutions could come from anywhere. At the same time, the lower start-up costs and operating costs allows less cost-effective or other marginal players to stay in business longer than would be the case in the real world.

Costs of knowledge codification and replication

As the Internet and other IT experiences have shown in the private sector, in this day and age, it is expensive to codify *tacit* knowledge, but once codified, new IT systems can easily replicate and disseminate these *explicit* forms of knowledge at low cost (Hansen, Noria and Tierney, 1999). Since the education sector involves codification and dissemination activities, this suggests a similar situation would be true. This is being borne out by the experiences of many universities and course developers. It also raises a more complicated question: what is the cost of delivering or engaging students in the creation of *tacit* knowledge, and what are the costs of using various technologies to deliver the various kinds of knowledge as new learning needs demand?

Scale issues

A second issue has to do with scale. Three of the most important advantages in "scale" that the Internet lends to a virtual university or a virtual education program are: (1) *scalability*, that is, the technological ability to accommodate larger numbers of students; (2) the related *economies of scale*, defined as lower per student costs (with increasing numbers of students); and (3) "diversities of scale", meaning, accessibility to a greater diversity of students and sources of knowledge.

These scale effects suggest that the virtual "classroom" environment can offer some advantages over traditional classrooms, both across different dimensions such as cultures and geography (i.e. interaction of students from widespread places), and in terms of access to varieties of knowledge on the Internet. However, many of these advantages can now also be replicated in traditional classrooms, such as the use of the Internet to connect students in a traditional classroom to other students in far away places.

Issues of convergence (or the blurring of boundaries)

Finally, it is useful to discuss in detail one of the main points running throughout most of our analysis. This is the idea that technology is causing the blurring of boundaries on a number of fronts, including:

- Blurring of traditional and virtual institution's capabilities—technology will provide both traditional and virtual universities with the same playing field. For instance, traditional universities could take advantage of technology to create virtual learning functions like digital libraries.
- Blurring of school and workplace functions. What is learnt or needed at school and workplaces is becoming more similar. This is partly because:

  (a) schools have to respond to rapidly changing workplace needs by mimicking them
  (b) technology now allows schools to offer virtual workplace simulations and environments similar to those of the workplaces themselves. For example, students can easily (and inexpensively) set up Internet businesses online through commercial websites for use in class projects.

- Blurring in the functions of educational institutions—the boundaries that define institutions are collapsing because of low costs and the ease of connection within virtual environments. Different virtual institutions are now able to offer the same educational activities. On the information highway, schools, digital libraries and other places for learning are no longer housed separately, but they can blend together, mutually support each another, or even compete against one another.
- Blurring across boundaries in knowledge—as knowledge is fragmented, the structure between different topics or areas becomes more fuzzy, sometimes being less hierarchically-structured, while sometimes, new structure is added (as in the creation of interdisciplinary fields). This promotes interchangeability between different environments. For instance, students now have access (from the Web) to business, scientific and other previously inaccessible information. However, the opportunities for plagiarism are also equally high because teachers cannot monitor such large amounts of information.

Constructivism and convergence

In the past, different types of institutions specialized in different types of learning and knowledge. For instance, in work environments, practical knowledge was important, and learning by doing was essential. Conversely, academic environments traditionally focused on theory and concepts rather than on practice. In the age of machines and physical products, in part because universities could not afford to recreate the workplace environment and infrastructure, they were largely confined to teaching theory, and the practical training of employees was left up to the workplace.

At the same time, we can also say that learning in workplaces was traditionally more task- and project-oriented, and skills had to be absorbed, while learning within schools was more structured, and subject-oriented, and learning in communities was unstructured and situation-oriented.

All this is changing with the Web and with the advent of newer ideas on learning, such as the notion of communities of practice (which describes how learning takes place within a community). While in some ways, theory can still be regarded as detached from practice, the workplace and academic learning environment are converging in many ways. The Web is a highly constructivist environment, meaning that meanings and solutions are constructed less from theory or single disciplines, and more in practical and interdisciplinary ways. Results are accepted as long as they are demonstrated to work, not simply because they meet the limits of some existing theory.

We can now recreate certain types of virtual workplaces within virtual universities at low marginal cost (after the initial fixed costs). An example is the simulation software used to teach stock market trading. Thus, technology makes it increasingly possible to teach both theories and practice in virtual environments. The pressure on schools to offer these is aided by workplace expectations for employees to possess more practical and constructivist skills.

All this leads to important questions such as: how do we integrate workplaces and communities into virtual universities' learning environments? And, how much integration is necessary? We will attempt to shed light on the first part of the question, but we will leave the more challenging second question to social commentators and philosophers of education.

Characterizing the Emergence of Virtual Educational Institutions

In this section, we will illustrate how educational institutions like virtual universities can meet the new needs for learning, and the comparative advantages of different institutions. As part of this, we will illustrate how the concepts of knowledge and learning identified in the last section may help.

Imagine that a university is comprised of at least three basic functions: teaching, libraries, and research. In the virtual setting, a variety of institutions could provide these functions:

- For *teaching*: traditional and virtual universities, publishers, corporate universities, software developers (see Mason's chapter for more details on some of these).
- For *digital libraries*: public libraries, universities, publishers using electronic media, professional associations (e.g. the Association for Computing Machinery's digital library).
- For *virtual research opportunities* such as collaboratories, which use Internet technology to allow research equipment at one site to be used by users from distant sites (see Chapter 8 by Mark for a discussion of collaboratories): universities, community-based organizations.

Some of these institutions are just now emerging to meet the demand for educational activities. However, the market is very much in flux because of factors such as the rapid change in technology, the experimental nature, lack of standards, and so on.

We might expect the emergence of virtual universities incorporating some or all of the various educational activities. However, many traditional universities can make use of the same technologies to provide similar functions. Furthermore, students can now participate in each educational activity in different places, such as a virtual classroom, a workplace or even at home.

Focusing on the teaching function as an illustrative example, some of the issues involved include:

- Who are the educational providers?—Will teachers become part of a disembodied educational market, and will various types of markets and market layers develop, e.g. markets for technology, courses, programs, knowledge, teaching, other educational functions (such as digital libraries)? As a result, how fragmented will knowledge be?
- What are the comparative advantages (i.e. costs and advantages) of different models and various institutions such as traditional and virtual institutions, for providing for these activities?
- What are the benefits to learners—Will virtual education eventually provide the same quality of socialization and classroom/campus learning as traditional campuses? What are the cost vs. quality tradeoffs? How much of the new student learning needs can each type of education provide?

First, we will examine some of the players and their advantages. Then we will develop a simple framework for analyzing benefits in terms of the knowledge and learning concepts shown earlier. This will be used to evaluate scenarios for the educational "providers".

The players in virtual education

The emerging market for virtual education involves a range of players. (A detailed discussion of selected universities is provided in Part III of this book). The traditional distance education programs are still well represented, as shown in the current Peterson's Guide, which lists about 900 colleges in the U.S. servicing seven million students. By the end of 1999, one expert expects as many as 1000 traditional universities to be offering online courses.[4] In this setting, new "virtual" universities like the Western Governor's University (WGU) and the California Virtual University (CVU) have been emerging, which offer programs of study comprised of offerings from dozens of traditional universities within the network.[5] Virtual educational providers like the University of Phoenix also cater to open learning needs such as individuals' self-development. Non-traditional actors such as corporate universities, professional associations and textbook publishers are starting to offer courses, ranging from the upgrading of employee skills to full university degrees. The number of corporate universities grew from 400 in 1988 to about 1600 in 1998, and they could "threaten" traditional educational providers should they leave the confines of companies, seek accreditation, and actively solicit students.[6]

The market for educational products is also highly competitive, particularly in the U.S, where corporate initiatives like IBM's *Global Campus* are signing agreements with many traditional universities to enhance their learning environments with everything from networks and laptops to software like Lotus *Notes* and *Learning Space*. Corporations such as Japan's NTT—which provides web-based training programs in state-of-the-art environments—also sell their products on the market (this is discussed in more detail in Chapter 6).

The combination of all these trends is creating markets in which the fate of players is unclear, often depending on each others' advantages and market strategies. Technological and other standards are also indeterminate at this current state. While the market is still very much in an experimental or early innovative phase, some experts expect that the market will become more well-defined in the next few years, but even that remains to be seen.

In addition to the pressure from virtual institutions and outside parties like corporate universities, increasing competition to traditional educational institutions is coming from institutions in other countries. For instance, educational institutions in many countries, including traditional and virtual institutions in developed countries, are filling their student pools by actively soliciting students in developing countries, thereby putting pressure on traditional universities in developing countries.

In the midst of this, many teachers will also see revisions in their roles, and possibly in their contracts. The question of whether teachers' knowledge can be adequately "captured" and infinitely replicated in virtual environments has been cause for some concern, and certainly the "technologically enabled" teachers have felt more at ease in their ability to stay ahead of the learning curve. Another perspective suggests that technological means of "commodifying knowledge" is leading to the disempowerment of faculty (Noble, 1998).

Comparative advantages of different teaching institutions

The variety of entrants entering the market for virtual education is mindboggling at best. There are many questions, such as whether traditional institutions will be able to use their comparative advantages to compete with new entrants. Will the WGU compete more directly with traditional distance education universities like the UK Open University or with traditional campus-based institutions? Will institutions like the WGU reap the advantages of being comprised of traditional institutions (and their teaching pool), as well as the advantages of offering the flexibility to operate on the Internet as "one-stop shops"? In the longer run, we may also ask whether the current types of virtual universities will be a dominant form, or whether they will eventually be succeeded by forms yet to be seen.

The issues can be broken into two types: issues of students' expectations (demand side issues), and issues of institutional resources and strategies (supply side issues). On the demand side are the traditional questions of cost, convenience, quality of the education, name branding, and other benefits that students seek, including the value and reputation of a virtual degree, versus that of a virtual name brand degree, a traditional degree, or a traditional name brand degree.

The supply side focuses on institutions and their capabilities. While some of these issues have been addressed in the literature, some of the underlying assumptions are either simplistic or reductionist, as discussed in Box 2.1.

Generally, the literature (particularly in the West) on virtual education has tended to view higher education as increasingly market-driven, or to assume that markets are a preferred mechanism for organizing the industry. There are numerous variances on this theme. One view suggests that the ideal, needs-driven market model is one where the market will provide student-consumers with greater options to select from (Twigg and Oblinger, 1996). Another view sees educational services and markets becoming fragmented along with their knowledge, requiring universities to operate in a free trade environment (Graves, 1997). Yet another possibility is one where educational brokers will help allocate these fragmented teaching resources in markets (Hamalainen, Whinston and Vishik, 1996). At one extreme, there is the view that the technology-induced advantages of virtual education could even eliminate campus-based universities (Noam, 1995). Another variant of this view suggests that technology and virtual education will provide such competition that institutional players will eventually reduce to one dominant "Virtual Harvard" which dominates the market through prestige and lower costs.

However, all these views do not rely on well-reasoned analyses of market structure and opportunities in virtual education, and are posited on fairly simplistic assumptions. Most of them assume that knowledge or its provision are easily separable, decomposable and resaleable in the Internet environment, that is, have commodity-like characteristics, and that coupled with the lower costs attained, this will be ultimately better for students. In fact, some of these assumptions are similar to those posed in the broader literature on electronic commerce and the economics of the Internet (see for example, Evans and Wurster (1997)). The problem is that in these analyses, teaching and learning are assumed to be simplistic processes that can be easily codified and duplicated on computers. There are other arguments that counter this, such as the view that learning is a predominantly social process, and that efforts at virtual education thus far downplay the important social and community environments that exist in schools (Brown and Duguid, 1995).

continued

Another analysis that has been popularly cited as a rationale for investing in IT is one that suggested how academic productivity might be enhanced, and that IT adopting institutions might put pressure on institutions that do not adopt IT (Massy and Zemsky, 1997). Whether this is true with regards to Internet technologies is another matter, as the issue is not simply one of cost, but one of offering a fuller-fledged learning environment with access to and the ability to use extensive sources of knowledge. As to which type of institution can best offer these is yet to be seen.

BOX 2.1
Virtual Educational Scenarios and Their Assumptions About Costs, Benefits and Market Structure

The advantages of some traditional campuses (and VUs based on them, like WGU) includes on-campus research capabilities and lower cost labor (e.g. on-campus students); possibilities to synthesize from a wide variety of existing academic specialties; and longer-term prospects for faculty development. However, traditional institutions may also be hamstrung by their slowness and inflexibility, which contrasts with new market entrants who may have virtual organizations that can restructure rapidly in response to changing markets).

Some observers believe that weaker colleges, already under pressure, may be further pressured by the emergence of virtual universities that broadcast lectures or build intimate online communities using powerful Internet technologies. Ironically, it may be the larger "mass production" university with their lower quality (e.g. 500-seat lecture halls) that could feel the pressure from virtual universities, particularly if the latter can be accredited.[7] On the other hand, if smaller institutions such as the four-year liberal arts colleges are able to maintain their comparative advantages—small class size and group and community learning—they could still provide a quality education, while expanding their students' access to online information. On the other hand, large universities will have to learn how to combine their research capabilities in practical fields with more customized learning opportunities.

Another more serious possibility stems from the perspective that comes from operating in a Web environment. As suggested earlier, certain conditions can favor the virtual university that operates as a totally virtual entity. In the Web environment, online environments that operate by the Internet's different rules of competition and economics could start to look more like virtual businesses, as shown in the following box.

In the most radical sense, universities that enter the online or virtual environment may decide to completely reform their systems according to the "rules" of a competitive Internet market and environment. Athabascar University in Canada epitomized this view when it developed the first online MBA program in Canada. Terrence Morrison, the university's president at the time, noted that the strategy was based on the (at the time) radical principle that such an online degree would look nothing like a traditional university nor rely on the resources of a traditional university, but rather would be run like a virtual business. This model views the production and distribution (delivery) of this knowledge to be important, and so the focus was on systems that can accomplish these tasks, much as with the Internet or retailing business models of Amazon.com or L.L. Bean respectively. In the Athabascar program, knowledge was considered to be easily bought or accessed from distributed sources. In other words, the focus was on network resources. The Athabascar distance MBA involved independently contracting with the best-qualified teachers to design courses, and coupled with a unique financing scheme offered by a bank to allow any student to afford a loan based on the MBA's postgraduate earning potential. The high quality instruction itself became the driver of enrollment and growth. Economies of scale allowed the program to grow rapidly and become self-sustaining, with the number of students reaching about 1000 students after the first three years of operation.

BOX 2.2
Running Distance Education on an Internet Business Model

Assessing the "delivery" of knowledge: scenarios using a knowledge-learning perspective

The different types of institutions and factors in the virtual educational marketplace makes it very difficult to predict which institutions are likely to succeed. The specific technology used, and economics, benefits and costs will all play a role. In the discussion in the previous section, these questions were left open.

Here, we will provide a qualitative set of scenarios based on cost and benefits, where benefits are defined as the type of knowledge and learning acquired. (A more detailed discussion of conventional benefits and costs is provided in the last section.) The issue of benefits involves how effectively learning occurs, which is in turn dependent on how interested students are, and on how useful that knowledge is (which is not analyzable in this framework). Using the knowledge and learning framework developed earlier, we will focus on how three representative technologies, for which we look at the (assumed) effects on deep or surface learning (as defined earlier), and the costs of codification and replication of that learning process. Three potential scenarios emerge for each technology:

Scenario one—In-class videoconferencing : Deep Learning at High Cost

In this conventional scenario, on-campus universities rely on techniques such as interactive in-class videoconferencing to supplement the education of their residential students, but *do not* codify this knowledge. This will provide greater learning opportunities for the same numbers of students. This is currently the scenario that many traditional universities follow. It does not allow them to service non-traditional students' needs easily, but it may add enough value to their program to ensure that it has unique learning opportunities.

Scenario two—Online text or directly codified video: Surface Learning at Low Cost

In this scenario, universities codify "explicit" knowledge without much alteration from the original (text or lecture) form. Examples include videotapes, online text material, and other materials (which may be on the Internet). The lack of intelligent feedback or interaction with the material will tend to generate more surface than deep learning. These universities may appeal to a mass audience in search of basic "credentials". This may be the scenario that many online "universities" currently follow.

Scenario three—Intelligent course delivery system: Deep Learning at High Initial/Low Repeated Costs

One possible future scenario is that universities will codify knowledge and/or use more sophisticated course delivery systems with more intelligence, i.e. interactive materials or feedback. This could promote more deep learning than in Scenario two (see Chapter 6 by Brusilovsky and Miller for a review of systems). This type of system may have high codification costs, including costs of software development or purchase, infrastructure, and costs of codifying the knowledge in forms for learning, but lower replication or reuse costs. This scenario could become increasingly more likely as tools become more sophisticated and more widely available, and as universities learn to couple content with sophisticated technologies.

Other scenarios could be based on this framework. For instance, Internet materials that are less structured but "interactive" could generate surface learning. Furthermore, while it is more than likely that institutions will use a combination of these new forms of teaching, some may focus on one form as their mainstay.

A Theoretical Framework for Designing Virtual Learning Environments

In this section, we will build a conceptual framework to try to answer the broader problem of *how to design learning environments.* We will use a systems approach to address this in two parts: conceiving of how to *organize* these learning environments, and *measuring* their effectiveness.

This should in principal also address the following issues:

- The need to identify a broader set of opportunities for learning, and the broader objectives involved in learning.
- The need to measure the effectiveness of learning within a larger system of institutions and individuals within society.
- The need to include broader processes of organizing, based on cooperative and social mechanisms (e.g. communities), and not just market-oriented processes.
- The need to recognize the possibilities for convergence in institutions and their functions, and the blurring (i.e. increasingly seamless nature) of boundaries between institutions in the Web. Box 2.3 provides a more detailed discussion of convergence.

Institutions can sometimes provide a false sense of boundaries, particularly in a fluid, "costless" environment like the Web. As noted earlier, the Internet is transforming and connecting not just educational environments, but all types of environments, including workplaces and communities. In addition, the nascent nature of virtual universities, coupled with the development of institutions like independent digital libraries and virtual communities, makes it useful to try to keep the larger environment in mind. In general, each of these "virtual" institutions can fulfill a variety of functions. That is, some institutions may fulfill similar purposes. For instance, independent digital libraries could also develop teaching and collaborative research opportunities for patrons. Imagine a digital library that offers an online database of public lectures as video on demand. A patron could also "get up" and offer public lectures that he or she is an expert or enthusiast in. Technology now allows digital libraries to easily set up environments for people to meet, form virtual "study groups", access search tools to do research, converse, share ideas, and drink virtual coffee. New technologies will provide the social cues needed to help users to feel at ease, as if they were in a regular library.

Overlapping functions can occur between virtual learning institutions because of a number of reasons. As noted earlier, in a virtual environment like the Internet, there are no physical barriers to a single institution's ability to link various learning resources together with software. The costs of this kind of vertical integration in a virtual institution may be less than those of a traditional campus. Furthermore, in the Internet environment, the boundaries of virtual institutions are not only circumscribed by walls and the human resources within them, but by the ability of that institution to access and process a wider set of information resources within that organization's environment.

At the same time, we should also recognize that each type of virtual institution may still have a native advantage or core function in meeting the learning requirements of students and other users. For instance, students may learn concepts in a more structured way within a "classroom"; while unstructured learning and research may be more properly performed in a digital library; and practical problem-solving in a virtual research environment.

continued

Thus, in a virtual university, the "core" function of teaching may be provided by certified "courseware" and instructor/mentors. But to provide library and research functions (i.e. a full-fledged learning environment), the virtual university may have to either design its own digital libraries and collaborative research environments, or it may have to "outsource" those functions to independent digital libraries and other institutions.

BOX 2.3.
Convergence (Blurring of Boundaries) Between Virtual Institutions

A *learning environment* can be at many levels, such as a classroom within an institution, an institution itself, or even multiple institutions. In order to simplify the framework, we do not directly address the issue of individual learning, but restrict ourselves to the "institutional" or social level, and focus on the general interaction among institutions and individuals, via incentives and other mechanisms.

A virtual learning environment may consist of an assortment of virtual institutions and functions, and there are no fixed rules for "designing" such an environment. There are, however, two important concepts that we will use to illuminate some of the critical choices that can be made. These are: the organizational structures for organizing different institutions into an effective learning environment, and measures of effectiveness.

Organizational structures: basic elements for designing learning environments

Organizational structures are a fundamental concept for understanding how to organize resources, virtual institutions and their relationships to meet the various needs within a learning environment. This involves both incentives, disincentives, enabling conditions, and constraints. Structure may arise from the bottom-up, as when different individuals in a community self-organize to provide services, or it may be imposed from the top-down, as in how the early Internet was planned and funded by the government. In these schemes, different types of organizational structures can be used to connect virtual organizations to each other and to ensure their proper operation as a whole.

We focus on three basic types of structures: markets, networks and communities.[8] While markets are self-explanatory, networks and communities are terms seeing increasing usage these days in the academic literature. Both involve *cooperative* forms of organization. We define the term *network* narrowly to describe how institutions form medium- to- long-term relationships for reducing their transaction costs (i.e. costs that are hard to price, e.g. search costs), increase reliability of supply, or create conditions for cooperation on task-based work. In our definition, communities are more informal mechanisms involving individuals who come together to fulfill some common purpose, e.g. sharing information or coordinating work tasks. While networks of institutions may have some aspects of communities, and some communities may involve networks, not all communities are networks. Said another way, networks of institutions may have embedded within them communities of individuals, as well as networks of individuals.

The Appendix provides a more rigorous distinction of these three structures by way of theoretically-derived perspectives.

These types of structures can be used to meet learning needs such as the following:

- Access to knowledge (or dissemination)—individuals and organizations need to exchange information or to tap into wider sources of information. On a larger scale, this becomes a knowledge diffusion process.
- Innovation, or the creation of knowledge—individuals require incentives and access to resources in order to innovate. Innovation could involve new ideas, products or practices.
- Allocation of knowledge—this involves allocating resources and various kinds of output fairly and efficiently. An example of "allocation" is the provision of virtual education to students.[9]

Some examples of organizational structures

In the last section, we discussed three types of organizational structures that could be used to organize virtual learning environments. In this section, we examine how three types of virtual institutions could be organized using those structures. Table 2.1 illustrates examples of each situation. All three—markets, networks and communities—can be developed as either top-down planned designs or bottom-up self-organizing systems. As the table shows, the structures need not be competitive, but complementary.

TABLE 2.1.
Some Examples of Virtual Institutions and Organizational Structures

	Virtual Education Institution	Digital Library	Collaborative Research Group
Economic Market Structures	Fragmented educational markets, institutional and inter-industry competition	Pricing/subsidy mechanisms for allocating information to users (according to needs, ability to pay)	Allocating collaboratory time via markets or auctions
"Designed" Network Structures	Institutional networks or umbrella forms of virtual universities, e.g. Western Governor's Univ.	Coordinated digital libraries (sharing resources via common interfaces)	Collaborating research institutions
Self-organized Community Systems	Learning via "Social Web" communities (see Chapter 8 by Mark)	Community-based digital libraries	Scientific communities (may be similar to networks)

(1) Market Structures

As discussed in the second section, market-based organizational structures have been very popular ideas in recent years, being thought to improve efficiency and stimulate innovation. As with "real-world" markets, "virtual" markets can provide information or allocate resources through prices and other mechanisms.

We earlier covered various types of markets for teaching, including educational brokers, and inter-institutional competition amongst virtual and traditional universities. It has also been suggested that market mechanisms could allocate time on expensive research equipment in *collaboratories*. In this application, markets can be used to aid the discovery, reservation and brokering of resources (Johnston and Agarwal, 1997). As with any economic system, prices only go so far in allocating resources, and it is also important to consider a public goods perspective on redistribution, that is, providing users with access on the basis of scientific merit and need.

Markets can also assist digital libraries in pricing information and making them self-supporting. However, since knowledge gained in a single transaction will be easily reproducible and reusable, it is vital to reward the producer with some kind of protection for intellectual property. There are a number of possible ways to deal with this, but this requires further study. The field of public finance will have to find ways for dealing with virtual environments. Different tax schemes can be devised to compensate producers of knowledge by taxing different stages of processing, such as transactions, e.g. packages of knowledge, or information value added, e.g. searching and compiling. The issue of equity is also important here, that is, providing those who are less able to pay with access to information, either through subsidies or special access programs.

(2) Organizational networks

Networks are useful structures for institutions which wish to pool their limited resources so as to form a bigger organization, or to reduce the risk of experimenting with new technologies and practices. Their more formal arrangements and lower transaction costs (such as the costs of searching for and hooking up with partners) makes them useful for medium- to- long-term relationships. In a systems environment, where every piece of the system creates new value added in terms of functionality, networks of producers will also provide *economies of scope*, or complementarities. For example, a type of network in which each institution provides additional information processing functions will make the network much stronger as a whole.

Some examples of "designed" networks include the "umbrella" virtual universities, such as the Western Governors University, which offers courses produced by a variety of universities located in 16 western U.S. states. Another example is meta-digital libraries which might employ a common interface for data search and other functions, thereby ensuring seamless access to multiple, if not most, digital libraries. Networks such as those formed in scientific societies are meant for sharing knowledge in a less structured social setting, and indeed, in cases, they may even be considered as "communities". The concept of "weak ties" is useful for explaining the value of scientific communities. Weak ties describe the broader professional or superficial relationships that allow individuals such as researchers to cast a broad net in their search for information and other resources (Granovetter, 1973).

(3) Community systems

Community-based systems are perhaps the least well-understood form of organization. Generally-speaking, attributes of communities—such as roles based on mutual needs for information access and exchange, flexibility of rearranging the structure, and the public nature of knowledge—may fit well in the Internet environment, and in particular, may help in the distribution of knowledge. This was the basis behind the Apple Computer Corporation's Educational Object Economy (EOE)[10], which seeks to create communities of educational program developers and teachers who share *applets*, which are programs written in Sun's Java programming language to be run over the Web. In fact, the sheer size and diversity of applets in the EOE suggests that communities can be as productive as markets at stimulating innovation and diversity. The market watchers on the virtual education scene might wish to take lessons from this experience.

One can distinguish between two types of virtual communities—a virtual community that is based on a physical community, e.g. a town's community information system, and virtual communities of interest that are physically dispersed with no connecting physical location, e.g. a mailing list (Blanchard and Horan, 1998). One important issue is of how to replicate the feelings of a physical community within virtual communities. By assuming that social capital underlies social relations, we are suggesting that virtual environments which seek to enhance *physical presence-based* social capital will lead to more feelings of "community" in either the virtual *or* associated physical community (Blanchard and Horan, 1998). This may be done by stimulating community participation, and strengthening norms of reciprocity and trust. This suggests that research should be directed at making virtual environments more amenable to the exchange of information that increases the prevalence of these norms.

One important implication of this logic is that incorporating aspects of communities into virtual universities (e.g. virtual student unions and organizations), and digital library user groups, can enhance the "community" feelings within them. The question then is: how can physical community-based environments and the relationships involved in them be duplicated in virtual communities?

Another set of issues involves learning. If we take as a precept the idea that learning is inherently a social process, then the support of community characteristics would be essential to the growth of virtual learning environments. In this regard, some important questions will have to be answered, including, what types of learning occur in virtual communities, and how does this learning take place?

Design issues concerning interconnection

The interconnection of institutions and individuals within the broader social environment involves addressing several issues:

- Making virtual institutions meaningful—just as traditional universities are wrestling with the need to supply *practical relevance* in the education they provide, virtual universities and institutions must also consider their relevance. This could involve providing different types of learning, or better linking virtual institutions with workplaces and communities.
- Interfaces between institutions and individuals—interfaces are needed for students in one institution to access services in another. For example, it will be important for students in virtual universities to have easy access to digital libraries.
- Interfaces between institutions and society—"social interfaces" are needed to support the links between virtual institutions and the physical society at large. This considers whether various *societal* needs are met, including the needs of a variety of cultures, and the issue of how society may be affected by virtual organizations.
- Interoperability of different systems and infrastructures—interoperability will become important from a physical infrastructure point of view, e.g. providing an open, common architecture infrastructure for the interoperability of e-commerce and distance learning (Schutzer, 1996), as well as from institutions' functional point of view (i.e. determining which institutions should hold which functions).
- Scalability—we define scalability here to be a broad issue of how the infrastructure can support increasing numbers of users and their needs, greater volumes of information, and higher productivity.

All these are by no means exhaustive, as similar issues have also been raised in other contexts.[11]

Effectiveness of design: costs, benefits and system measures

Just as the design of a product needs to meet certain design parameters like the cost of production and speed, we must also consider what design parameters are important for a learning environment. The most important thing is to get educational institutions working effectively in order to meet the student's needs for different types of learning and knowledge. Effectiveness can be measured in a number of ways, including: (1) standard approaches such as benefit–cost analysis; (2) modifications of benefit–cost concepts to consider risks and other factors; and

(3) other indicators, e.g. indicators of the effectiveness of learning, and indicators at the systems level, e.g. speed and spread of information diffusion. These will be examined in the following sections.

The benefit–cost analysis of technology-enhanced virtual education

Benefits and costs are fundamental to understanding the value of virtual education and the likelihood of its success. The analyses have ranged from simple cost calculations to more complex benefit–cost frameworks. In examining benefits and costs, it is important to identify or at least keep in mind to *whom* the benefits and costs accrue to. For instance, some costs are borne by institutions, but many end up being passed onto learners. In this section, we will try to take a more general systems approach, which can be specified for target groups later.

A commonly held assumption is that the costs of virtual educational programs will be lower than that of a traditional educational program. Various approximations have been given. One study estimated the cost of a "virtual university" servicing 2,000 students to be $15 million, that being for computer equipment, personnel (including faculty) and the "physical campus" or location (Turoff, 1997). The cost of developing an individual virtual course at a traditional university has ranged widely, with one estimate being at about $6,000 per course (Helford, Fritz and DeVries, 1998)[12], while two other estimates peg the cost of an advanced system at between $100,000 to $250,000[13], and between $50,000 and $500,000 respectively.[14] At Berkeley, extension courses ranged from $10,000 to $20,000 after an initial infrastructure investment of $500,000.[15] These are obviously very rough guidelines, because the basis for calculating costs varies a great deal, and technological change makes it even harder to estimate costs for the future (we deal with some of these issues in the next section on extensions). Some factors that can cause variances in the cost basis are:

- size of the course (e.g. number of hours of teaching time)
- the degree with which technology has to be developed or bought "off-the-shelf"
- the degree of advanced technology and supporting course materials involved—some courses may be very basic, while others may involve very sophisticated materials (e.g. multimedia)
- whether scale economies or reduction of time for future learning (after the initial learning curves) can be taken advantage of, e.g. for repeating the initial course's layout and structure for other courses.
- whether hidden costs and subsidies such as infrastructure costs were included—the funding for the Internet and Internet2 originally came from the government, making the actual costs seen by end users and software providers lower, hence the costs of "distributing" information appear lower than in reality.
- whether or not institutional support for developing software is included, e.g. lower-wage student labor can be a form of "subsidy" to projects.

While the above discussion addresses only the cost side, benefit–cost analysis offers a broader framework for determining the value of virtual education. As noted by Cukier (1997), benefit–cost analyses of virtual and traditional forms of education have been studied extensively, and various methodologies have been developed using standard economic approaches, including mathematical models, comparative approaches, and return on investment models.

One standard classification for costs includes categories for fixed and variable costs, average costs (needed for program justification), and marginal costs (significant for program administration). Benefits traditionally include cost savings, opportunity costs and learning outcomes (Cukier, 1997). Examples of fixed costs are infrastructure costs, e.g. buildings, Internet connectivity, hardware, while examples of variable costs include human resource costs e.g. training and operating.[16]

Whereas in much of the literature, the benefits considered are really the cost-savings, Cukier describes a project which takes a broader perspective on cost–benefit analysis including "value added" benefits such as "value-driven" benefits (e.g. quality of teaching), and "indirect" benefits like community economic development. Some have also noted that it is "almost impossible" to measure the monetary worth of education and its benefits, making it an imprecise exercise (Moonen, 1997).

Technology also has the potential to increase organizational productivity. This can be reflected in either the cost or benefit (additional revenues) side. The belief in the productivity of IT in the education sector has been influenced by Massy and Zemsky's white paper in 1995 for Educom, which provided a framework for examining the costs of using IT in education, and which outlined a set of scenarios for the future of U.S. higher education institutions (Massy and Zemsky, 1995). They developed a model which suggested that academic productivity could be improved by IT in two ways: by increasing the economies of scale (i.e. a form of increasing returns), and by allowing mass customization of the educational experience to individual learners. However, the paper notes that barriers to change also existed in traditional institutions, and suggested that the structure of the U.S. educational market might change as a result of the use of IT and the differing adaptability of institutions to change.

Extensions to the benefit–cost framework

We will look at two types of extensions to the framework: benefits and costs to individuals and society, and secondly, to the institution.

The first extension involves the intangible benefits of social impacts or individual learning. The issue of individual learning and benefits was briefly addressed in the fourth section, with a particular focus on knowledge and learning in the Internet environment. The issue of social impacts includes various types of learning effects on other actors, or society at large. This may include benefits to society at large, such as the social stability deriving from a more well-informed and responsible public. These are not necessarily part of a benefit–cost analysis done from an educational institution's point of view.

At the same time, there are risks to students, such as the risk that students may learn incorrect or unimportant knowledge, because of a lack of standards in a diverse environment such as the Internet.[17] In part, the lack of attention to these may be due to the difficulty of measuring these outcomes. Although we do not directly address this second type of extension, we provide later a few related observations on general considerations at the systems level.

The second type of extension—that of calculating benefits and costs to institutions—can also be complicated. We will focus on just one aspect to illustrate the complexity: the aspect of understanding of the benefits and risks of choices in a dynamic economic environment, e.g. an educational market under rapid technological change. A typical benefit–cost analysis would ignore strategic considerations to an institution such as the following:

- Dynamic effects in technology adoption—some technological options will provide greater flexibility and expandability in the long-run. One example is the Internet's ability to provide access to more knowledge, which provides more possibilities for learning. This is more uncertain and difficult to quantify in monetary terms for a benefit–cost analysis, but instead, non-monetary measures can be examined, particularly in the case of comparative analyses between two or more technologies.
- Risks of loss or technological failure—risk can be considered as an avoided cost to the institution. The cost equation can be adjusted to account for the amount of *risk* that an institution is willing to take on, where the risk can be one of the different variants. One type of risk is that of choosing technologies that do not end up as the "market standard". To counter this, an institution might experiment in more than one technology in order to "spread" its risk and potential reward (of staying in "new" markets). Experimentation helps an organization move up along a *learning curve* for a technology, which also provides a future benefit. On the other hand, the power of IT and the Internet to reproduce information and virtual environments costlessly has major implications for institutional strategy, as it suggests that institutions may also consider cooperating with other *competing* institutions that are investing in technologies, to save on their resources.
- Sunk investments and the loss of future options—the opposite risk occurs when the institution has committed too much of its resources to a particular technology. Those committed resources are "sunk" investments, and cannot be used later to purchase other options which may have proven to have been better.
- Economies of scope—an institution's development of networks like connections to other institutions' services, or ensuring of system compatibility, will to ensure a greater market presence. This has been shown to be the case for Web search engines, in which the companies add to their popularity by offering

more services to customers, such as free email or information for specific communities.

- Transaction costs—traditional cost analyses often do not include many other types of costs, such as transaction costs, such as the cost of the time spent searching for information or partners, and in negotiating contracts. Amongst other things, transaction costs are important for understanding why institutions prefer to form networks rather than to operate in markets, since networks establish stable environments for repeated transactions between partners, thus reducing the need to search for new partners.

In general, the more rapid turnover of knowledge on the Internet creates greater uncertainty for both students and institutions. Thus, it may be appropriate to think in terms of reducing uncertainty where possible (i.e. diversifying a technology base), and maintaining a "no-regrets" policy or a minimum standard of safety, such as a policy to ensure that minimum levels of standard learning and experimental learning both occur.

In a virtual university, there will be interesting questions of how to factor in the cost of providing educational activities. A full-fledged virtual university based on traditional institutions, such as those typified by the Western Governor's University, seems well-positioned to add collaborative research functions and digital libraries (once they overcome the nascent stage) to their traditional functions. However, virtual research and library functions may currently be out of reach of start-up virtual universities, because of costs and other institutional barriers such as the lack of research resources. In this case, they will have to develop network strategies together with other institutions.

Measures of effectiveness at the systems level

The incorporation of individual learning, social learning and broader learning environments into the benefit–cost analysis complicates matters, and alternative indicators of effectiveness may have to be defined. Examples of broader learning environments are organizations, or larger systems that involve multiple institutions. In this case, the criteria for effectiveness in one institution may not necessarily be the same as for the whole system, i.e. what is cost effective for one university may not be so for a region or a group of universities. In these cases, we can use a systems approach to help untangle the issues at stake. To measure effectiveness at a systems level, a systems view of knowledge involving concepts such as the knowledge lifecycle or characteristics of learning may be used.

Some parameters that may be useful for characterizing "learning" effectiveness in larger systems such as organizations include:

- The rate of diffusion of knowledge—this is the rate at which knowledge is transferred from one part of the system to another. Whereas normal diffusion processes only consider a knowledge transfer model, in education, another important issue is that of how well the knowledge diffused *and* learnt in each place it diffuses to. We can compare the effectiveness in diffusion of different mechanisms, e.g. communities versus markets.

- The degree of innovation—the issue of innovation is a complicated issue, as witnessed in the large literature on technological innovation. The ability of an organization to innovate depends on many factors, including external resources, internal ability, and incentives. One measure of innovation, such as number of successful ideas (where success is based on some specified degree of diffusion). This is similar to the patents measure used in the economics of innovation.
- The ability of organizations to "unlearn" old knowledge—while we have focused largely on learning, the ability of an organization to unlearn bad or outmoded practices can also be important. This is related to the organization's ability to capitalize on innovations (i.e. diffuse ideas successfully throughout the organization).

As with any system, there are a number of potential problems. One such problem is the issue of where to draw boundaries around the system or institutions. This is critical because some systems can be characterized by more than one boundary, and this could cause the values of measures to be very different. An example is a "virtual university" that has both teaching and library functions. Do we measure learning effectiveness within the classroom and library separately, or do we measure it across the learning environment as a whole?

Systems for global access

It is now widely recognized by international development agencies that knowledge is a critical resource and commodity of development (World Bank, 1998; Mansell and Wehn, 1998). All the educational activities discussed here may need to be accessed by users across the globe. However, the appropriateness of organizational structures need to be examined from the viewpoint of the very diverse needs of the global population. In particular, this includes considering issues of access, including financial affordability and technological infrastructure, and differences in language and culture. In a virtual environment, each type of organizational structure can be made global in scope, but the best means for doing so remains less clear. The design of global learning environments to improves the opportunities for learning on a worldwide basis is nothing less than essential.

Conclusions

The knowledge-based economy demands new skills from employees, and new virtual learning environments from educational institutions. In general, we have observed that technology is changing the economics of virtual educational activities by way of lowering the cost of reproducing knowledge and virtual environments. The Internet is also causing convergence in virtual institutions and the functions they provide. As a result, different types of institutions can offer the same functions, and educational institutions may start to duplicate settings such as workplaces, if not link directly to them. At the same time, more institutions can afford to offer learning opportunities, thanks to the increased functionality and low costs facilitated by IT.

We have introduced some concepts to help understand the emergence of virtual learning environments and virtual universities, and to consider in their design. The foremost considerations in both predicting emergence and doing design is the type of learning and knowledge provided, and their costs and other benefits. Learning efficacy can only be determined through further study of the effectiveness of various technologies and practices, both at the classroom level as well as at the community or institutional level.

At the institutional level, while costs and benefits are typically used to measure the economic worth of a learning approach, there are much broader issues at stake. These include how learning opportunities can be provided within the broader environment, and the need for institutions to take into account the dynamic characteristics of technological change, their own learning capabilities, and market dynamics.

Despite the predominance of markets in the discussion of virtual education in the U.S., there are numerous alternative ways in which government, institutions and communities can organize resources to organize information resources and "design" learning environments so as to provide educational opportunities.

Appendix

Table 2.2 illustrates the distinguishing characteristics of the three types of organizational structures that we consider.[18] The table consists of several broader categories that help distinguish among the structures: *purposes*, or the reasons for using a particular structure; *incentives*, which illustrate the individuals' motivations; *relations*, which illustrate how the structure operates*;* *output characteristics* (especially for knowledge), and *structure* of the connections.

While networks can be composed of either individuals or organizations, our definition here is based more on the type of *cooperative network of organizations* found in the organizational network literature. On the other hand, our concept of a community is based on individuals (although community "organizations" or self-organized collectives can also exist). Our definition of community is informed by an assorted, but not necessarily consistent, literature. A large part of it is based on the idea of *social capital*, which describes the formation and maintenance of social relationships.[19] Our use of the terms *social capital* and *communities* here is focused more on groups in which members acquire knowledge for self-improvement or mutually-led interests.

Structures like *markets* and *organizational networks* are used to make transactions and to allocate resources. On the other hand, *communities* are means of meeting the needs of a *pool* of people, e.g. people who share similar interests, specific interests or a broader range of interests (including, in the case of physical communities, needs for living and mental well-being). For the individual members, networks and communities also share the same purposes of meeting members' needs for access to widespread information, exchanging information and mutual learning. All three structures could be used to foster innovation, such as the development of new learning technologies or practices. In the "design" of the broader societal learning environment, it will be useful to identify how they can play complementary roles.

In our "model", all three structures can be driven by economic or other rational incentives. The economic drivers in networks are transaction costs, i.e. costs of search, evaluation and so on, rather than prices. Mutual interest also drives community formation and sustenance, but these may be due less to costs, and more to a variety of (still rationally-based) interests that are stronger when shared (e.g. unique interests, intellectual satisfaction, etc.).

The nature of relationships can be important for achieving certain learning objectives. For example, cooperative longer-term relationships (such as those found in networks and communities) may be necessary for developing joint projects based on the participants' pooling of resources.

TABLE 2.2.
Characteristics of Three Types of Organizational Structures

Main Category	Characteristic	Market (neoclassical)	Network	Community
-	Unit of analysis	Organization (firm)	Organization (firm)	Individual
Purpose	Overall purpose	Resource allocation (providing forum for transactions	Fulfills long-term needs; advances cooperative's interests; allows pooling of resources	Sustains a cooperative's (specific or wide) range of interests; allows the pooling of resources
	Information purpose (to individuals)	Price signals overcome individuals' limits on gathering information	Wide-spread information access (through weak links)	Wide-spread access to (usually) situation-specific information
Incentives	Economic or rational incentives	Price mechanism (costs, revenues)	Minimizing transaction costs	Maintaining social capital
	Role and responsibility resolution	Price-led, market norms	Mutual needs, repeated contracts, negotiated outcomes	Mutual needs, social norms

Main Category	Characteristic	Market (neoclassical)	Network	Community
	Incentive level	High	Medium to high	Low to high
Relations	Type of relationship	Competitive	Cooperative	Cooperative
	Length and nature of relationships	Short-term (spot contracts)	Medium to long-term relationship	Medium to long-term relationships
	Level of trust	Low	Medium to high	High
Output Characteristics	Type of information/Knowledge exchanged	Knowledge is commoditized (both structured or unstructured)	Knowledge structured by situations	Knowledge structured or unstructured by situations
	Output (product) "variation"	Low to high	More customized	Highly customized by "spot" relationships
	Nature of knowledge[20]	Private (public aspects held outside of market)	Public (freely available to the network)	Public (within the community or select networks)
Structure	Boundary, topology	Flexibly re-arranged and lack of structure	Somewhat less flexible structure	Highly flexible networks (free-floating but circumscribed)
	Dynamics (of internal structure)	Dynamically-changing structure	Mutual relationships, stable over time	Mutual relationships, may change over time
	Robustness (i.e. stability when members leave)	Low (due to low entry cost)	Low to high (depends on player's importance)	Low to high (depends on player's importance)

The characteristics of knowledge "outputs" may also be a consideration in the design of a learning environment. In markets, we may get customized or uncustomized knowledge (depending on what is paid for) with a wide range of product variation. An example of the provision of customized knowledge (structured according to the situation) is given by law and consulting firms, which tends to disburse customized advice to clients. Knowledge disbursed in networks and communities tends to be more customized on average. Knowledge in networks also tends to be "structured", i.e. catered to the particular situation of the member requesting it, whereas in communities, both structured and unstructured advice can be given, where the latter may be one member's simple broadcast announcement to the whole community.

In terms of their boundaries and flexibility of rearrangement (of internal structure), communities can consist of either a densely connected network (as in a social network), or free-floating networks (as in groups or cliques within a pool of people). Markets are even more flexible because of their "one-time" relationships. Networks represent the least flexible situation, because of the longer-term contracts that may have to be signed.

Notes

1 By *complex*, we mean the distributed "phenomena" occurring across the Internet (e.g. websites and information-related processes occurring all over), and the aggregation of those into complicated patterns which could not have been determined from examination of the phenomena separately.
2 The URL (or Universal Resource Locator) is a link within a website. Thus, websites may have more than one URL. While the number of websites is a useful indicator from the standpoint of counting people and institutions who create a "complete" site, URLs are more commonly used as indices by search engine companies.
3 In economics, knowledge is usually considered to be either a private good (i.e. privately held) or a public good which is freely available to all. Economists also distinguish between knowledge and conventional economic goods by saying that knowledge is "non-rival". This means that knowledge can be infinitely reproduced (without increases in per unit cost or decay in quality), and as many people can make use of that knowledge as necessary, including the seller (this is related to the increasing returns to scale assumption).
4 *Business 2.0*, July 1999.
5 At the time of writing, the CVU was suffering a lack of commitments to provide start-up funding from the universities themselves. The WGU only had a hundred-odd students signed up by late 1999.
6 *Financial Times*, 18 June 1998.
7 This is assuming that the lecture quality between an online on-demand lecture and a 500-seat class in a traditional campus is approximately the same.
8 In the literature, the discussion of these organizational structures is often loose and disconnected from each other.
9 We should note that due to the economies of scale in virtual products, it is quite possible that certain situations could end in a "winner takes all" result. While we do not have the space here to illuminate this further, suffice to say that this is not necessarily the case as long as virtual learning is constrained by the lack of real-world characteristics (e.g. socialization), or the need to include costly real-world features, like online tutors.
10 This is technically more a community than an economy, since exchange of information is done on a mutual basis rather than being price-led as in a market. However, in the EOE, other typical characteristics of communities, such as the ability of individuals to recognize each other, are not necessarily satisfied. It is also notable that many such "communities" exist in software programming environments, where programmers worldwide share "objects" (i.e. programs written in object-oriented code), to help each another save time and effort.
11 One report detailing similar issues suggested that the evolution of digital libraries needs to consider systems issues like: interoperability and scalability (of various kinds), adaptability and durability, and support for collaboration; collection-centered issues such as archiving needs and intellectual property rights; and user-centered issues like designing for a variety of users, and providing multicultural interfaces (National Science Foundation, 1997).
12 This is probably a very basic course. It includes 60 to 100 hours of teacher effort, 80 hours technological support staff time and 80 hours of student labor.
13 Rough estimate by Phil Miller of Carnegie Mellon University, based on the development of an Oracle database platform that could be "reused" for multiple courses, and which focused especially on student assessment. This estimate included faculty time and programming assistance.
14 Rough estimate given by Green (1998) in the electronic newsletter *On the Horizon*, University of N. Carolina at Chapel Hill.
15 *Business 2.0*, July 1999.
16 One could argue that the fixed and variable costs distinction may be meaningless in some situations, for instance, while computers may look like fixed costs, if they need to be continually replaced because of rapid technological advances, in a compressed time scale, they could actually be considered to be variable costs.

17 Setting student assessment standards may help alleviate this, as will the use of technology such as filtering software to ensure that student searching is more targeted to their needs.

18 We adapted several of the categories in the table, and their information for markets and networks, from a review of the literature by Van Alstyne (1997). The characteristics of communities and knowledge are drawn from wider literatures. While we examine three such structures: economic markets, networks and communities, others may exist. Normally, there are multiple types of organizations, and the distinction is commonly made between "networks", "markets" and "hierarchies", the latter being one type of organization (see Van Alstyne [1997]). However, we will not consider hierarchies, since they are less suited for describing virtual environments, given the latter's flatter organizational structures. The concept of a community, loose as it may be, also appears to offer an all-embracing alternative suitable for describing open learning systems in non-traditional environments.

19 See for example, Putnam (1995), as well as the literature on computer-mediated communication and on virtual communities (see ACM article). Putnam used the concept of *social capital* to capture the social relations between citizens, social participation and government (or at least the rationalistic aspects of the relations), which is how communities sustain themselves. Social capital includes the "features of social organization such as networks, norms and social trust that facilitate coordination and cooperation for mutual benefit" (Putnam, 1995). Putnam's well-cited work stated that the decline of neighborhood bowling leagues in America suggested a decline in societal social capital. One typical counter-argument suggests that the rise of Internet communities represents an opposite effect.

20 This is based on the economic notion of knowledge being private or public goods. Recently however, Arrow (1994) suggested another notion that within a network, knowledge can have both private and public characteristics.

References

Arrow K. J. (1994) Methodological Individualism and Social Knowledge, Richard T. Ely Lecture. *AEA Papers and Proceedings*, Vol. **84**(2), May 1994.

Blanchard, A. and Horan, T. (1998) "Virtual communities and social capital," *Social Science Computer Review*, **16**(3), 293–307, Fall 1998.

Brown J. S. and Duguid P. (1995) Universities in the Digital Age. *Change*. http://www.parc.xerox.com/ops/members/brown/papers/university.html.

Brynjolfsson E. and Hitt L. (1998) Beyond the Productivity Paradox. *Communications of the ACM*, August 1998.

Carter A. P. (1996) Measuring the Performance of a Knowledge-based Economy. In Foray D. and Lundvall B. (eds.) *Employment and Growth in the Knowledge-based Economy*, OECD, Paris.

Cukier J. (1997) Cost–benefit Analysis of Telelearning: Developing a Methodology Framework. *Distance Education,* Vol. **18**(1): 137–152.

Dewan S. and Kraemer K. L. (1998) International Dimensions of the Productivity Paradox. *Communications of the ACM*, August 1998.

Evans P. E. and Wurster T. S. (1997) Strategy and the New Economics of Information. *Harvard Business Review*, Sept–Oct 1997: 71–82.

Foray D. and Lundvall B. (1996) The Knowledge-based Economy: From the Economics of Knowledge to the Learning Economy. In Foray, D. and Lundvall B. (eds.): *Employment and Growth in the Knowledge-based Economy*, OECD, Paris.

Granovetter M. (1973) The Strength of Weak Ties. *American Journal of Sociology*, **78**:1360–1380.

Graves W. H. (1997) "Free Trade" in Higher Education: The Meta University. *Journal of Asynchronous Learning Networks*, Vol. **1**(1), March 1997. http://www.aln.org/index.htm.

Green K. C. (1998) Money, Technology, and Distance Education. *On the Horizon*, 5(6). http://horizon.unc.edu/horizon/online/html/5/6/.

Hamalainen M., Whinston A. B. and Vishik S. (1996) Electronic Markets for Learning: Education Brokerages on the Internet. *Communications of the ACM*, June 1996, Vol. **39**(6): 51–58.

Hansen M. T., Nohria N. and Tierney T. (1999) What's Your Strategy for Managing Knowledge? *Harvard Business Review*, **77**(2): 106–116, Mar/Apr 1999.

Helford P., Fritz M. and DeVries J. (1998) Beyond Courses to Curriculum: Building Web-based Programs. Conference on *Universities in a Digital Era: Transformation, Innovation and Tradition*, European Distance Education Network (EDEN), Bologna, 1998.

Johnston W.E. and Agarwal D. (1997) Issues in Architectures and Technologies for On-line Global-Scale Scientific Environments. White Paper for the *America in the Age of Information* Forum, E. O. Lawrence Berkeley National Laboratory. http://www-itg.lbl.gov/~johnston/CIC.Paper.1.fm.html.

Mansell R. and Wehn U. (1998) *Knowledge Societies: Information Technology for Sustainable Development*. Oxford University Press, Oxford. For the United Nations Commission on Science and Technology.

Marchese T. J. (1997) The New Conversations About Learning: Insights from Neuroscience and Anthropology, Cognitive Science and Workplace Studies. In *Assessing Impact: Evidence and Action*, American Association for Higher Education, Washington, DC. Also at: http://www.aahe.org/pubs/TM-essay.htm.

Massy W. F. and Zemsky R. (1995) Using Information Technology to Enhance Academic Productivity. White Paper, *Educom Conference on Enhancing Academic Productivity*, June 1997.

Moonen J. (1997) The Efficiency of Telelearning. *Journal of Asynchronous Learning Networks*, **1**(2), August 1997. At: http://www.aln.org/index.htm.

Noam E. (1995) Electronics and the Dim Future of the University. *Science,* October 13, 1995. **270**: 247–249. http://www.vii.org/papers/citinoa3.htm.

Noble D. (1998) Technology in Education: the Fight for the Future. *Educom Review*, **33**(3), May/Jun 1998.

Nonaka I. and H. Takeuchi (1995) *The Knowledge-creating Company: How Japanese Companies Create the Dynamics of Innovation.* Oxford University Press, New York.

Polanyi, M. (1948) *Personal Knowledge: Towards a Post-Critical Philosophy*, Chicago: University of Chicago Press.

Putnam R. D. 1995 Bowling Alone: America's Declining Social Capital. *Journal of Democracy*, **6**: 65–78.

Schutzer D. (1996) A Need for a Common Infrastructure. *D-Lib Magazine.* April 1996. http://www.dlib.org/dlib/april96/04schutzer.html.

Tsoukas H. (1996) The Firm as a Distributed Knowledge System: A Constructionist Approach. *Strategic Management Journal*, **17**(S2): 11–25.

Turoff M. (1997) Costs for the Development of a Virtual University. *Journal of Asynchronous Learning Networks*, **1**(1), March 1997.

Twigg C. A. and Oblinger D. G. (1996). *The Virtual University.* Report from the Joint Educom/IBM Roundtable. Educom, Nov. 1996.

Van Alstyne M. (1997) The State of Network Organizations: A Survey in Three Frameworks. *Journal of Organizational Computing*, **7**(3).

World Bank (1998) *World Development Report 1998/99. Knowledge for Development.* Oxford University Press, New York. Published for The World Bank.

3

Learning Environments: The Virtual University and Beyond

MICHAEL L. DILLINGER[1]

It is impossible to cross a chasm with a thousand tiny steps.

Chinese proverb

On the eve of the 21st century, large segments of society are rushing headlong to leap across the chasm between yesterday's capitalism and tomorrow's post-capitalist Age of Information (Drucker, 1993), toward a Culture of Knowledge (Duderstadt, 1997). The ponderous University trails behind the crowd, lumbering to the chasm between today's classrooms and tomorrow's learning environments, to consider its options.

In the course of this transition, the learning industry's importance is skyrocketing, and will clearly come to assume a pivotal role in all walks of life and all activities of tomorrow's society. For this industry in particular, the ability to cross the chasm is defined in terms of the kinds of learning environments today's institutions can offer tomorrow's society.

The pressures that these trends have brought to bear on the University are clear: there is a growing market of non-traditional students, a need for providing students with more flexible and more relevant skills, increasing budgetary difficulties and dwindling subsidies, increasing dissatisfaction with the out-of-date information offered in some programs, and a growing number of more adaptive competing service providers.

It has become just as clear, however, that these external pressures have not been sufficient to stimulate substantial change. The traditional classroom learning environment has attained the status of a rationalized myth: on the basis of long-standing and deep-seated consensual beliefs, it is regarded as the single best and necessary means for promoting learning (Jaffee, 1998, p. 26), even in the absence of systematic empirical verification. One option that the University has, therefore, is to preserve this valued institution and as many of its characteristics as possible in this moment of transition. The classroom institution has centralized power and influence in the hands of the instructor; it is natural, then, that instructors, the main agents of transition, will tend to favor this option (Jaffee, 1998).

Another option is to change. Developed institutions change significantly, according to Toffler (1985), only when three conditions are met: first, there must be enormous external pressures; these have just been observed. Second, there must be people inside the institution who are strongly dissatisfied with the existing order. And third, there must be a coherent alternative embodied in a plan, a model or a vision.

Unfortunately, the second condition is clearly not met: "Despite examples of increased experimentation (...), however, most universities have not challenged traditional assumptions and approaches with respect to learning, students and processes" (Hanna, 1998, p. 75). There is no vociferous group of young Turks within the university clamoring for radical change; in general they have been excluded, leaving a conservative teaching staff reluctant to change. In a detailed and informative review of seven different emerging organizational models for the university in the digital age, Hanna (1998) provides the data, but does not call attention to the fact that *all* of them continue to use the lecture format (with greater or fewer technological enhancements) as their primary "learning" technology. And Jaffee (1998) argues that changes to traditional teaching practices will be opposed by the vast majority of faculty, because they define their professional identity in terms of their role in the traditional classroom setting. Based on a model of diffusion of innovation, the part of this group that is only somewhat receptive to change would make up some 70% of university faculty, and the strongly opposed, another 15% (Geoghegan, 1994 cited in Jaffee, 1998).

One kind of change is to attempt to *reform* the classrooom environment and enhance it with information technology, adapting it to the new means increasingly available. Most discussions of the nature of the Virtual University seem to be carried out under the assumptions of this option, perhaps in the search for a politically and socially viable path for initiating at least some innovation. This view is associated with faculty who are comfortable with the new technologies as well as already somewhat dissatisfied with the disengaged, alienated student behavior that the classroom seems to foster. For Geoghegan (cited in Jaffee, 1998), this group of early adopters and visionaries would make up only 10 to 12% of the faculty.

But will the Virtual University be able to carry us over the chasm by trying to put the University into the computer and onto the Web, with digital lectures, multiple-choice examinations and fixed curricula? Certainly not, although an alarmingly large number of discussions seem to suggest that the technology itself will be sufficient. We will very clearly need information technology to make the difficult transition from one scenario to the other, and lots of it (see e.g., Bourne, 1998 on asynchronous learning networks and Fletcher, 1995, 1996 on computer-based instruction). But technology is merely a tool for multiplying the effects of our actions. The Internet is just as effective for promoting knowledge and understanding as it is for increasing ignorance, prejudice and bigotry, as good for peace as for terrorism, as good for democracy as for dictatorship. In order to understand the Virtual University, then, we have to turn away from the technology and focus on what is to be done with it (see Brown & Duguid, 1995; Perelman, 1997; Wurman, 1997).

This takes us to Toffler's third condition for change, here rephrased by Hanna (1998, p. 93):

> "Leaders of all institutions and programs, to be effective in this era of digital competition, need a strong rationale and framework for organizational change. This rationale will provide a foundation for organizational adjustments and even transformations necessary to respond to the opportunities and risks presented by increasing world-wide demand for learning, advancing learning technologies, and growing competition among multiple providers, all seeking to gain competitive advantage."

In essence, the crux of the problem is redefining the mission of the university for the next millenium (Graves, 1994a). Note that here Hanna uses "need," not "have." A coherent vision of the future is missing, although elements of it are easy to find. Sherron & Boettcher (1997) provide an excellent overview of practically all the main issues and options that have arisen. Katz &West (1992) and Hanna (1998) provide useful perspectives on administrative issues (best read together with Jaffee (1998) and Brown & Duguid (1995)), the National Learning Infrastructure Initiative has offered many interesting contributions[2], Jaffee (1998) contributes valuable sociological perspectives and Duderstadt (1997) and Brown & Duguid (1995) give thoughtful and insightful discussions of directions and goals for the Virtual University, just to offer a few examples. Options abound, but a coherent vision has been elusive.

Another kind of change for the University is to use information technology and the first-rate knowledge workers in its ranks to leverage a metamorphosis into the de facto leader of a new learning-centered society (Duderstadt, 1997), rather than a marginalized vestige of days past. Perelman (1992, 1997), however, makes a strong case that today's universities are singularly unsuited to such a leadership scenario and that aggressive, adaptive, accountable commercial institutions will hasten what he sees as the already iminent demise of traditional schooling. The main problem? Entrenched traditions and mindsets that make adaptation and change nearly impossible within a socially relevant timeframe (as analyzed in Jaffee, 1998). The focus of change? The nature of the learning environment (see also Talbott, 1999).

Conceptions of learning environments

What underlies all these options are the ability of today's university faculty to redefine the way they view their core professional identity in terms of different learning environments and the ability of today's university leadership to articulate and promote a coherent view of the university's role in tomorrow's society. The central issue of the Virtual University, then, is one of change of vision, in particular a vision of learning environments that are adequate to the needs of tomorrow's demands.

This chapter considers three broad conceptions of the kinds of learning environments the University can aspire to design, implement and foster (see Denning, 1999 and Brown & Duguid, 1995 for parallel analyses). The first conception focuses on the needs and characteristics of the institution that hosts learning, in particular its resources and its researchers. On this view, better learning environments are those that have better resources and better researchers, taking the term "environment" rather literally. The section below on the information-dissemination economy focuses on the institution as the learning environment par excellence.

The second kind of learning environment focuses on the needs and characteristics of individual learners, and, indirectly, their future employers. On this view, better learning environments are more individualized, promote a more active role for the learner and focus on skills that will be palpably relevant in the learner's professional activity outside the university. The section below, about philosophical and psychological perspectives on learning, focuses discussion on the individual's mind as *the* learning environment.

The third conception decouples the notion of learning environment from that of a physical place and puts it in the realm of collaborative group action. On this view, better learning environments are social groups that are more focused on more specific goals or products and that foster more synergy and greater mutual empowerment in varying real-life situations. The section below on communities for self-empowerment redirects the discussion to focus on integration of individuals' cognition in collective activity and on innovation as a driving force for learning.

Implementing any of these conceptions on a significant scale at the Virtual University will tax all the resources of today's institutions to the breaking point, and vast sums are already being spent (Hanna, 1998). Just as an example, Sherron & Boettcher (1997) estimate the cost for implementing a single, Web-based, stand-alone, three-credit course to be between $500,000 and $1 million dollars, which doesn't include resources dedicated to planning new curricula, evaluation schemes, administrative changes, marketing, etc. With investments being made on such a scale in the context of ill-articulated goals, it is no surprise, then, that members of the business community are either predicting the imminent demise of the university (Perelman, 1992) or, in very many cases, founding their own, alternative, institutions (with an increase of 400% over the last 10 years, as cited in Talbott, 1999).

Focus on the Institution: The Information-Dissemination Economy

Today's society revolves around the flow of information just as much as it does around the flow of money: major institutions have come about that are dedicated to producing, packaging and delivering information.

The key concern of the information-dissemination economy is delivery of information and the basic goal is to keep information, whatever its characteristics, flowing in society, from producers to consumers: information is unilaterally disseminated or broadcast. Educational institutions, the media, publishing, and computer industries, along with research centers, the human resources development industry and the arts, can all be seen as major players in the information-dissemination economy.

In this economy, information is viewed much as any other commodity: something to be produced, packaged, distributed and consumed. It is "facts"—news, weather, geography, politics, dates, times, statistics, definitions, indicators, classifieds, etc.—small, well-delimited informational units that are independent of most other cultural context. Because control over information is so important in this context, privacy, secrecy and intellectual property rights (different facets of ownership of information) are central concepts for the institutions involved, along with market share.

Information also has value and in this kind of economy the value attributed to facts is determined to the greatest extent by their producers, packagers or distributors, either explicitly by disseminating value judgments as facts, implicitly through frequency of dissemination, or by other means. Relevance for the consumer, reliability and quality of content (by whatever measure) are not systematic indicators of the value attributed to information; timeliness, appeal (or entertainment value) and popularity seem to be better predictors. The "need to know" is defined externally to the consumer in this economy. Thus, the relation between the content or value of the facts disseminated and the consumer's individual needs or purposes is often difficult to ascertain. In other words, few of the facts transmitted in the context of the news or the classroom can be seen as affecting directly the course of a given individual's day, month or year. Hence, learners' insightful favorite question: "why do I have to study this?"

This is the framework that underlies most of the information economy, although we are already starting to show signs of progress beyond this model. The benefits of a vast information distribution network, however, also raise the questions of the quality and relevance of the information that it is used for, and these questions are even more relevant for the university than for the media.

Producing information

Information is produced on an economically relevant scale by a small segment of society: scientists, researchers, artists, government and the media. The distinction between producers and consumers of information is asymmetrically defined:

everyone consumes information, but "producers" are a special kind of consumer: they consume information in order to produce even more information, whereas "consumers" only consume it, often with unknown objectives or benefits. Distinct kinds of information are produced for other producers ("technical" information) and for non-producers ("non-technical" information). For those who produce information (researchers, journalists, etc.), dissemination to other producers is a high-prestige activity, whereas dissemination to non-producers is lower-prestige. Among researchers, this is known as the difference between "publishing" and "popularization". This state of affairs both reflects and reinforces the clear separation of functions in society between producers and consumers of information.

Packaging information

Given the vast amounts of information already available and new information being produced each day, industries for packaging it have become an important part of the economy: portal sites on the Web, wire services and publishers are among the most important; editors, authors, professors, curriculum developers and film producers have this function at an individual level. The packaging sector takes indeterminate quantities of raw (or partially processed) information and packages them into units of a variety of sizes and degrees of complexity: news items, Web pages, instructional units, articles, books, commercial and educational TV programs, radio "spots", etc., so that information can be distributed in convenient units.

The central question about the activities of the packaging sector reduces to how they choose what information to include and what information to exclude for the mass market. The two most important strategies for information choice are clearly those of maximizing degree of consensus and minimizing consumer requirements; this can be called "market-driven" information choice. Consensual information is favored over controversial information, so as not to alienate potential consumers. One-sided or biased information is also selected for different market segments, for example, textbooks about the same topic presented from the perspectives of rival approaches are based on very different bibliographies and present different (usually complementary) information. Simpler information is preferred over more complex information in order to minimize the cultural and intellectual demands on the consumer's part and maximize comprehension of content. Again, part of this strategy is systematically presenting topics from only a single point of view, without treatment of alternative views. Although they have been termed "dumbing down" by critics, these have proven to be very effective strategies for increasing market share.

Delivery of information

With increased industrialization, unbundling of production, packaging and delivery of information has increased, giving rise to an independent delivery sector and information workers with more specialized functions. Delivery of information is a widely discussed issue, especially as information-technology infrastructure improves. In information-dissemination economies, discussions of delivery technologies (bandwidth, storage, infrastructure, etc.) are much more prevalent than discussions of how these means are best used.[3] In its narrowest interpretation (in fact quite a common one), the Virtual University is simply an alternative delivery system for traditional academic content.

Barriers to information delivery are currently a major focus of discussion, because they drive developments in infrastructure and the key concern of the whole economy is assuring delivery of information. The time barrier shows up as speed of access and as time-of-day at which information is available (Bourne, et al., 1997; Bourne, 1998). Interestingly enough, another notion of time—the time needed to deliver or find necessary content ("time-to-fact")—has not been an important topic of discussion, although it poses significant problems to the consumer. Geographic distance is often discussed, but "distance" or degree of similarity between content areas or specific facts has not been an important topic of discussion, although it presents significant difficulties in information retrieval. The need to "do more with less" is often present in discussions, but the cost to the consumer of dealing with ever greater amounts of information is not addressed.

Consuming information

In information-dissemination economies, the focus of major institutions is on dissemination of information: its production and especially its packaging and delivery. There are no major institutions of information consumers and only a few experiments in which consumers contribute directly to information programming, such as community TV.

Consumer functions are simple: consumers pay to access information, they are expected to recognize it (hence the term "recognition value") and sometimes to store variable amounts of it in memory. They are not required or stimulated to use, manipulate or react to most kinds of content in any way. The main exceptions to this are commercial content (advertising) to which they are supposed to react to by spending money and educational content which they are supposed to react to by reproducing it on demand (see Postman, 1985 for more analysis).

Finally, consumers are seen as rarely, if ever, needing the skills to participate in the production, packaging or delivery of information. This is part-and-parcel of the rationale behind mass education: to make education more efficient and available to a greater segment of the population, teach only what the students will effectively need to continue as (passive) *consumers*. This includes basic, introductory concepts of a variety of subject areas, basic literacy and mathematical skills—what Resnick & Resnick (1977) call the "low-literacy" tradition of education.

The Information-Dissemination University

In keeping with the dominant focus on the institutions of production, packaging and delivery of information in the economy at large, rather than the social setting or consumer benefit, the mission of the university focuses on "creating, preserving, integrating, transmitting and applying knowledge" (Duderstadt, 1997, p. 79). These activities in an educational context are all activities done by professors, hence the system is often and accurately called "teacher centered". Concepts of quality of education in most universities focus on "measuring *inputs* to the instructional process, such as (...) curricula, faculty experience, student quality [upon admission]. The concept behind this approach is that, taken together, these inputs (...) *in*directly measure *anticipated* student learning." (Hanna, 1998, p. 69, emphasis added). That is, quality of education is judged by production, packaging and delivery rather than by consumer benefit or usefulness in a wider social setting.

The question that this observation raises, of course, is why there are so few efforts to *directly* measure the *actual* benefits of university education in order to define quality. There is no question that the technology to do so can be developed; on the other hand, there is no political entity to defend consumers' interests and in fact carry this out. As commercial educational institutions increasingly demonstrate the direct measures of actual benefits (with respect to skills that learners and the workplace deem relevant), the university's indirect and anticipated benefits will wane in the public eye.

Production

In educational institutions, information production is a much higher-prestige activity than delivering it, as reflected in the strong negative correlation between salary and teaching hours among university professors (Brandt, 1997). In other words, consumers or learners are of secondary importance. Production of information, "research", is *the* prestige activity in the university, particularly if it is to be disseminated to other information producers, in particular government and industry, who reciprocate by underwriting further research. As an environment for the production of information, the university is particularly interesting in that it imposes very few constraints on the content of the information produced, particularly when outside funding is obtained. These researchers generally have much more freedom to pursue intuitions and specific topics than those in government or industry. The independence of the processes of producing and delivering information is emphasized here, as elsewhere in the economy.

Packaging

Packaging information in the university setting is uniformly carried out by professors, who choose what the consumer/learner "needs" to study and when.

The participation of consumers in packaging information is limited to choosing the ordering of information packages (as constrained by professor-imposed prerequisites) and choosing elective courses which are considered irrelevant to the course of study.

In university contexts, packaging of information is still a craft or "cottage industry": there is no specific training for it, it is usually carried out individually and with little systematic feedback, there is little re-use of others' previously acquired expertise, there are no standards for evaluation and it is carried out on a very small scale, usually a single course at a time. This emphasizes the fragmentation of knowledge and makes it much more difficult for the learner to see coherence in the curriculum. With the growth of the textbook industry, packaging has become more professionalized, and packaging functions of professors have diminished, since for most topics existing textbooks are adequate and may only need to be supplemented with multimedia material and/or more up-to-date additional reading.

Delivery

Information is delivered on-campus in print and orally, with small but systematic increases in non-linguistic media and alternative delivery media. It is also delivered off-campus (in lower information-density packages) by all the media available, from postal service to interactive Web-based applications. The focus on "course delivery" might lead to calling this the "FedEx model" of education: the teacher's responsibility for learning practically ends when the material is laid at the consumer's doorstep. The institution's role is to provide information; learning something useful from it is up to the consumer. This is seen in concepts like "...the fundamental essence of learning [is] to have *access* to experience and knowledge..." from Microsoft's Online Institute (Watson, 1996, emphasis added).

Professors as agents of information delivery are in transition: divested, in many cases, of the functions of producing and packaging the content they deliver, they are losing prestige. The unbundling of packaging and delivery is illustrated by the professors of previous generations who used the texts they themselves wrote, as compared to the newer generation that uses textbooks prepared by others. It is a curiosity of the university model that professors in most cases have no training in information delivery: they are content experts, not experts in delivery, although their functions require expertise in delivery as well.

Consumers

In keeping with the focus of information-dissemination economies on the activities of non-consumers, educators have developed an extensive body of information about professor activities and little accumulated information about how students learn. This is exemplified by schoolteachers' perennial interest in and possession of relatively large amounts of information about teaching techniques (teacher actions in the educational setting), politics of teaching (policies, salary issues and parent reactions) and disciplinary problems of students (student actions in the educational setting). Correspondingly, there is little information in the community about how instructors plan delivery or adapt instructional goals to variable groups of students, nor about the cognitive processes of student learning. Since university professors have no specific training in pedagogy and are not evaluated systematically for teaching effectiveness, they have little incentive or opportunity to revise and update their teaching skills.

This teacher focus is also reflected in the locus of control of relevant decisions in the classroom: deciding what is to be taught, when is it to be taught, how it is to be taught and how it is to be evaluated are all teacher activities, planned, executed and evaluated by the professor. Student involvement in these issues is most often unwelcome, even though they are supposed to be the main beneficiaries and in fact benefit more when they can control some parameters of the learning environment (Doherty, 1998).

On this view, as well, the differences between teaching and learning are emphasized, starting from the fact that teaching is what professors, not students do and learning is what students and not professors do. Not only are they different activities, but learning is dependent on teaching, and not vice-versa. Thus, the logical dependency of the processes mirrors the social dependency of the participants and the directionality of information flow in the economy. The prevailing assumption is that without professors, students cannot learn.

Professors reinforce this dependence in their conceptions and actions. One example is professorial conceptions of reading, which is generally thought of as a process by which the reader extracts a single, pre-existing message ("*the* meaning") that is stated to be "in the text". Meaning is therefore dependent on things outside the reader: the text, the author or producer of information. In those (very common) situations where a single meaning for the text is not clear, readers are dependent on the professor for finding out which is the correct message to be extracted from the text. The teacher interprets the text for them. The interpretation of texts and/or events is dependent on the producers and deliverers of information, relegating the consumer to an unimportant and passive role in the interpretation process as well as creating the belief that consumers cannot process information without assistance.

Conclusion

The information-dissemination economy has made significant contributions. The greatest is to make information a valuable commodity and give it a central role in modern culture, to establish a vast infrastructure for the production, packaging and delivery of information, and to make available an amount of information of stunning breadth and detail about literally millions of topics. It is inevitable that such impressive amounts of such a valued commodity would lead to sporadic, then systematic, questions about how to put all of it to use in the effort to universally improve quality of living. This itself can be seen as yet another contribution: it made such questions both possible and necessary.

Will expansion into cyberspace improve the information-dissemination university? The evaluation of educational methods and tools (such as multimedia systems) is of necessity carried out in a particular historical, political and situational context (Kintsch et al., 1996, p. 57). In the context of the information-dissemination university looking for new market share, the Internet only makes sense as an additional delivery technology, and multimedia as a marketing strategy to add glitz to what is often seen as out-dated, irrelevant information. This is basically a strategy of putting old wine in new bottles, and not very sound from the outset. Notice that multimedia and Internet-based teaching are often cited as "no worse than" face-to-face classroom teaching (Moonen, 1997; but see Fletcher, 1995, 1996); investing in them, therefore, only makes sense if they are seen as delivery technologies or marketing ploys. If universities were looking for tools for improving *learning*, then multimedia would immediately be identified as a bad strategy right now: the initial investments and resistance to innovation are very high and there is *no* reliable, qualitative research about its effectiveness in constituting learning environments (Kintsch, et al, 1996; Moonen, 1997). Moreover, multimedia is a "bolt-on" technology that only increases overall costs without improving cost-effectiveness (Twigg, 1994; Williamson, 1996; Hiltz, 1997; Moonen, 1997; Martín Pérez, this volume).

Besides making significant contributions to human understanding, this model has created universities built on teachers without training in teaching or in the cognitive processes of learning, as well as on curricula that are responsive to teachers' expertise rather than to learner's needs and desires or to the demands of a modern society. In classrooms and lecture halls, information is transmitted rather than cultivated. Of course the standard caveats apply to the effect that there are many gifted, nurturing professors who are doing excellent work with their students; this, however, is by accident, not by design. No departments base their selection of professors on teaching excellence, but on quality and productivity of research; rarely will teaching excellence warrant tenure in a prestigious university. In the name of tradition and prestige, accountability of the university has become a secondary concern, hence its responsiveness to external change has diminished proportionately. The focus is squarely on the needs of the institution and its teachers, rather than on the needs of individual and corporate consumers. It is important to note that although we have long since stopped *talking about* education in these terms, they are still deeply entrenched as the basis for *acting* on education in the classroom.

This model of teacher-centered information dissemination is what has brought the university to its present crisis. Porting it to cyberspace will do nothing but push the university into the chasm.

Focus on the Individual: Philosophical and Psychological Perspectives

Information-rich ignorance is what today's educational institutions produce, on an industrial scale: very well-informed graduates who don't know what to do with what they've learned and have few skills that the marketplace requires. What use is a Ph.D. in linguistics or medieval history outside the university? Even many computer science students leave the university without the basic skills of software engineering and collaborative development that the industry so sorely needs (Thayer (Ed.), 1997).

Part of the challenge universities face is to develop an alternative model—a plan for action that overcomes the shortcomings of the institution-centered framework—and to implement it, both on campus and on line. The current intuition is that the key is to make universities "learner centered", so the concept of what the learner is doing is crucial (see Twigg, 1994). Where information dissemination focused on objective facts, the learner-centered view focuses on the learner's cognitive skills; where the old model focused on teacher activities such as teaching and research, the new focuses on student activities such as reasoning and learning.

This intuition and the clamor for change in education at all levels is at root based on the same conceptual change that has been wrought in the sciences, at different rates and with different degrees of success, during the course of this century. It is a change in *Zeitgeist* that has drastically altered our ways of doing science and now this same change is, as it were, howling at the gates of academe. Most fundamentally, it is a change in how we view knowledge and the learner, a change of epistemology, not a change in technology or administration.

In philosophical circles, this change is described as a shift from the empiricist epistemology of the positivism of the first days of the century to the more rationalist conception of the closing hours of the millennium. It is a shift from a focus on the world objectively observed to one on constructing our own view of the world around us.

In specific sciences, the same change has been described with different names: in psychology, the shift was from behaviorism to cognitivism (Bindra, 1984; Chomsky, 1995), in linguistics, from descriptivism to generativism, and so on in many different fields of research. In pedagogy, the shift is one from teacher-centered pedagogy to learner-centered pedagogy and from facts outside the learner to understanding happening within. I point this out to say that this change is already familiar; it is a natural consequence of the other shifts already consecrated or in progress, so we can look to other fields to understand the transition that is happening in university pedagogy as well.

It is important to point out that these two perspectives on knowledge have been vying for dominance since before the ancient Greeks, and equally important to note that as history progresses, each time one point of view predominates again, it does so with more characteristics of and concessions to the opposite view. Knowledge has always been conceived of as involving *both* experiences of the world and our ways of understanding and organizing these experiences. In simplest terms, we have shifted (in the sciences) and are shifting (in pedagogy) from a view that strongly emphasizes experiences of the world, or information and facts, to a view that strongly emphasizes how we understand and organize them, or thinking and theory. The shift is from a focus on the learner's environment and its institutions to the individual's internal cognitive processes; from observation to reasoning as principal determinants of knowledge. It is of fundamental importance to formulate this as a question of strong differences in *emphasis* rather than a dichotomic revolution: the learner is present in both views, as is his environment. In other words, a careful analysis shows that in fact the "opposite" views are simply different mixes of principles from both positions and the historical trend is toward a balanced synthesis of the two extremes (see Bunge, 1974–1989, vols. 5 and 6 for a detailed analysis of the resulting synthesis in epistemology). We seek, then, such a synthesis of the most useful aspects of both positions.

For a conceptual change such as this to come to fruition, we will need a well-developed theory of the learner to guide the transition. For this we cannot look to teachers; theirs is a specialized subculture that focuses on the activities of teachers, not learners. A careful reading of educational theory from around the world shows that it accords far more importance to social and environmental factors than to anything cognitive and individual. The politics, philosophy and sociology of education receive much more attention than the learning processes of the individual.

Four other subcultures, however, focus their activities on knowledge and the learner: epistemologists, scientists, cognitive scientists specialized in human learners and computer scientists specialized in machine learning. "Epistemology" is just a specialized name for the study of knowledge and the learner, that uses the philosophical tools of conceptual analysis. Cognitive scientists study the learner's activities with the names "knowledge acquisition," "inference," or "problem solving," often emphasizing the mental abilities and tasks of school-aged children and university undergraduates. "Science" and "research" are jargon terms for a particular kind of learner: the scientist who uses the particularly refined tools of the scientific method. Some computer scientists have also made great progress in studying how we can get machines to mimic human learners' behavior. Researchers in each of these areas can provide us with different and complementary perspectives on how learning happens.

Cognitive scientists have studied the learner in laboratory and school settings for more than 20 years, producing a large body of qualitative results about the cognitive processes that occur in learners. This body of research has been used as a systematic basis for proposing educational reform in primary and secondary schools (Bruer, 1993), and has been conveniently summarized in the form of a list of "principles for active learning" (Kintsch et al., 1996, p. 10–11). It is important to note that these are not the intuitive principles teachers derive from practice; the principles cited here are backed up by abundant research with a qualitative focus.

In this research, the learner's activities have been observed to be most effective when thought of and organized according to the following specifications, which have been paraphrased and adapted to better reflect the point-of-view of the learner.

1 Learning is a learner activity, a process that happens in the learner, thus the intelligence and prior knowledge of the learner need to be engaged and exploited during all of its phases. Independent learners require skills in: setting and developing meaningful goals for themselves (choosing topics and subtopics for study, decomposing complex problems, planning, determining and assigning subtasks, etc.); constructing meaning for themselves; monitoring understanding and progress; repairing breakdowns and errors in learning; and designing and implementing assessment measures.
2 The learner's prior knowledge is the starting point of any learning activity: assessment, use and development of prior knowledge are key components of learning.
3 The goal is for learners to develop more usable knowledge, that is knowledge that can be easily accessed and applied in novel situations. Hence, their processes are directed toward building accurate and detailed mental models in a domain.
4 Learners are most efficient when engaging in activities whose value and purpose are evident to them, activities that are most often, from the learner's standpoint, of moderate difficulty.
5 Learners work best in domains that are realistic and meaningful to them.
6 Social construction of knowledge is an important component of the learner's activity, which requires skills in dialogue, cooperation, organization and dealing with multiple points of view.
7 Learners' progress emphasizes conceptual understanding of mechanisms rather than rote learning of observable phenomena.
8 Learners' progress is optimized by developing awareness of their own thinking processes and styles and socializing these processes for analysis and assessment.
9 Learners require support for higher level reasoning, such as visualization tools.
10 Cognitive skill development, unlike rote learning, requires flexibility and adaptability of cognitive strategies: transfer of reasoning from one domain to another, comparing, contrasting and synthesizing multiple perspectives on a single topic, exploration of analogies. Iterative processing—multiple analyses of the same problem to revise and reconsider one's solutions and those of others—is a central component.
11 Learners progress best with qualitative, relevant and timely feedback at optimal frequencies.
12 Identifying and analyzing errors and difficulties are important opportunities for learner development.

13 Individual differences in interests, motivation, knowledge, point-of-view, abilities and style are essential for generating possible solutions for problems in learner-centered environments. Diverse perceptions, conceptions and solutions provide the basis for discussion of cognitive processes.
14 Assessment of conceptual change is the basis of systematic learner development.

This is essentially the same position defended by Bruer (1993) in the context of motivating reform of primary and secondary education, and in many of the contributions to the Educom Review in the context of higher education reform.[4] Using such a rich notion of the learner to guide instruction, enhanced by the details of the research that these principles summarize, will lead to major changes with substantial and desirable results.

Notice that this rich view of the learner focuses almost exclusively on cognitive processes, on understanding alone, with little mention made of cognition in group settings and none of the interaction of cognition with observation. This is in no way a shortcoming of the research, since the explicit goal was to understand individual processes of a certain type. However, for the purposes of rethinking the university, we can see it is incomplete.

Transition to learner-centered environments

As can easily be seen from the concept of learning developed above, most classrooms and lecture halls are not learning environments. They are *teaching* environments and will have to be bulldozed or recycled as quickly as possible if promoting students' understanding is taken seriously as the mission of the university. If this seems extreme, go back and consider how well the modern lecture hall with 500 students is suited to promoting the activities in the list given above.

The space available to a student in such an environment is less than one square meter, his mobility is severely constrained, and the resources at his disposal are very, very limited: his desktop (with notebook and pen and, increasingly, his DiskMan CD player or crossword puzzle), and the material provided during teaching events. It is difficult to check other materials, for lack of space; it is not possible to consult with colleagues, for lack of permission; the means for learning are strictly limited to passive observation of the teaching event. The teaching event is organized by someone who is very familiar with the content, hence no longer remembers the difficulties encountered in coming to understand it. The material is adapted to the professor's interests, goals, abilities and cognitive style; as is the temporal organization of teaching. Student goals are to "assimilate" the material and reproduce it in the form the professor stipulates to pass the course and fulfill a requirement for graduation. The professor is of course not doing this on purpose or with ill intent; he is simply a cog in a machine that sanctions and requires this sort of behavior, capitalizing on his lack of training in teaching and learning.

Perhaps the most expedient method for provoking change at universities, then, would be to lock all the classrooms and lecture halls. Try it as a thought experiment: imagine a university's administration in fact locked all of its classrooms and lecture halls, prohibiting their use (Penn State actually did something similar on a very limited scale in 1995 (Sherron & Boettcher, 1997, p. 24). How would activities proceed? Initially, amid much dissatisfaction, classes would be held in the corridors, the cafeterias, and on the lawns. Thus far in the experiment we see only the immediate chaos and dissatisfaction.

But what would happen next, if the administration held out? Many professors would start to adapt. They would brainstorm and negotiate with the students alternate arrangements: smaller groups, different schedules. Repeating the same thing to ten different groups would bore the professor so much that she would start looking for alternatives. Very quickly lectures would start to become more infrequent and mechanisms for awarding credit for competency instead of classroom hours would develop. Teacher contact hours would probably evolve into small-group discussion periods and evaluation activities. Thus far in the experiment, we see that students are already responsible for mastery of content and that lectures offer little additional value. Web-based teaching is already proceeding along these lines. Geographical and temporal concerns have, as it were, made lecture halls impractical.

Loss of precious research time would inspire more imaginative professors to find ways of doing research at the same time as they satisfied student contact-hours, by involving students in their research. This would have the added advantage that the research would proceed more productively with more assistance and without the added effort of having to find funding for research assistants. Professors used to having one or two research assistants would now have 30 or 40 at their disposal for a semester or more. Because of the semesterly turnover, they would develop new materials for training (not teaching) the students to help them. Core training would be in planning research projects based on study of existing bibliography, collecting and analyzing data, presenting results and writing reports—reading, writing and arithmetic in a new guise and with new relevance. With all this help, publication rates would soar. Soon, because of the increasing need for laboratory space, professors would pressure the university administration to renovate the lecture halls as laboratories and re-open them. Student satisfaction would increase drastically and employer satisfaction with graduates even more. At this stage of the experiment, people would start to see advantages of the new model, in which learners play a more active role.

At this point would anyone, professor or student, want to go back to the previous, lecture-centered model of teaching? Probably only the least successful adapters. Would this be the ideal scenario for all disciplines and universities? Surely not. But it provides one plausible scenario for change and a clear illustration of the difference between teaching environments (classrooms) and learning environments (laboratories and study groups). I leave as an exercise for the reader to continue the thought experiment by locking the *gates* of the university and imagining what might happen (see, for example, Perelman, 1992; Brown & Duguid, 1995). The point is simply that there are greatly improved alternatives to the teacher-centered lecture model that tradition has saddled us with.

This scenario is consistent with a different view of the university's mission, now seen as producing knowledge and promoting understanding, avoiding the "false dichotomy between research and teaching" (Graves, 1994a; see also Mason, 1998) and the confusion of teaching with learning. Evolving this far would be a great improvement and a minor miracle. Great strides would be made in developing students' cognitive skills and increasing professors' productivity.

Let us consider in more detail one kind of learner who will be seeking out the Virtual University. The university freshmen of the year 2010 are seven years old today, in 1999. More than half of their parents are knowledge workers, to take the case of North America, and in most cases both work. Almost all these children have already had intense exposure to television and telephones, to video games and movies, to radio, CD players and a wide variety of complex gadgets in the home and at pre-school. They have just entered mainstream schooling, and they have already learned in pre-school how to read aloud, some simple math, socializing skills, world knowledge about traffic conventions, holidays, history. Not to speak of their idiosyncratic interests, about which they already know amazing amounts (cars, dinosaurs, TV characters, etc.). Printed matter abounds. They are frequently familiar with multiple social and physical environments: home, preschool, parents' workplace, vacation locations, libraries, shopping centers. Their adolescent older brothers and sisters talk incessantly about the Web, e-mail, chat rooms and other computer-related topics, as well as issues of social interaction: romance, betrayal, sex, seduction, fashion. Nearly half already have computers at home and they have increasing supervised, and sometimes unsupervised, access to them. In many families, they already have responsibilities such as feeding pets or cleaning their room. These kids may be only seven years old, but they have an amazing amount of experience and knowledge that they take along to their first day of elementary school.

Over the next 11 years, the typical young adult will have accumulated more than 50,000 additional waking hours of experience with information and the world: thousands of hours of passive experience with every kind and type of human happening from all over the globe and from every corner of the entertainment world, as well as thousands of hours of independent interaction with a wide variety of other people, topics and problems to solve, including all of his primary and secondary schooling and many, many hours of computer and Internet time.

How will such a person react to being treated as an empty vessel to be filled with knowledge? Will such a freshman worry about how well ideas are strung together, if so much of her experience is with fragmented information? How will she react to information that is artificially organized in an extremely coherent fashion, if she's rarely experienced that before? How will she react to a text-only document that is supposed to be read in isolation from others, from beginning to end, when all of her experience is with intuiting coherence from multiple sources? How will she react to being forced to think about topics that seem irrelevant, "because it's a required course", when she has become so accustomed shaping her own information environment?

It seems clear that for such a generation, coherence will not be as important as relevance and relevance as defined in terms of usefulness for solving and preventing real, not academic, problems. Subservience to authority will be out of the question; they will put a very high value on autonomy and individuality, since their world is so full of variability and their heroes are idiosyncratic individuals. In the 21st century the workplace is predicted to be organized around the intellectual capital of individual consultants, to inform short-term specific projects rather than to fill long-term positions, so the workplace will value autonomy and individuality, as well (Perelman, 1992). Relevance, autonomy and individuality will be very important in such a market, as will be social skills and the ability to deal with multiple perspectives on the same problem. It is quite clear that today's teaching environments will become increasingly irrelevant.

Conclusion

This learner-centered model has already begun to create nurturing, challenging environments for cognitive growth and has signaled fundamental changes in the design of learning environments. It is the best way we know of to ensure relevant learning environments that are adaptive to learners' changing needs.

There are major challenges, however. Such an adaptive, individualized apprenticeship-like model requires a great deal of quality contact hours focused on individual learners' developmental paths and so seems doomed to be a commodity only for the on-campus elite. Another challenge the university faces is making these advances available to a wider audience; there are neither facilities nor staff to provide such environments for today's 14 million university students (Software Publishers Association, 1998), much less for the increasingly broader segments of society that are beginning to look to the university for their cognitive needs. A third challenge is adapting the advantages of this sort of approach to cyberspace, something about which there is apparently very little discussion.

This model of learner-centered development of cognition can help lead the university out of its present crisis. However, focusing on individual cognitive skill, although it will be fundamental to education, will not be enough. The cognition part is right, but the sociology is wrong: the dependency on the institution for learning continues, and the problem of establishing a new, more dynamic and more essential place for the university integrated into modern society remains. Reinventing teaching will not be enough; the university has to reinvent the role of learning in society.

Focus on Integration and Innovation: Communities for Self-Empowerment

Reconsider our notion of the learner. Epistemology provided the basic framework in which to formulate our questions, setting up the parameters of observation and reason. Earlier psychological views of the learner equated observation and reason, teaching and learning and attributed little complexity to mental processes beyond uniform and automatic "assimilation". Even so, focus on the learner's environment revealed important contributions of that environment and its organization, regardless of the learner's actions. Cognitive science provided a rich and well-founded focus on the individual's cognitive processes in classroom learning, emphasizing reason over observation. Focus on the learner's cognition revealed very important contributions of learner actions to the acquisition of information and skill, often independently of the environment. With this, the goal of education has begun to evolve from absorbing information to acquiring the skill of understanding. But a look at additional perspectives on learning shows that there are other important dimensions to take into consideration.

One characteristic common to these perspectives on the learner is that what is to be learned focuses on the solutions that our predecessors have found for the problems they chose to solve. Instruction is about known solutions to known problems, in fact practically *all* of our education is about the past. This is an interesting and curious cultural bias that deserves more systematic analysis and discussion. We teach much, much more about how science has been done (history of science), for example, than about how to *do* science. In all domains, theory for us has become inextricably entwined with past history, rather than with current limitations or future problems. Learners have been given detailed information about, then led to in-depth understanding of, past solutions to known problems, while rarely being provided the experience or tools for evaluating the usefulness of these solutions for *other* problems, current or imaginable. But the solutions of the past may not be relevant to the problems of the future; all too often entirely new solutions have to be devised, and our educational system simply does not prepare them for this possibility to any significant extent, with the possible exception of some parts of professional degree programs. We produce reactive rather than proactive learners; consumers rather than creators of knowledge.

Machine learning research and analyses of scientific practice, in contrast, reveal a focus on developing the ability to deal with the open problems of the present and the unknown problems of the future (Ram & Leake, 1995). The computer scientist seeks to empower the computer with its own adaptability, its own ability to interact and learn with the environment so that it can deal with unforeseen problems (Hunter, 1995; Michalski & Ram, 1995). Doing science also emphasizes the development of empirical and theoretical methods for constantly dealing with previously unknown problems of measurement and understanding. Unlike teaching in educational settings, the focus in these contexts is on learning to be able to solve *present or future* problems with *unknown* solutions. This future-oriented kind of cognitive development can best be called "empowerment" rather than learning; it focuses not only on assimilating information or understanding it, but also on knowing how to *apply* it to unforeseen situations.

Scientific research adds still another dimension: scientists emphasize communication, systematicity and explicitness to facilitate *sharing* of methods. Sharing methods of acquiring reliable data, methods of analyzing it and methods of thinking about the world (hypotheses and theories) enables them to empower *each other*. This key characteristic of scientific research is unlike any other kind of learning: science is *self*-empowering, and most often explicitly so. Thus the kind of learning we call scientific research is the hallmark of the modern mind (Donald, 1991) and so unique that it bears closer consideration.

Scientific research is the most highly refined way we have of learning about what we observe, that is, of building mental models for observable phenomena based on what we know. It is the embodiment in practice of the synthetic view of knowledge as being built from equal attention to experience, observation and description of the environment on the one hand and reason, thinking or theorization, on the other. The general principles of learning in science can be seen as derived from a very simple observation: the individual learner has finite (and in fact, quite limited) capacities of perception, reasoning and memory. Hundreds of studies in cognitive psychology have confirmed these limitations and teased out many of the qualitative and quantitative details, including the fact that the specific limitations are very variable from individual to individual.

Although these limitations may seem somewhat obvious and uncontroversial, a few examples will help remember why they seem so clear. There are, for instance, many kinds and bandwidths of energy that we cannot, because of our organism's make up, perceive directly—radio waves, microwaves, infrared and ultraviolet light, sounds about or below certain thresholds of amplitude and frequency, and objects smaller or larger than certain dimensions are just some examples. Absence of information flow (sensory deprivation) or excess (sleep deprivation, information overload, etc.) can lead to mental breakdown and hallucinations. Moreover, a film or speech that is otherwise perfectly normal becomes pure, unintelligible noise at very high or very low speeds: there are limits to the rate at which humans can process information, independently of the kind of information. We can easily visualize a mathematical space or system of coordinates in two or three dimensions, but not in five or 20. We can understand the influence of two or three factors on a given process, but not the simultaneous interactions and influences of 30 factors. Many humans can understand a text in one language while they produce and verify a translation of it in another (simultaneous translation), but they cannot produce a text on a different topic at the same time or understand simultaneously three texts on three topics.

These examples illustrate that there are kinds of information and rates of information flow that humans simply cannot deal with: however wonderful the human cognitive apparatus is, it has very clear limits. Besides this, the existence of distractions, illusions and hallucinations shows that this limited capacity is also limited in reliability, and, as memory and reports of subjective experience show, they are often unreliable in unpredictable ways. Finally, the concept of selective attention shows that not all of the information available from the environment is in fact used.

The inevitable conclusion is that the fundamental tools that humans have for producing knowledge—sensation, perception, reason, and memory—are both imperfect and severely limited. Obtaining reliable knowledge about a very complex environment therefore requires the development of special strategies for circumventing these limitations. These strategies constitute the cornerstones of systematic learning, or what is also called "the scientific method".

The first consequence of this conclusion is that, for all these reasons, observation, however detailed and systematic it may be, does not constitute a sufficient base for the production of reliable knowledge. Observation is *not* knowledge, nor is it understanding. The second consequence is that by all the same arguments, reason, however enlightened it may be, does not constitute a sufficient base for the production of reliable knowledge. Thought is *not* knowledge, nor does it require understanding of the world.

In sum, learning—producing knowledge—requires a synthesis of observation with reason, which when systematized is called theorization. How we understand something (our reasoning) guides our observation of it and what we in fact see modifies how we understand it; theorization in effect directs observation and observation restricts theorization. This shows up as much in our use of social stereotypes (based on race, gender or ethnicity) as in scientific research. Part of the "magic" of science is this discovery that neither observation nor reason alone are enough, but the synthesis of the two has proven to be the best strategy of producing reliable knowledge (see Jacob, 1982, pp. 362–3).

Key pieces of the synthesis of observation and reason are induction and deduction. Inductive generalization is the leap of faith that the learner takes when, after studying 30 people reading two different texts, he proposes that what he observed will be true for any other literate person or any other text. It is by leaps of this kind that knowledge and theories are built, little by little: based on partial and limited observations, the learner proposes a universal hypothesis. The other key component of the synthesis of reason and observation is deduction: from the general hypothesis, one can make a series of predictions (deductions) about what should be found in relevant observations of specific situations.

Perhaps the most important principle for guiding the synthesis of observation with reason is to search for hidden mechanisms as the causes of observable phenomena. This is a consequence or corollary of the interaction between reason and observation, since occult mechanisms have to be imagined or hypothesized and therefore are in the realm of reason. Always looking for them guarantees reasoning about the observations available.

With the coupled activity of induction of hypotheses and deduction of predictions, we generate a view of scientific investigation as a virtuous circle: a cycle of activities that mutually feed, stimulate and constrain each other (Bunge, 1967). Learning in scientific contexts is intensely self-reflexive: we start from current knowledge and systematically build on it. We imagine something (make hypotheses), deduce from there what things should be like in the world (make predictions), see if that's how things in fact are in the world (make observations), re-evaluate our original hypotheses, then think of our object of study in a different way and try again. Progress is measured by the confidence (accumulation of evidence and arguments) we have in hypotheses and by the detail with which they are formulated. As learning progresses, we accumulate more and more confidence in hypotheses that are more and more detailed.

Idealization

The fact that reason and observation are both inherently limited leads to another important conclusion: the simplification or idealization of the object of study is inevitable. Humans can only study or systematically observe a few characteristics of something at any one time, and we thus say that we "abstract away" the existence of other aspects that are not of immediate interest. Idealization or abstraction of the object of study, studying it a part at a time, therefore, is a crucial strategy of learning in general, and of the scientific method in particular. It also means that it is only the *sum* of these partial observations that will eventually offer a more complete view of the object of study. This is important: it means that no one has the whole truth about even the most specialized topic; it means that every learner or researcher has only a part of the whole story; it means that no one is "right" and that collaboration is essential. The issue of idealization is often contentious in research because in practice how much you focus on or ignore is most often based on a subjective assessment about which is the best strategy for learning at a given point in time. Exaggeration in either direction ("wallowing in the data" or "pie-in-the-sky theorizing") impedes the progress of learning and fragments the learners into groups that have a harder time collaborating.

Error

Human limitations lead to yet another fundamental conclusion: error is inevitable, both in observation and in theorization. Since we can only observe a few of the factors that determine the occurrence of a given phenomenon, all the observations, measurements and hypotheses we make are, by definition, approximate and partial. In other words, we have no access to the absolute, complete Truth, only to an approximate, partial and temporary Best Guess. This is why in the empirical sciences researchers avoid talking about proofs or proving things: "right" and "wrong" are useless; we only have better and worse ways of viewing the world. The strategy that scientific research has adopted to deal with omnipresent error is to always seek to detect and minimize it, without becoming obsessive about eliminating it entirely. Many statistical techniques were developed to provide means for evaluating how much error is present in a set of measurements and therefore how reliable they are.

Since knowledge is always incomplete and always partially wrong, we can relax about making "mistakes": knowledge can always be improved, and in fact always has to be improved. No matter how important a theory or piece of research, there will always be a way to improve it and expand on its results. The professional learner is comfortable with this situation and with the inevitability of making mistakes. The incompleteness of knowledge and inevitability of mistakes also mean that an infinite supply of problems is available for study. Finding a solution to a problem raises new problems for study: where are the errors? what is missing? how can the solution be made more detailed? It also means that all past problems are open for rediscussion; knowledge is never final.

Collaborative learning

Another important strategy devised to overcome the limitations of individual observation is collaboration (see Brown and Duguid, 1995; Mark, this volume, on emerging technologies to make this possible over the Web). Knowledge is produced by whole communities (Salomon, 1994; Resnick, 1996; Papert, 1993); the role of the individual is to provide the input (data, hypotheses, arguments, theories) for this process. The objectivity or "intersubjectivity" of scientific hypotheses that Popper (1959) placed so much importance on is only one aspect of this strategy. Comparison of the observations of different researchers can contribute to determining which aspects observed are likely to be subjective or individual and which are not. The same argument is doubly true for collective hypotheses and theories, which are initially produced from individual intuitions, imagination and interpretations of available data.

The key to collaboration in learning and research, then, is knowing how to deal with multiple perspectives: how to compare and contrast them, how to evaluate their relative merits, how to accept none as the truth but each as a distinct and valid contribution. Multiple perspective taking leads to more complex and complete ways of looking at things and it also expands the kinds of observations and data that will be useful for checking these perspectives. This is a major improvement over individual learning. The collective evaluation of hypotheses and theories is more complete and more demanding than individual evaluation, ensuring greater reliability of those hypotheses that are accepted. This is just a more systematic reformulation of the prosaic "two heads are better than one" (and two thousand better still).

This necessity for collaborative research is what underlies the importance that the scientific community accords to publication, since this is the main channel of communication and therefore most important tool for the evaluation of ideas. An unpublished hypothesis or theory is worthless because the parts of it that are true do not contribute to the progress of the community and the parts of it that are false go undetected. Similar considerations hold for non-professional learners: "peer teaching" and communication among learners allow the individuals to adjust their understanding to reflect the group's consensus, thus building collective knowledge. Moreover, in collaborative learning, one learner's insight advances several learners' understanding simultaneously. Collaborative research also compensates for the fragmenting nature of idealization cited above. Collaboration and communication are needed to sum existing studies, with their differing simplifications and different perspectives. Only this sum and not the work of an individual is what can be evaluated in terms of completeness, coverage and progress of understanding.

Assessment and diagnosis

One simplifying assumption common to non-scientific perspectives on learning is that there is some absolute standard of correctness for evaluating learning: Truth, true facts, correct use of skills, use of the correct skills, etc. This golden standard is external to the learner and focused on the solutions that our predecessors have found for the problems they chose to solve. This has led to an all-or-none strategy for evaluation which only rewards blind perfection. However, this golden standard approach is both rigidly context-insensitive and inadequate in that correctness is not judged in terms of the specifics of a particular learner or particular problem and it is rarely revised. The golden standard approach also fails to recognize the inevitability of idealization and error. Scientific learning, by contrast, suggests that absolute correctness undermines learning and that basing learning on problem- and context-dependent notions of correctness ("best available solution", "partial truth" or "satisficing") leads to best results.

A final, important strategy used in science is frequent and detailed evaluation: evaluating a learning situation and discovering why it is that way. At each step of learning, assessment of how good a particular way of reasoning works is central: the quality of content is a primary concern. Professional learners evaluate their decisions every day, every time a problem arises or new information appears. They continually ask: How does this idea or hypothesis fit with the data we have? With other data available? How does it fit in with other ideas that we are using? With ideas that we are not using? The more explicit the criteria and methods of evaluation, the better. Checking often and in detail how coherent a given set of ideas and observations are, and asking frequently "what did I learn today?" are among the professional learner's most valuable tools. By way of continual and detailed assessment, they use knowledge to improve knowledge and to improve the methods they use as well; efficient learning is essentially self-reflexive.

The Web, too, is becoming increasingly self-reflexive. *All* of the top 10 web sites in traffic cited in a September, 1998 report were what can be called "metasites": sites that point to and organize other sites, providing information that makes it more efficient to find further information (Yahoo! Internet Life, 1998,

p. 44). The metasites had a reach (in percent of Web population) between three and 18 *times* larger than that of information-dissemination sites (such as the Weather Channel or ESPN SportsZone).[5]

This discussion of science as learning shows that learning can be non-empowering, empowering or self-empowering. Absorbing information or even understanding it are non-empowering forms of learning, in that they do not develop the skills of applying what was learned to unforeseen problems. The learner's successes are restricted to problems already encountered and his options limited to re-applying previous solutions to newly encountered problems. Learning for empowerment, by focusing on potential, future problems, as machine learning researchers were quick to find out, provides much more flexibility and greater success in problem solving.

Science adds a further insight: even greater success is possible when learners mutually empower each other. This is self-empowerment from the point-of-view of the group whose members continuously foster the learning of other members. Once we observe that almost all of our education focuses on how *other* people have understood our world, on their understanding of what exists, and even then in a one-sided fashion, we can see that the next step is to focus on how we create *our own* understanding of the world, on how we create what doesn't exist (answers to questions, solutions to puzzles, objects with novel properties, new theories, etc.). This is epistemological *self*-empowerment, from the point-of-view of the individual, because the key concept is autonomously learning about how best to use and adapt your own personal style of understanding. Instead of a focus on facts or cognitive skills, it is a focus on cognitive tools, on learning to use as well as to make tools or strategies for your own learning. This is not a recycling of an old emphasis on study skills, but a new focus on *designing* one's own customized problem-solving environments. The explicit goal is to promote autonomy and innovation as a learner, rather than dependence on external authority or traditional processes.

Analysis of the strategies that scientists have developed as professional, active, learners, shows them to be essentially identical to the principles of active learning that cognitive scientists have identified, although they are the result of entirely independent research efforts. The importance of collaborative learning, the use of prior knowledge and new information to develop reasoning, the key roles of error, feedback and assessment, the systematic planning of learning as a function of meaningful goals, all are components of both. The analysis of the origins of these strategies in science also suggests why these principles should be important, and the convergence of the two very different lines of thought encourages us to believe even more strongly in the validity of these principles.

Epistemological self-empowerment, then, takes learning a step further, cognitively, socially and culturally. Cognitively, we advance from having understood what we have experienced to being able to create an understanding for that which we can only imagine. Socially, we advance from understanding as being dependent on external individuals and institutions to understanding as being socially autonomous but based on consensual criteria for evaluation. Culturally, we advance from valuing a single point of view to valuing multiple points of view, and from an unconscious focus on the past to a goal-directed synthesis of past, present and future learning. When learning environments are designed in such a way that one person's learning fosters others' learning and vice-versa, the groundwork is laid for exponential growth of knowledge. In sum, scientific learning environments suggest that empowerment is essential and self-empowerment leads to best results.

Using this approach, science has revolutionized human understanding. It may well be that with this approach, we can revolutionize human education, as well. Is this, however, the conceptual learning environment that is created in the lecture hall? Clearly not. Is this the core curriculum for the learning environment of the Virtual University? Hopefully it will become a part of it.

One important and well-articulated view that is consistent with this position is that of the National Learning Infrastructure Initiative (Graves, 1994b; Twigg, 1994; Twigg & Oblinger, 1996; Heterick et al., 1997). The NLII reports identify very concisely the range of problems and pressures that the university faces and articulate a cogent series of principles for orienting change toward a "technology-mediated environment in which the learner can thrive" (Twigg, 1994; see also Brown & Duguid, 1995). This chapter can be seen to complement the NLII view of learn*er*-centered environments with a broader discussion of, and focus on, learn*ing*-centered environments so as to lessen the chance that others might take NLII's well-placed emphasis on the learner to less-productive extremes, as well as to seed their discussion with additional views on how virtual learning environments might be constituted.

Learning Environments Revisited

We usually think of learning environments as places people go to for the specific task of learning. But think of a learning environment not as a place, like a Web site or a classroom, but as a problem awaiting solution, a puzzle or a challenge. Learning environments, then, can be seen not as geographic or physical, but cognitive entities. It is within the environment of a specific problem or puzzle that learning occurs. A social group coalesces around a given problem and *the group* learns together in that environment, each individual contributing and benefiting to the extent of his or her abilities. This is the same in the laboratory, at a construction site or in a product development group. Learning environments, from the individual's perspective, include a systematic statement of both the initial and goal states that characterize the problem, as well as the known constraints on how it can be solved. From a group perspective, learning environments have the same components, but a consensual understanding of each has to be negotiated by the group.

Consider an example scenario. A group of 18-year-olds is engaged to help design and execute a Web site for their peers about, say, information visualization, in collaboration with more experienced designers, programmers and content experts. The 18-year-olds are experts in their peers; the others are experts in design, programming or content. Who are the teachers and who are the learners? Are the adolescents hiring the more experienced people to be their "teachers" or are the more experienced people hiring *them* as consultants? Who pays and who gets paid? The ambiguity or plurality of roles of *all* the participants is a hallmark of this sort of scenario: the focus is the problem or product.

Add to this training and coaching for all the participants about how to learn the most from all of the others, as well as how to enhance the productivity of the group as a whole by leading all to teach the others, and you have an example of a learning environment designed to maximize epistemological self-empowerment. The success of a very similar approach has been demonstrated in the ThinkQuest competitions for high-school students [www.thinkquest.org], which have resulted in excellent instructional websites (much better than many at virtual universities) idealized and executed by students.

The learning environment is relevant and challenging, everyone is learning from and teaching everyone else, each is valued for his respective competencies, and something of value is being produced during the learning experience. Collaborative and problem-solving skills are being enhanced, multiple perspective-taking is part of day-to-day life and the collective learning activity is goal-directed so more effective. The experience mixes the best of internships, apprenticeships and consulting. As the variety of such experiences increases, so does the individual's content knowledge about an increasing array of domains, and, more importantly, his ability to learn how to learn and to learn in different settings. Finally, the focus is on using what we have learned in the past to build something new for the future.

Understanding that the "learning community" is not restricted to a community of students is key: learning happens throughout society. Moreover, viewing learning as directed toward constructing products (rather than for its own sake) makes this view more compatible with the productive sectors of society and theories of goal-directed learning. On this view, then, education becomes integrated with the productive sectors of the economy, rather than marginalized and largely alienated. It eventually becomes invisible and ubiquitous, since the learning environment can happen on the Internet, in an industrial research lab, at some corporate headquarters or wherever there are problems to solve. This is parallel to the approach of ubiquitous computing (Weiser, 1991): instead of putting reality into the computer/learning (the virtual reality approach), the idea is to fit the computer/learning naturally into reality. On this view, education is valuable and valued, and an integrated part of everyone's lives. The main market changes from 18- to-22-year-olds to the whole society, including everyone from Nobel laureates to illiterates. No one knows everything so everyone has something to learn.

What if *all* higher education was like this? This would be the Virtual University in its broadest sense, not just the traditional university broadcasting over the Internet. A large part of the university's function would be to design these experiences, put the relevant people together, coach them before during and after the event and charge them all for the benefits obtained, as well as a percentage of the profits, without anyone necessarily stepping onto anything like a campus. It is clear that if universities don't fulfill this role of *learning brokers*, in a timely, relevant and responsive way, then a dynamic, aggressive, well-focused industry will appear to address this need outside of the university. If the leaders of higher education do not take initiative to make a radically new educational system happen, then the university as we know it may disappear in two senses, not just one. We might not only lose the campus, but the institution itself.

Skills and Metaskills for Learning Environments

Where before we had education for information, and now we are building education for understanding, the insights gained from scientific practice make clear that for the future we need education for integration and innovation. What are the kinds of skills that the freshmen of 2010 will need to function effectively in social groups organized around learning environments like those above?

As a first approximation, we can group the kinds of skills used in professional, innovation-oriented learning into three groups:

- Planning: Skills for designing learning environments or formulating problems;
- Management: Skills for traversing learning environments or solving problems;
- Communication: Skills for communicating about learning environments/problems.

Planning: Skills for designing learning environments or formulating problems.

Skills for designing learning environments are those used for identifying and delimiting a problem to be solved, as well as evaluating its importance. They are the skills used in industrial environments to develop specifications for new products and assessments of the costs and benefits of producing them, and the same skills used in laboratory environments for writing grant proposals and detailing experimental designs. In essence, they are the skills for mapping out a group's position before taking on a given problem: specifying prior knowledge, the set of available skills and resources, a specification of where exactly they want to arrive and an evaluation of costs and benefits. Well developed business plans and research proposals have all these elements.

Problems or learning environments can be seen as paths in an imaginary space: the problem of getting from some initial situation or state to some final state or goal. The initial state and the final state, along with any restrictions on the paths that can be taken, define the problem to be solved or delimit the boundaries of the learning environment (Newell & Simon, in Rosenbloom et al. (Eds.), 1993).

"Reviewing the bibliography" and "market research" are activities for determining the initial state of a problem. These activities require skill in gathering information, determining what is and is not relevant, synthesizing it, identifying the hidden mechanisms that make the important processes work, identifying the resources available and the skills at hand. A given problem has to be contextualized by relating it with existing, related problems, in order to identify and assess the currently available options. This, in turn, requires attention to multiple perspectives on the problem to generate a maximum of relevant options. Determining the initial state of a problem is a very complex and intellectually challenging activity for which virtually no one receives explicit instruction, in spite of its great importance in so many domains. This is one of the skills that has to be brought to the fore to generate a focus on innovation.

Pinpointing the research questions to be answered and developing detailed specifications for a product are both activities for determining the final state of a problem. This is another, even more challenging activity, that plays a central role in learning in research and commercial activities. It requires the same analytic and synthetic goals as above and in addition imagination and creativity. It requires skill in predicting the effects and interactions of known factors in unknown situations; a skill most often ignored in today's curricula. It requires a kind of emotional intelligence in being able to identify and detail one's own vision and goals; in turn necessitating training in creativity, imagination and self-confidence. This activity leads to the identification of several different kinds of possible benefits derived from actually solving the problem or traversing the learning environment: theoretical benefits stated in terms of new knowledge to be produced, methodological benefits stated in terms of new skills, empirical benefits stated in terms of new information that will become available and practical benefits stated in terms of other non-cognitive problems that can be solved. These skills of imagination and projection are another set that has to be developed explicitly to underpin a focus on innovation.

Skills in planning and developing complex structures of interacting goals— A fundamental skill for planning is the ability to decompose a general goal into its more specific parts and organize them according to which are requirements for achieving another. These are skills used in project management, and less often in academic research, to produce a workflow diagram for estimating the resources necessary for executing a given project. Experience and domain knowledge play an important role in developing detailed and accurate breakdowns of goals into subgoals, which are the basis for much of project management.

Another part of defining a problem is identifying what is known or hypothesized about the relations between the initial state and the goal state, so that an adequate inventory of possible paths can be developed. Time and resources limit the number and types of different paths between initial state and goal state than can in fact be explored. Prior knowledge can as well, thereby focusing problem solving activity on relevant options. One manufacturer has determined that making product X with a given plastic is too expensive or makes it too brittle; this identifies one path in the problem of manufacturing X that can be more systematically evaluated and probably rejected. This activity, again, requires highly developed skills of evaluating relevance and judging the relative plausibility of different options.

Skill in evaluation or relating the learning environment to multiple goals so as to evaluate coherence (relations with known factors) and importance (relations with unknown, future factors)—This requires consideration of the predicted costs and benefits of actually traversing the learning environment, in terms of new information obtained (empirical contributions), new ideas obtained (theoretical contributions), new methods and techniques generated for or during the traversal (methodological contributions) and practical problems that may be solved with all these benefits (practical contributions). Moreover, political and economic costs and benefits of traversing a learning environment are also important to take into consideration.

Finally, formulating problems requires the ability to evaluate precisely and in detail the costs and benefits of attempting to solve them, during the planning phase. Grant proposals and business plans both raise these crucial issues, but here the domains are quite different. Where costs and benefits are rarely a focus of discussion in academic activities, and never a topic of explicit instruction, in business they are the key to success or failure and as such are the focus of most discussion. This one difference seems to account for much of the common impression that academics are alienated. Once more, this ability is not trivial and cannot be developed systematically when people are expected to "pick it up as they go". Just as much in academic as in commercial activities, these are fundamental skills that have to be developed explicitly.

Designing learning environments is the same as formulating problems or defining projects, in whatever domain. Learning occurs in solving the problem or traversing the learning environment, to the same extent that it occurs in designing the learning environment itself. By having learners only solve problems (often only trivial ones) rather than also formulate them for themselves, education has not allowed learners to develop many of these important skills.

Management: Skills for traversing learning environments or solving problems

Skills for managing learning environments are those used for organizing, staffing, directing and controlling the groups that are solving problems, in essence the basic skills of project management (see Mackenzie, reprinted in Thayer (Ed.), 1997).

Organizing a learning environment entails establishing roles for participants and defining relationships, as well as establishing requirements for the people to fill each role. In a very simple case, skills in this aspect of learning management would make it possible for supervisors and students to determine who has which responsibilities and capacities; thus avoiding misunderstandings and unnecessary friction.

Staffing a learning environment entails selecting, orienting, training and developing human resources for the different roles required by the specific problem to be solved. The people are chosen not to fill stereotyped roles of "teacher" and "student", but to fill adaptive roles in specific problem-solving environments. Students and supervisors, workers and managers all need to be trained in these skills so that their responsibilities are clear, and important orientation and training is not overlooked with a corresponding decrease in productivity. In university environments, students are expected to know how to write technical papers, how to present them, how to develop research plans, etc.; however, the university often neglects to offer appropriate directed opportunities for developing these aspects of their training and development.

Directing a learning environment or problem-solving activity entails delegating responsibility, motivating people to take action, coordinating the activities of group members, managing the differences in cognitive and social style and managing the group in the face of changing goals. These are not skills that are developed explicitly and systematically among the professors who today have these responsibilities in their roles as research supervisors and department heads. It is as unproductive to require these skills of the professors without providing them the opportunities for on-the-job training as it is unfair to offer such unstructured learning environments to the students who pay for access to effective learning opportunities.

Controlling learning environments entails establishing reporting systems for intermediate results, developing performance standards, measuring observable results, taking corrective action when the results are above or below stated goals, and providing positive and negative rewards in the appropriate doses and situations. This corresponds to today's evaluation and grading, but explicit training in these skills will lead to more systematic and productive practices than those in use today.

Communication: Skills for communicating about learning environments/problems

Communication is essential to collaboration in general and to planning and management in particular, which require collective understanding of a learning environment or problem to be solved, and collective planning and collective execution of the course of action agreed upon. Communicating (understanding and making others understand) requires skills such as the following.

Skill in identifying communicational goals, or intent and message, and based on them elaborating and evaluating communicational plans.

Skill in identifying audience characteristics, in particular prior knowledge and biases, and based on this predicting audience reactions and adapting communicational plans accordingly. This requires developing the ability to read from and write for different perspectives or viewpoints.

Skill in choosing and organizing linguistic elements according to communicational goals and audience characteristics.

Skill in effectively acquiring, evaluating and incorporating feedback about text adequacy to communicational goals and audience characteristics.

Any professional learner (research scientist, industrial engineer or product development manager, for example) will attest to the necessity of having all of these skills. With some extra effort, it would be possible to show how these skills are used in even the most menial tasks. The point here is that they provide a framework for organizing what is to be learned and why it is to be learned.

This is the richest notion of learning environments that seems possible to piece together at this time. There is a vast amount and a wide variety of research that can contribute to enriching further this model of learning and learning environments. But, because it is from a range of disciplines, few stop to bring it together. Philosophers and sociologists of science have been analyzing the behavior of scientists for many, many years. The view here is based on such a metascientific perspective. Researchers in machine learning contribute to enriching this model by elaborating further the special role of goals (Hunter, 1995; Ram & Hunter, 1995; Ram & Leake, 1995) in learning, and cognitive scientists who focus on human problem solving contribute their more detailed understanding of how people understand and formulate for themselves the problems they are asked to understand (Rosenbloom, et al. (Eds.), 1993). Sociologists, clinical psychologists and motivational experts can bring to bear their experience on creating motivation and fostering empowerment, and management science can contribute its experience in organizing and fostering effective collaboration. Other cognitive scientists can provide much relevant experience about how people best understand texts or images and how they plan writing about them. Once a research field dedicated to "cognitive ergonomics" (or the study of the principles of mental work) arises, much of this research can be tied together in a single framework. Note that this will be very different from the field of pedagogy, which focuses on *teacher* activities in a learning environment.

The Best Way to Predict the Future is to Construct it (Kay, 1993)

In this chapter it has been argued that the top priority for the University is articulating its mission for the next millennium, a mission based on a clear and coherent conception of the learning environments that will be at its conceptual center and how these environments meet the needs of its constituents and stakeholders. In so doing, it seems very clear that instead of meekly following the changes happening in the rest of society, universities will have to reinvent themselves and the market in time to lead society into a new Age of Innovation. Instead of putting more reality into the learning it fosters, universities will need to put more learning into reality.

For this to happen, the first step is to forget about reform. As the word itself shows, reform is about giving the university a new *form*, rather than new content. The crucial issue is to re-*conceptualize* the university, or redefine its goals and content. Form will follow naturally from there, as will the technology to be used.

At the center of the effort of redefining the goals and content of the University will be the notion of learning environment, which has been our focus of discussion here. Three broad conceptions were discussed, to organize some of the most important issues at stake.

One conception emphasized the university and its faculty as mostly autonomous—the "ivory tower", with weak input from students and society in general. The kinds of learning environments developed under this conception helped develop and concentrate knowledge resources on a vast scale, but suffered from a lack of responsiveness to the needs of students and society.

Another conception emphasized students and their cognitive learning processes, still with weak input from society in general. The kinds of learning environments developed under this conception are beginning to help produce greatly improved instructional materials and active, autonomous learners, but still suffer from a lack of responsiveness to the needs of society, in particular to the demands being made of today's knowledge workers.

The third conception of learning environment that was raised for discussion deinstitutionalized learning environments, generalizing the notion to almost any goal-oriented group activity. This ascribed to the physical environment of the University a secondary importance, in keeping with the current trend toward increasing use of internships, distance education, asynchronous learning networks, on-the-job training and learning brokering. The kinds of learning environments developed under this conception are beginning to address the issues of market relevance of skills, diversity of learning times, places and styles, as well as bringing to the fore traditionally non-academic skills (such as self-empowerment and planning and management of collaboration) for explicit instruction. Accountability to student and marketplace needs are emphasized. Brown & Duguid (1995) explore a variety of organizational options for developing this kind of learning environment.

The term "University" has the same root as "universe", suggesting an attempt to squeeze the universe into a campus bounded by physical time and physical space. "Virtual University" focuses on an effort to free that universe from these physical boundaries, by moving it into the realms of cybertime and cyberspace built to human scale. The next step will be to recognize that this universe is the individual and that the relevant boundaries are only those of collective human organization and creativity.

Implementing any mix of these conceptions provides a wide variety of fascinating administrative and technical challenges.

The language barrier

One stumbling block to more efficient and pervasive learning environments is the fact that most knowledge and information is still very parochial in nature:

restricted groups of people accumulate certain kinds of information (doctors and lawyers, but also restaurant critics and glass makers, for example) and build barriers to this information with jargon and languages. If laws and legal procedures were written in plain English, as they are in Canada for example, much of the market for lawyers' routine advice would disappear. Similarly, many of the visits to the doctor's office are for complaints that any well informed person could deal with by him or herself. Technical jargon, then, hides certain kinds of information, making it accessible only to the initiated.

On a global scale, this problem is much more visible: some countries invest more money in research but one has to learn their language to take advantage of it. More prosaically, a tourist in a foreign country can't even access information about where to find a bathroom or a restaurant without knowing the language that happens to be spoken in that particular place. The problem in these cases and many others is the same: differences in ways of using language (jargon, dialect or language) create barriers to the exchange and use of knowledge and experience, on all scales, small and large. From what's the best brand of diaper to what's the best technique for performing coronary bypass surgery, once we overstep the bounds of our usual jargon and local language, words and sentences become a problem rather than a solution.

The information barrier

With the great strides in infrastructure that the Internet has provided, thousands of times more information is available than any single individual could possibly make use of, even if he did not face any linguistic problems. Perhaps the clearest scenario to illustrate that this is problematic is that of a surgeon in the middle of an operation: he opens up the patient and discovers that the patient has a different or additional problem that was unforeseen. All of the information accumulated over centuries of medical practice is just down the hall in the library and much of it can be accessed through the computer on his desktop. Even if there were a computer right in the operating room would it be of any use? Very clearly, no: none at all. The volume of information available is such that by the time the surgeon found the 30 different techniques he might want to consider, the details of how they are applied and data about their advantages and disadvantages, the patient would be long dead. This is not to say that there are no marvelous computers full of important and relevant information in operating rooms: there certainly are. But they are not Internet access points, and that is the point: broadband access to knowledge is simply useless without the tools to locate it in a reasonable time frame. The question is not only one of speed of access, but of speed of relevance.

Space probes, for example, transmit terabytes of data per day (Fayyad et al., 1996). The Internet increases by an average compound rate of 30% every six months [see www.nw.com]. The global amount of published information doubles every seven years. Music, images and video information is increasingly available.

More information is being collected every day about more people, more businesses, more opinions and made available to anyone who happens to be interested or happens to find it by chance. This situation leads to a grave paradox: the more information is made available and the easier and broader the access to this information is, the less useful the information becomes. The knowledge is only available in principle and not in fact. This is still simply information-rich *ignorance*.

Solutions to the language barrier and the information barrier are beginning to appear. One promising solution is to devise a universal content markup language, just as we already have markup languages for text and hypertext formatting. This would be an international standard for exchanging not text but knowledge, for describing what is meant rather than (or as well as) what was said.[6]

The other major challenge I will cite, although there are many more, is the multifaceted challenge of leadership and "retooling" the university. Current universities, by almost all accounts, have at their head a fund-raiser, a promoter and a manager: the president. It has been said that management is doing things right, but leadership is doing the right things. The decentralized nature of university activities has the advantages of autonomy and freedom, but the disadvantages of lack of cohesion and a weakening of leadership. Moreover, very strong roots in tradition and a focus on theoretical change that leaves little time for thinking about social change have led to a kind of conservatism by inertia. Application of the principles of research to the organization of the university would yield obvious conclusions about adaptability and change, but designing a new university has not been proposed as a collective research problem, that is, in terms that professors are accustomed to thinking about. Management cannot promote a culture of change, but leadership can. In particular, leadership focused on applying the university's vast intellectual capital to reinventing itself.

The path ahead requires major retooling of the University: rebuilding it around an explicit mission, tallying and assembling the resources available for the assault on the marketplace and identifying and obtaining the additional resources needed are major tasks. Vast relevant resources are already available: most of the faculty is already well-trained in designing learning environments ("research design" in its many guises) and they already have experience in content design and delivery. Large amounts of quality research on the cognitive science of learning, machine learning, computer-based instruction, instructional design, the nature of scientific research, the cognition of research, the social psychology of empowerment, group dynamics, language processing, knowledge representation and hundreds of other relevant topics is available to support such an effort. The information infrastructure, in terms of libraries, computers, networks, etc. is enormous. Students' learning experience with the university can be conceived of as contributing to this retooling, so that there are 14 million knowledge workers to assist in the project. Funding of university activities is high, some $250 million per year (Software Publishers Association, 1998), and almost all of the retooling effort can be presented as either research or teaching, to satisfy the demand of this or the other funding agency. Jaffee (1998) provides a useful analysis of the institutionalized resistance to change that characterizes retooling efforts having to do with information technology.

With vision, leadership and management, today's University can become the pivot of tomorrow's Culture of Learning. The cognitive, intellectual and technical resources necessary abound; the specific, technical information needed for drafting and executing a detailed plan of action can be found nowhere else; the challenge of articulating and executing a coherent and audacious plan lies ahead. What seems clear, though, is that the greatest threat to today's university is reform. At the edge of a chasm, a tiny step can spell disaster.

Notes

1 The author is currently affiliated with Logos Corporation (Rockaway, NJ, U.S.A.) and can be reached at mike.dillinger@pobox.com.

2 www.educause.edu/nlii/nliiHome.html.

3 See, for example, the (vast) amounts of press accorded the National Information Infrastructure [nii.nist.gov] and Global Information Infrastructure [nii.nist.gov/gii/whatgii.html] vis-à-vis the (tiny amount of) attention given to debate about what kinds of information should be disseminated with them and how exactly they should be used.

4 www.educause.edu/pub/er/review/ teachLearnIndex.html.

5 FAQ [frequently asked questions] pages number in the thousands and, in my view, are another kind of self-reflexive knowledge; one of the most important parts of the Web: they provide information that can be leveraged into other information. As an indicator of this kind of information, a search with Altavista [www.altavista.com] on "how to" returned some 5 million hits, a search on "tutorial" returned a million and a half, and "FAQ" returned almost 18 million hits, not including the *.answers newsgroups.

6 Automated systems for "translating" text into this standard or interlingua and back again form the backbone for making this a viable solution. Moreover, search engines can explore knowledge bases expressed in this interlingua to find information no matter how it is expressed, with different synonyms or paraphrases, different jargon or different languages. The most advanced project developing this sort of solution is the Universal Networking Language Project (Dillinger, 1997; UNL, 1996) run out of the Institute for Advanced Studies of the United Nations University in Tokyo. Since 1996, groups around the globe have been working to develop the markup language and the language-specific systems necessary for English, French, Spanish, Portuguese, Russian, German, Indonesian, Hindi, Chinese, Japanese, Swahili, Mongolian, Latvian, Arabic and Italian. The importance of these efforts for the development of truly global markets for information, learning, leisure and commerce is difficult to overestimate.

References

Bindra, D. (1984) Cognition: its origin and future in psychology. In: J. Royce & L. Mos (Eds.), *Annals of Theoretical Psychology*, vol. **1** (pp. 1–29). New York, Plenum Press.

Bourne, J. 1998. Net-learning: Strategies for on-campus and off-campus network-enabled learning. *Journal of Asynchronous Learning Networks,* **2**(2): 70–88.

Bourne, J., McMaster, E., Rieger, J. and Campbell, C. 1997. Paradigms for on-line learning. *Journal of Asynchronous Learning Networks,* **1**(2): 38–56.

Brandt, N. 1997. Executive summary, Realizing the potential: Improving postsecondary teaching, learning and assessment. [http://www.ed.gov/offices/ OERI/PLLI/norm2.html]

Brown, J. S. & Duguid, P. 1995. Universities in the digital age. [http://www.parc.xerox.com/ops/members/brown/papers/university.html]

Bruer, J. 1993. *Schools for Thought: A Science of Learning in the Classroom.* Cambridge, MA: MIT Press.
Bunge, M. 1967. *Scientific Research*, 2 volumes. New York, Springer.
Bunge, M. 1974–1989. *Treatise in Basic Philosophy*, 7 volumes. Dordrecht, D. Reidel.
Chomsky, N. 1995. *The Minimalist Program*. Cambridge, MA, MIT Press.
Denning, B. 1999. Teaching as a social process. *Educom Review*, **34**(3).
Dillinger, M. 1997. The Universal Networking Language Project: Principles, Perspectives and Current Work. In: *Proceedings of the UNL-Brazil Workshop*. Rio de Janeiro, UFRJ–Núcleo de Computação [ftp://ftp.nce.ufrj.br/pub/reltec/UNLRio97.zip]
Doherty, P. 1998. Learner control in asynchronous learning environments. *ALN Magazine*, **2**(2), October, 1998.
Donald, M. 1991. *Origins of the Modern Mind: Three Stages in the Evolution of Culture and Cognition*. Cambridge, MA, Harvard University Press.
Drucker, P. 1993. *Post-capitalist Society*. New York, Harper/Collins.
Duderstadt, J. 1997. The future of the university in an Age of Knowledge. *Journal of Asynchronous Learning Networks*, **1**(2): 78–88.
Fayyad, U., Shapiro, G., Smyth, P. & Uthurusamy, R. (Eds.) 1996. *Advances in Data Discovery and Mining*. New York, AAAI Press.
Fletcher, J. 1995. Advanced technologies applied to training design: What have we learned about computer-based instruction in military training? [http://www.ott.navy.mil/1_2/nato.htm]
Fletcher, J. 1996. Does this stuff work? Some findings from applications of technology to education and training. [http://www.ott.navy.mil /1_2/salt.htm]
Graves, W. 1994a. The Learning Society. *Educom Review,* **29**(1). [www.educause.edu/pub/er/review/teachLearnIndex.html]
Graves, W. 1994b. Toward a National Learning Infrastructure. *Educom Review,* **29**(2).
Hanna, D. 1998. Higher education in an era of digital competition: Emerging organizational models. *Journal of Asynchronous Learning Networks,* **1**(2): 78–88.
Heterick, R., Mingle, J. and Twigg, C. 1997. The public policy implications of a global learning infrastructure. [http://www.educom.edu/program/nlii/ keydocs/polícty.html]
Hiltz, S. 1997. Impacts of college-level courses via Asynchronous Learning Networks: Some preliminary results. *Journal of Asynchronous Learning Networks,* **1**(2): 1–19.
Hunter, L. 1995. Planning to learn. In: A. Ram & D. Leake (Eds.) 1995. *Goal-driven learning*. Cambridge, MA, MIT Press.
Jacob, F. 1982. *The Possible and the Actual*. Harmondsworth, England, Penguin Books.
Jaffee, D. 1998. Institutionalized resistance to asynchronous learning networks. *Journal of Asynchronous Learning Networks*, **2**(2): 21–32.
Katz, R. & West, R., 1992. *Sustaining excellence in the 21st Century: A Vision and Strategies for College and University Administration*. Boulder, CO: CAUSE, Educause Professional Papers, number 3008. [http://www.educause.edu/pub/profess.html]
Kay, A. 1993. *The Best Way to Predict the Future is to Invent It*. Xerox PARC Forum, 23 September, 1993, cited in Sherron & Boettcher, 1997.
Kintsch, E., Franske, M. & Kintsch, W. 1996. *Principles of Learning in Multimedia Educational Systems*. Technical Report 96–01, Institute of Cognitive Science, University of Colorado, Boulder.
Martín Perez, M. [this volume]. The ITESM Virtual University: Towards a transformation of higher education.
Mark, G. [this volume]. Social foundations for collaboration in virtual environments.
Mason, R. 1998. Models of online courses. *ALN Magazine*, **2**(2), October, 1998.
Michalski, R. & Ram, A. 1995. Learning as goal-driven inference. In: A. Ram & D. Leake (Eds.) 1995. *Goal-driven Learning*. Cambridge, MA, MIT Press.
Moonen, J. 1997. The efficiency of telelearning. *Journal of Asynchronous Learning Networks,* **1** (2): 68–77.
Pappert, S. 1993. *The Children's Machine: Rethinking School in the Age of the Computer*. New York: Basic Books.
Perelman, L. 1992. *School's Out: A radical new formula for the revitalization of America's educational system*. New York, Avon Books.
Perelman, L. 1997. Barnstorming with Lewis Perelman. *Educom Review*, **32**(2).
Popper, K. 1959. *The Logic of Scientific Discovery*. London, Hutchinson.

Postman, N. 1985. *Amusing Ourselves to Death: Public Discourse in the Age of Show Business*. New York, Penguin.
Ram, A. & Hunter, L. 1995. The use of explicit goals for knowledge to guide inference and learning. In: A. Ram & D. Leake (Eds.) 1995. *Goal-driven Learning*. Cambridge, MA, MIT Press.
Ram, A. & Leake, D. (Eds.) 1995. *Goal-driven Learning*. Cambridge, MA, MIT Press.
Resnick, D. & Resnick, L. 1977. The nature of literacy: An historical exploration. *Harvard Educational Review*, **47**, 370–385.
Resnick, M. 1996. Distributed constructionism. [http://el.www.media.mit.edu/ groups/el/Papers/mres/Distrib-Construc/Distrib-Construc.html]
Rosenbloom, P, Laird, J. & Newell, A. (Eds.) 1993. *The SOAR Papers: Research on Integrated Intelligence.* Cambrigde, MA: MIT Press.
Salomon, G. (Ed.) 1994. *Distributed Cognition*. Cambridge, UK: Cambridge University Press.
Sherron, G. & Boettcher, J. 1997. *Distance Learning: The Shift to Interactivity.* Boulder, CO: CAUSE Professional Paper Series, # 17.
Software Publishers Association. 1998. Executive Summary, *SPA's 1998 Education Market Report: Post-secondary.* [http:www.spa.org/project/edu_pub/ 98EMRPSEXEC.HTM]
Talbott, S. 1999. Who's killing higher education? Corporations and students: the unusual suspects. *Educom Review*, **34**(2): 26–33.
Thayer, R. (Ed.) 1997. *Software Engineering: Project Management.* Los Alamitos, CA: IEEE Computer Society.
Toffler, A. 1985. *The Adaptive Corporation.* New York, McGraw-Hill.
Twigg, C. & Oblinger, D. 1996. The Virtual University. [http://www.educom.edu /nlii/VU.html]
Twigg, C. 1994. The need for a national learning infrastructure. *Educom Review,* **29**(4–6).
UNL. 1996. *UNL: The Universal Networking Language.* An eletronic language for communication, understanding and collaboaration. Tokyo: United Nations University Institute for Advanced Studies.
Watson, B. 1996. Tricks & Traps: Lessons the Microsoft Online Institute has learned. [http://www.uvm.edu/~hag/naweb96/zwatson.html]
Weiser, M. 1991. Computers for the 21st century. *Scientific American*, September, 1991, 94–105.
Williamson, S. 1996. When change is the only constant: Liberal education in the Age of Technology. *Educom Review,* **31**(6).
Wurman, R. 1997. Technology and learning: Celebrating the connections. [Interview with Richard Saul Wurman]. *Educom Review,* **32**(6).
Yahoo! Internet Life. 1988. p. 44.

Webliography

distancelearn.miningco.com – about.com (formerly the Mining Company)'s distance learning GuideSite.
nii.nist.gov – National Information Infrastructure ("the information superhighway") official homepage.
nii.nist.gov/gii/whatgii.html – official Global Information Infrastructure page.
promo.net/pg – Project Gutenberg, an important electronic text initiative.
sunsite.unc.edu/nii/NII-Table-of-Contents.html – another National Information Infrastructure site.
web.ansi.org/public/iisp – the American National Standards Institute's Information Infrastructure Standards Panel site.
www.adec.edu/virtual.html – the ADEC Distance Education Consortium.
www.aln.org – Asynchronous Learning Networks site.
www.aln.org/alnweb/journal/jaln.htm – Journal of Asynchronous Learning Networks.
www.altavista.com – An important search engine and metasite.
www.cudenver.edu/~mryder/itc_data/net_teach.html – Metasite on "Teaching and learning on the Internet" by Martin Ryder.
www.educause.edu/nlii/nliiHome.html – National Learning Infrastructure Initiative homepage.
www.educause.edu/pub/er/review/teachLearnIndex.html – Educom Review's Teaching and learning index.
www.emtec.net/links/construc.htm – Metasite on "Constructivism and related sites".
www.gii.org/index.html – the Global Information Infrastructure Commission's site.
www.about.com – formerly the Mining Company, an important metasite.

www.sil.org/linguistics/etext.html – The Summer Institute of Linguistics' page of links to digital libraries and electronic text projects.
www.spa.org – Software Publishers Association.
www.thinkquest.org – the Thinkquest competition.
www.virtualuniversity.com – an Association of Virtual Universities.
www.yahoo.com – Yahoo!, an important metasite.

4

Reforming the Educational Knowledge Base: Course Content and Skills in the Internet Age

JANUSZ SZCZYPULA, TED TSCHANG and OM VIKAS

Introduction

We are living in an era in which technological and societal changes are rapid, virtual social and business environments are increasingly common, and vast information resources are an integral part of the current work environment. IT and the Internet allow timely and comprehensive access to huge amounts of distributed information and knowledge. In this technology- and knowledge-rich environment, both institutions and individuals have to constantly and quickly deal with complex situations requiring strategic decisions. The time for appropriate responses (i.e. the decision-making time) is significantly reduced, while at the same time, there is the need to process huge amounts of information for these decisions. This places an overwhelming cognitive load on humans as well as a computational load on machines. As a consequence, new thinking abilities are required on the part of workers, such as the ability to process diverse forms of information into knowledge (where we define knowledge as the innate ability of one to make sense of information, or to cater information to specific situations). Other obstacles to working effectively in these new environments include poor "connectivity" (broadly defined as the degree of isolation or disconnection of people from information), and from the increasing rate of change in the environment, which makes keeping up with trends much more difficult.

To better prepare future professionals for rapidly changing and knowledge intensive work environments, new curricula, programs and learning paradigms are being developed by universities and other educational institutions. Information technology (IT) curricula have been available for a number of years, particularly in business and other professional schools. However, the development of the

Internet, and its profound effect on the way in which society and business operate, is causing a qualitative change in the curricula of almost every field of study. The Internet influences traditional fields as well as brings about the creation of new fields and programs (e.g. Internet commerce) emphasizing the retrieval and manipulation of information, including specialist fields such as information security management.

Whereas traditional thinking treats curricula as well-defined subjects, and assumes that students are "empty vessels, into which knowledge is poured," some of the current U.S. educational thought is evolving towards curricula revisions based on constructivist and related ideas. In fact, some even believe that the traditional notion of curricula has run its course (Gehl, 1996). Constructivism focuses on problem solving, integration of diverse knowledge, the introduction of students to diverse sources of knowledge, collaborative (group) learning, and student-centered learning (Dowling, 1996). Furthermore, in information-rich environments, it is often advocated that teachers should become facilitators and guides, rather than remain in their traditional roles as instructors.

In general, while on-campus curricular content has been changing significantly to reflect the Internet's influence on the society and economy, other, more subtle changes in thinking and learning abilities, such as the ability to think in constructivist and independent fashions, are also increasingly visible. In fact, many universities have already been teaching some of these skills in conventional ways, e.g. through group seminars. The question is whether Internet-based learning environments can duplicate these same teaching/learning pedagogies. There is no doubt that the Internet is itself a constructed environment which creates the impetus for change, if not provides the tools to do so. New learning pedagogies may be supported by many new tools being introduced on the Internet, ranging from browsing tools to Web-based simulation software. Whether these changes are sufficient to carry traditional universities through the wrenching societal changes to come is another open issue.

This chapter examines what we call the educational knowledge base of a university, which we define as the composition of three areas that students need to know in order to function in new workplaces:

- new curricula content and subjects
- new learning and thinking abilities, and the pedagogies for fostering them
- new skills, such as the ability to use IT and the Internet.

While our study is mainly focused on curricula content, we will provide a discussion of some of the main kinds of learning and thinking abilities needed, as well as the uses and advantages of Internet-based technologies in learning.

Our study focuses on two geographical perspectives: (1) that of traditional comprehensive campus-based universities in the U.S. which are developing new curricula to address the needs in information societies, especially those arising from the growth of the Internet, and (2) that of India's educational system, which provides an instance of a developing country that is positioning itself to respond to these technological forces.

In addressing these issues, we also recognize that a broader context exists. For one, the Internet trends, such as the demand for professionals and skills, are also part of a wider set of technological changes and imperatives. In the case of developing countries, a great need also exists to provide for more basic information technology skills and infrastructure.

These changes in education are also part of a trend in which schools have to be more relevant and responsive to societal needs, or at least cognizant of broader social trends brought about by technologies. In this context, education should be viewed as an open system that fulfills many objectives, both personal and societal. However, since our focus is on how curricula changes in traditional institutions address the needs of emerging knowledge-based economies, we will not be able to address other salient perspectives, such as the needs of individuals for lifelong learning, or for enriching their lives, e.g. developing appropriate personal philosophies towards these significant changes in their lives. Ultimately, these changes that we see reflected in individual courses or select programs of study could lead to overall changes in entire degree programs or university offerings.

In the first section, we motivate the treatment of the knowledge base by a discussion of the trends in the society and economy, seen in the increasing demand for IT professionals and skills in the U.S. In the second section, we examine the new Internet-oriented curricula being developed in American business, public policy, engineering, and other schools. This section examines the Internet's influence on curricula in terms of: course content, learning and thinking modes, and the utilization of IT tools. In focusing on the U.S. perspective, we recognize that, while it is not necessarily the dominant or preferred model for the rest of the world, the U.S. situation is nevertheless reflective of the most rapid changes of the information societies. In the third section, we will examine some of the critical issues pertaining to developing countries, with a particular focus on India, which has a rapidly emerging information technology sector, and is positioning its education systems to respond to the needs of this sector.

The New Human Resource Needs of the Knowledge-based Economy

One of the important roles of higher education systems is to support the changing needs of society and the economy. Universities are being hard pressed to track and respond to these changes, including the increasing global nature of society, the impacts of new technologies, an increase in informal networking, increased competition, and new needs for human resources (Duderstadt, 1997). In this section, we will emphasize the new needs of the knowledge-based economy, since those are driving many of the changes in professional education.

Employment needs

While the IT industry may be the primary sector in the knowledge-based economy, IT and other technologies influence all types of industries and society, making them knowledge-intensive as well. Despite the preponderance of technology, and its potential effect of displacing traditional occupations, people are still a critical resource in IT-based environments. Rapid advances in IT, the rise of the Internet, and the resulting new sources of growth in the economy are creating new occupations, and consequently, a high demand for skilled IT professionals.[1] However, universities with their traditional, somewhat inflexible, curricula haven't been able to train IT professionals in sufficient numbers or with adequate skills to fill the changing and growing labor needs.

The growth in demand for competent and skilled IT professionals in the past few years has been, and continues to be, overwhelming according to the Information Technology Association of America (ITAA). A survey of 104,000 IT and non-IT U.S. companies with more than 100 employees indicated that there have been about 346,000 unfilled IT positions in the United States in three core IT occupations: programmers, systems analysts, and computer scientists/engineers (ITAA 1998). The estimated number of vacancies in broadly defined IT professions was 606,000. Further, the total number of IT employees is estimated to be over 10 million, with projected growth within the next five years of over three million. These estimated numbers and the growth resulted from a variety of factors, including the rise of electronic commerce and the "year 2000 problem", as well as the reclassification of occupational categories, such as more programmer positions being reclassified as systems analysts. Some companies have adopted multiple strategies to deal with their needs for IT professionals. Outsourcing to countries such as India or Israel, with their large IT labor force, or hiring new immigrants are two major strategies (USA Today, 1997; Mosquera, 1998). However, the gap between the supply and demand of skilled IT professionals appears to be global, as indicated by reports from Canada, UK, and other countries (ITAA 1997).

While information services and other IT sectors are creating an increased need for IT workers, industries outside of IT will also grow and improve their productivity through IT. In particular, the Internet's role in raising the demand for IT professionals is profound. The Internet is leading to the creation of diverse new businesses, while at the same time transforming traditional businesses. According to Business Week, 760,000 persons were working at Internet-related companies in 1997.[2] However, the number may be greater if we include traditional (i.e. non-Internet) companies' in-house staff members who use the Internet in their work.

New types of professionals

The overall continuation of employment growth will depend critically upon the existence of a new workforce skilled in IT and possessing interdisciplinary knowledge of how the Internet creates new work environments, as well as changes in traditional work settings. Thus, workers will need to have some knowledge of IT and the mechanics of using the Internet and manipulating the information on it, as well as knowledge of the new economic, legal and other structures that affect the characteristics of information. In the new Internet economy, we could say that there are several breeds of "Internet inhabitants", such as new entrepreneurs and traditional businesses who use electronic commerce to reach larger numbers of customers or to better develop products and services; technocrats who manage the public aspects of the system's infrastructure or make policy for the Internet; media professionals and artists who use hypermedia tools as a work medium, to present their work in virtual settings, and to serve other inhabitants in their sophisticated presentation needs; and information specialists needed to retrieve, manipulate, and use widely distributed information. The growth of virtual organizations, such as virtual communities and companies relying on networks to tie together their worldwide operations, will further create new occupations. These jobs will require a mixture of skills and thinking abilities that goes beyond the expectations of a traditional occupation. Some job descriptions will have to be defined on the job, such as those involving the management of virtual communities, organizational knowledge and other new creations.

Restructuring educational systems to meet new needs

Curricula changes are being made quickly in U.S. universities in response to workplace needs, though perhaps not across all universities at the same pace. Some of these are cosmetic patches such as adding a course or two to help old programs look new, while others are more fundamental. U.S. higher education has become increasingly more needs driven, particularly with the rise in professional practices and professions. For instance, the ITAA has proposed that the current educational paradigm needs to be re-examined from the perspective of the knowledge-based economy. Further, the educational system needs to respond to the changing requirements for new professionals and produce world class work force for the Information Age.

All across the world, the role of education is being revised, but the degree with which education centers on technology varies from country to country. In the U.S., the more dominant strategy is the "technology-push" approach, which may be partly due to the strength of the technological advances, and partly because of how the federal government formulates science and technology policy (e.g. making IT in education the national priority). For example, both the public and private sectors fund research programs in many schools, including the Internet II, the National Science Foundation's digital library initiatives, the National Institute for Science and Technology's Web education program, and the Sloan Foundation's Asynchronous Learning program. These incentives cause institutions and faculty to invest heavily in technology, and to focus on technology-based approaches to teaching. Often, this ends up becoming "more of the same teaching", except technology-assisted. Nevertheless, countries need to invest first in information infrastructure, making the initial impetus at the national level a technology-push situation. However, it is critical for institutions to think critically about what does and does not work, and how to make it work. While technology is often portrayed as a savoir, it has been proven many times that a narrow focus on purely technology solutions does not work, and that an equal if not greater appreciation of content creation, pedagogies and other non-technological activities is necessary for achieving educational success.

Towards a framework for reshaping curricula to workplace needs

Our conception of curricula as three distinct, but related, components can be used as a basic framework to identify solutions to current workplace needs for skills. While many employers, particularly in IT workplaces, commonly request specific skills of new employees, they often cannot get those types, or have to work with fresh graduates who are inappropriately-skilled. Often times, graduates with the most general training and good thinking skills are the most readily retrainable. In this circumstance, while *technically* skilled employees may be the preferred ideal of employers, the reality is that *thinking and learning* skills are the most valuable for socializing into the organization and its work. Furthermore, given that the two are intertwined, it is likely that some balance of the two types of knowledge would be optimal for employers as well as for employees as a whole.

A second type of relationship between these aspects of the curricula involves the working context, which is highly critical for transforming conceptual knowledge into working knowledge. In particular, technological skills can only be effective when brought to bear within specific working or applied contexts. In our framework, *curricula content* addresses this by providing students with the necessary contextual knowledge to frame problems and develop solutions. Thus, content provides a context for the development and application of both thinking skills and technological skills, both in the classroom, as well as for work.

The changing nature of knowledge and curricula

University curricula traditionally consisted of didactic, well-defined or circumscribed, and somewhat immutable content (i.e. slow to change), all of which changed incrementally with scientific progress.[3] However, the present day's more rapid technological metabolism, including changes caused by the advent of the Internet, are making this model obsolete. Changes in modern day curricula are driven more by the rapid and fundamental shifts in the environment for information. New information and knowledge are being created in the Internet domain in quantities and at rates that may be faster than traditional universities can absorb, or teach. The nature of the Internet makes time-dependent information a valuable commodity, and at the same time makes the lifespan of knowledge (or the period of its relevance) much shorter. The knowledge constructed on the Internet is also highly distributed and therefore more difficult to find, more diverse (making the ability to integrate knowledge more crucial), potentially inconsistent, and of variable quality.

In the face of these changes, universities can either ignore the situation or confront it. Of the three components of the educational knowledge base, curricula contents are changing most rapidly (as appeared to us during our survey—discussed later on); changes in the other two being more subtle. Some changes to curricula may occur within disciplines while others require interdisciplinary approaches, or the fusion of different types of academic activities, e.g. the introduction of research into teaching environments.

Universities are for the most part still experimenting by adding courses and new programs of study. These are typically done by applying traditional methodologies and fields of study to understanding the Internet's impact on society and the economy. In some sense, this slow rate of change is due to a generational gap between faculty and the real world. An example of this is the field of Internet economics, which examines how traditional economic methodologies can be brought to bear on issues such as pricing access, services and information on the Internet.

The Influence of the Internet on the University's Educational Knowledge Base

In this section, we look at how the Internet and other associated ITs have influenced the university's knowledge base. In particular, we focus on the current curricula to students in traditional universities in the U.S., as they reflect the perceived needs of workplaces. However, one open question needs to be kept in mind: even as these changes help students become more employable, do they also fulfill more personal needs or other societal needs, e.g. learning how to learn (for lifelong learning), or how to reflect on changes in one's life and one's position in society?

Figure 4.1 illustrates one representative structure for the knowledge base. Our treatment of curricula considers course content as a part of a larger system of learning that also includes the learning and thinking modes and the tools of instruction. The Internet impacts each of these in profound ways, some of which we will examine in the following subsections.

Whereas traditional curricula tended to be organized by well-defined subjects and set according to standards of disciplines, nowadays it is increasingly common to find interdisciplinary approaches, studies addressing real-world problems, and a reliance on a wider variety of information sources. The Internet has had an even greater influence on this disintegration of information, and subsequently, on corresponding needs to reintegrate that information.

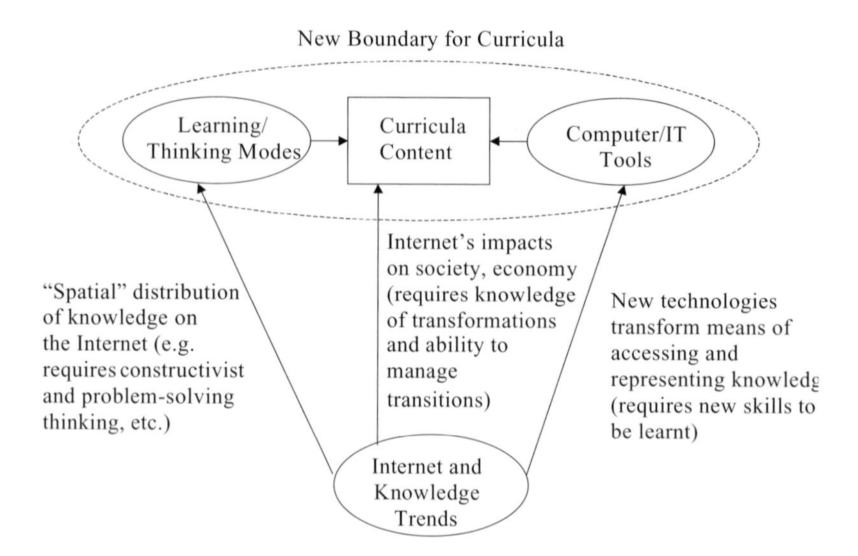

FIGURE 4.1
The Convergence of Trends in Curricula.

The new environment for knowledge

The new environments for creating and managing knowledge, particularly those involving the use of the Internet, can be characterized by several factors:

- Diverse sources of knowledge and large quantities of information, which complicate the task of retrieving and processing that information for decision-making, as well as using it in further knowledge creation. In areas such as business, the digital archival of information such as sales data provides a basis for supporting decisions.
- Rapid diffusion of information, which reduces the life span of many types of current information. For instance, fractions of a second can be critical in the accessing and use of financial information such as stock prices.
- Distributed nature of the system, which can cause the localization of information such as through privatization or the formation of network communities. This creates barriers to the access of information.

In general, the overall "system", comprising of the different sources and locations of knowledge, involves many individually-acting and at the same time interacting activities. For example, the effect of new IT is often not anticipated until it rapidly diffuses, catalyzing a transformation in the way a business is handled. Traditional business operations are being transformed by the Internet. Well-known examples are Amazon.com's book-sales over the Web, which has transformed the book selling business, and the transformation of the encyclopaedia business, like the forced revamping of Encyclopaedia Britannica's whole operation (Evans & Wurster, 1997). In addition, to facilitate access to information, organizations need to design information environments that are customer-friendly, as well as useful for them. Individuals also need to be able to access and use information, and to integrate it into their daily decision-making.

A knowledge-based economy also requires the devising of new forms of education to support new socio-economic functions, including the needs to manage information and virtual environments such as virtual marketplaces, and to formulate public policy. All this challenges schools to provide substantially different types of education and skills to students. At the same time, schools must recognize that content must change as rapidly as the business or social environment changes. The Internet-related course of today will be dated by tomorrow. Given this dynamic state of affairs, the review that follows should be taken only as an indicator of the direction that programs are headed in.

Curricula content: New courses and programs of study related to the Internet

The Internet is creating opportunities for both private and public enterprises to be more agile and knowledge-intensive. This requires their employees to have new skills and knowledge. Thus, universities are creating new fields of study as well as changing the curricula of traditional fields to focus on the Internet. To study the current trends, we surveyed leading schools for various fields of study and their courses related to the Internet (shown in Table 4.1)[4, 5] At the same time, we selectively sampled several emerging research areas, particularly in the engineering and the social sciences, in order to help frame the current trends in course development.

In the table, the *Topics and Trends* column illustrates the subcategories that courses can be classified under. In some cases, the column refers to trends on how the Internet is transforming the field. The *Examples of Curricula* column lists some of the non-conventional courses that have been added to curricula to reflect the transformation/trends. On the whole, we can see from the table that Internet technologies are altering many spheres of economy and society, and that activities in these spheres are increasingly overlapping.

In the following sections, we examine some of the broader curricula changes in each field of study, as determined from our survey of courses and a review of the academic literature to examine the new programs and the thinking in their fields of study.

TABLE 4.1
Changes in Curricula of Selected Fields of Study.

Field of Study	Topics and Trends	Example of Curricula
Business	IT and telecommunications	Telecommunications Management (CMU); Communications and Connectivity among Information Systems (MIT); Telecommunications Technology and Competitive Strategy (Penn); Internet Resources (UT)
	Internet and electronic commerce	Electronic Commerce (CMU); Electronic Commerce and Marketing (MIT); Information, Technology and Business on the Internet (Arizona)
	Specific topics in electronic commerce	Principles of Internet Marketing (Stanford); Seminar in Internet Marketing (Stanford)
	Competitive strategy, organizational change, and business transformation	Building Information Age Businesses (Harvard); Managing in an Information Age (Harvard); Managing Virtual Organizations (UT); Information Technology and Business Transformation (MIT); Competing in the Information Age (Harvard)
	Policy environment for business	Internet Business Strategy: Law & Policy (Harvard); Information: Industry Structure and Competitive Strategy (Penn); Technology in Global Markets: Corporate Strategy, National Policy (Penn)
	International business environment	Global Information Infrastructure and Business Strategy (CMU); International Dimensions of the Information Technologies (Arizona)
	Other	Social Issues of Computing (Arizona); Economics of Information (UCLA)
Public Policy	Technical	Information and Telecommunications Protocols and Policies (Harvard); Economics of Information Management (SUNY-Albany); Electronic Commerce (CMU); Internet Commerce: Technologies and Issues (Harvard)
	Social and organizational	Telecommunication & Society (Georgia Tech); Information Technology & Social Accountability (CMU); Ethics & Values in the Use of Information Technology (Michigan); Mass Communication Policy (Georgia Tech)

Field of Study	Topics and Trends	Example of Curricula
	Public Policy	Public Policy and the Internet (CMU); Government & Politics in the Cyber Age (Michigan); The Politics of Telecommunication (Georgia Tech); Information Security Management (CMU)
Law *	Trend: Regulation of the Internet, intellectual property, etc.	Internet Business Strategy: Law & Policy (Harvard); Law, Internet & Society (Harvard); Ethics and Law on the Electronic Frontier (MIT) jointly with The Law of Cyberspace-Social Protocols (Harvard); The Law in Cyberspace Seminar (Wayne State University)
Library and Information Science	Trend: Library functions in digital environments	Telecommunications Systems (UNC); Telecommunication and Information Network Technology (Syracuse); Network Security (Pitt); U.S. Telecommunications Policy (Pitt); Wireless Interactive Communications (Syracuse)
Schools of Information	Trend: General, interdisciplinary programs that deal with information acquisition, management, processing etc.	Web Security and Commerce (Michigan); Information Users and Society (Berkeley); Information in Society (Berkeley); Copyright and Community: The Future of the Information Society (Berkeley)
Engineering*	Trend: Transformation of engineering by the Internet and IT tools	Rapid prototyping courses (in manufacturing engineering); Agent-based engineering; Simulation and visualization tools (e.g. in chemistry)
Social Science*	Trend: Study of socio-economic phenomena on the Internet, or caused by it	Study of collaborative work; Organizational change; Internet economics; Virtual communities-Sociology of Cyberspace (UCLA)
Art, Media and Design*	Trend: Advent of new media for design and display (e.g. computer art, virtual museums)	Designing for Dynamic Interaction (CMU)

* based on partial survey

Business

In the field of business, IT has been transforming markets, products, and business processes. For instance, electronic marketplaces are being designed to enhance the World Wide Web's functionality for matching buyers and sellers;

facilitating the exchange of information, services, goods and payments; and providing institutional infrastructure such as legal and regulatory frameworks (Bakos, 1998). The Internet has the potential to transform business relationships, both between firms and their customers, and between firms themselves, e.g. buyer–supplier relationships. Business economics will be significantly altered, such as the elimination or lowering of transportation and transactions costs. At the strategic level, this means that the Internet has the potential to fragment traditional value chains, which could lead to the rise of new businesses (Evans & Wurster, 1997). Thus, businesses will need to be able to deal with a variety of issues, including information, virtual organizations (and other organizational change), new markets, policy environments, and competitive strategies. Businesses need to be highly adaptive, in their ability to use tools for controlling processes, and to deal with the wider environments.

Information can be considered as a key input to production or as an output "commodity". An example of the first aspect is businesses' use of information to improve their practices, such as the use of software "robots", statistical techniques and other means to gather and analyze consumer behavior from diverse sources of knowledge such as websites. One such example is the technology developed by the Firefly Network, which helped businesses develop consumer profiles and determine consumer behavior from Web search information.[6] At the same time, such technologies can also be used to help consumers specify their preferences in searching for information from websites. New techniques of data mining/warehousing also combine hardware and software to help companies to correlate huge, complex sets of data for analyzing their business operations, e.g. determining marketing and sales patterns, and thereby improving them.

Universities have recognized the significant influences of IT and the Internet by modifying their curricula to include courses that provide an understanding of several aspects of the IT and the Internet. Based on our survey of business school Web sites, we can classify the courses into several themes:

- *IT and telecommunications*—many schools are offering telecommunications courses that focus on telecommunications technology and its management. Telecommunications courses cover topics such as the various kinds of networks (e.g. data, voice, local area, or wide area). Other courses in IT cover a range of topics including hardware, software, and emerging technologies like heterogeneous and multimedia databases. Some courses also cover trends in key technologies, such as distributed, open computing, and the need for interoperability (or connectivity issues).
- *Internet commerce*—these are basic courses that many schools are offering. The courses cover basic principles on topics ranging from different models of electronic commerce, to general computing technologies. Other courses focus on selected topics like pricing and payment schemes, software agents, virtual communities and so on.

- *Specific topics in electronic commerce*—while "e-commerce" could appear to be a broad field, specific courses have also been developed to tackle specific issues in Internet marketing. These courses focus on topics such as engagement of consumers, conceptual business models and Internet marketing strategy, as well as practical aspects, e.g. design for e-commerce web sites. Some universities are already offering MBA concentrations in electronic commerce, and some are planning complete master's degrees (Davis et al., 1998).[7]
- *Competitive strategy, organizational change, and business transformation*—these courses bridge the gaps between traditional thinking and the broader implications of IT for competition. These courses examine topics in new management concepts, including emerging business models, the transformation of businesses, the competitive/strategic implications of IT, and IT as an integrating force and catalyst for organizational change.
- *Policy environment for business*—these courses examine the interactions between public policy and business. Topics include the current technology policies of the U.S. government and interactions with businesses and their strategies, the economics and technology of telecommunication markets, emerging public policy and market structures.
- *International business environment*—courses on international perspectives cover the international dimensions and comparisons of IT strategies and policies, at the firm, national and global levels.
- The general view that emerges from this sample of courses is that the influence of IT extends to encompass a much larger operating environment than simply business activities. The Internet essentially creates new interactions and considerations for business at all levels (particularly the national and global levels), and in many functions ranging from internal ones like organizational change to external ones like strategy and marketing. The most commonly found courses, perhaps deemed the most essential to programs, are general courses on technology, electronic commerce, and business strategy.

Public policy and law

Together with the growth in public networks, services and electronic commerce, a parallel need has emerged for more effective public policy towards the Internet domain. Public policy and law schools have started experimenting with curricula reform with varying degrees of emphases. From the sample shown in Table 4.1, we can see the changes in public policy and law school curricula taking at least three directions:

- *Technical*—courses that provide an understanding of the communications and information technologies, including the Internet, and their effective uses.
- *Social and organizational*—courses that examine how to manage the social, political, institutional, and organizational aspects of communications and information-related environments. Some courses examine issues affecting society

and individuals, including values and ethics, security, privacy, and government control.
- *Public policy*—courses which examine national and international policy-making on communication and information environments. Important issues include standards-setting, legal principles and regulation.

Some of these courses are similar in content to what is found in business school curricula. In fact, public policy schools like Carnegie Mellon's Heinz School and Harvard's Kennedy School have also started to offer electronic commerce courses. Increasingly, we are starting to see the fusion of Internet and information issues with other traditional areas of public policy, ranging from the conduct of international diplomacy, to national affairs such as media and government services. Furthermore, rising Web use by grassroots organizations such as human rights and environmental organizations will also inevitably spawn course content relating to them.

With the advent of a "global" Internet where "information wants to be free" but producers of knowledge want to be compensated, issues of intellectual property have become critical. Law schools and other programs are adding new course offerings on this topic, such as the *Internet Business Strategy: Law & Policy* course at Harvard, which treats legal issues of intellectual property protection, rights of privacy, content control, antitrust, and problems of jurisdiction. As the Internet environment starts to dominate or integrate with the conventional business and social environments, courses such as these will eventually have to become part of mainstream curricula on public policy and law.

Library and information science/information management (LIS/IM)

Specialized programs in information technology are now commonly embedded in Library and Information Science (LIS) curricula. Many LIS programs have traditionally focused on how the functions of libraries could be done electronically, including functions such as archiving, information storage and retrieval. However, LIS curricula are undergoing a fundamental transformation at many schools (Dalrymple, 1997). New breeds of schools, such as *schools of information*, sometimes partly reconstituted from LIS programs, are also emerging (Marcum, 1997). In contrast to the traditional LIS programs which specialize in libraries, these information management programs train graduate students for broader information environments, such as information management in private enterprises. Two of the most recent of this type of program are the University of California at Berkeley's School of Information Management and Systems (SIMS), and the University of Michigan's School of Information (SI). Both schools are interdisciplinary. Berkeley's SIMS seeks to educate information managers who are skilled at "locating, organizing, manipulating, filtering and presenting information".[8] This involves the management of technology, information and people; familiarity with law, economics, ethics and management; and training in computer science, cognitive science, business, law, library/information science, and communications.

In similar fashion, Michigan's SI has a heavy multidisciplinary emphasis. In particular, the core of the master's program involves academic disciplines such as the behavioral sciences, the organizational sciences, and the computer sciences. The curriculum is based on librarianship, information and computer science, business, organizational development, communication, and systems engineering. The program uses a human-centered approach to teaching and learning that stresses the collateral learning between technologists, end-users, educators, and others. The teaching and research are augmented by affiliated faculty from other colleges and schools. Practical experiences/internships are also offered at these schools as a way of bridging the conceptual and theoretical knowledge developed in programs, with industrial needs that are rapidly changing.

Based on our partial sample of programs, the courses at LIS and information management programs can be roughly classified as follows:

- *LIS or information management courses*—both the longer-standing LIS courses and the related information management courses look at information sources (e.g. databases), and administration and retrieval of information in its various forms, for different contexts ranging from libraries to business.
- *Technical courses*—courses describing the essential elements and concepts of information and communication systems, including Web-based applications and services. These are similar to the technical courses offered in policy and business schools but have somewhat different emphases.
- *Courses on policy and the social aspects of IT*—As in public policy and business schools, LIS/IM courses are developed to examine the social, legal (e.g. intellectual property), and policy issues pertaining to information systems and the Internet. These address the social and private needs to manage or regulate the growth of information and its uses.
- *Other topics.* IM and increasingly, even LIS programs, also offer courses on a variety of topics, including information system security, economics, cognitive-social learning, and software agents.

Many programs are multidisciplinary. The instruction within a particular course may also employ multiple paradigms and perspectives in the examination of a topic. For example, the *Copyright and Community* seminar at Berkeley's SIMS attempts to construct "an appropriate copyright policy for communities using digital documents in networked environments", by "exploring various social theories for information society" (including concepts from economics and electronic commerce), and examining what "conception of society" technologists are building into network technologies. In similar interdisciplinary fashion, the foundation course on *Choice and Learning* at the University of Michigan's School of Information uses both cognitive psychology and microeconomics to provide a basis for the planning and design of information systems. This type of course offers a strong theoretical training, but often has to be team-taught by specialists from the different disciplines.

Engineering

While the contributions of engineering and computer science to the Internet are quite obvious, the Internet has also changed the way in which engineering and product development is handled. Nowadays, in virtually connected environments, new product development often happens on "Internet time", meaning that product development life-cycle stages can overlap and are compressed in time (Iansiti & MacCormack, 1997). Similarly, complex products are now developed in "virtual enterprises", or enterprises that are networked and coordinated using the World Wide Web (Fielding et al., 1998).

Some of the learning experiences that the IT and the Internet helps facilitate include:

- *New simulation tools*—simulation is a mainstay of the engineering design process, and is increasingly used in science curricula such as chemistry. New tools, such as interactive simulation tutorials, are located at increasing numbers of websites, serving as important resources to students. An example is the tutorial developed at the University of Michigan/Carnegie Mellon on control engineering, using MATLAB mathematical modeling software. The tutorial is accessible through the Web, and has had thousands of accesses (Bollentin, 1998). In the sciences, the Internet is now becoming a home for simulation and visualization tools, including those for three-dimensional representations of complex chemical molecules, and other information (Krieger, 1996).
- *Remotely operated laboratories*—The Internet can allow scientists to remotely operate expensive scientific equipment, more commonly known as collaboratories (Kiernan, 1997). This allows the more cost-effective sharing of resources, as well as providing accessibility to larger bodies of students.
- *On-line learning networks*—these can provide a richer learning experience by allowing students to interact with their peers or with tutors at a distance (Bourne *et at.*, 1997). They may be used for anything from class discussion to group problem-solving and projects. In fact, these sorts of networks are more generally applicable to other fields of study.

More advanced industry trends are being reflected in engineering curricula. One such trend is that of virtual manufacturing. Manufacturing engineering programs at various schools teach virtual prototyping as a way of realizing concurrent engineering principles, which seek to make information flows, design processes and other aspects of engineering seamless and integrated. A typical course on rapid engineering design, prototyping and fabrication techniques allows students at one campus to design parts in virtual form, then to "send" them to other campuses for rework and/or fabrication. The experience integrates the use of simulation and other IT tools for design. The goals are to encourage students to experience the design in new ways that "should not only be about the success of the final product, but about learning to experiment, fail, converse, share, reflect, and create. Learning these skills may require unlearning some of the skills acquired in a standard undergraduate engineering education." (Finger et al., 1996).

Some research areas are so new that they have yet to be seen in courses, but that is a matter of time. One such area is agent-based engineering, which enhances manufacturing design capabilities with software agents. Software agents can be used for automating almost anything on the Web, from searching for and brokering information to the exchanging of engineering design parameters (Petrie, 1996). One typical configuration in engineering has multiple agents assisting users in gathering information on design parameters, then sharing and reconciling the parameters via electronic brokers, for eventual use by computer-based design tools.[9]

Social science

The social sciences have been relatively slower at researching and teaching about how the Internet has impacted society or the economy, but curricula are gradually being revised. A number of research projects have been done on the Internet's impacts, and faculty and researchers are increasingly feeding their ongoing research into both introductory and graduate level courses. Some examples of new research topics in social science fields are:

- *Internet economics*—the field of Internet economics looks at how economics and economic mechanisms can be brought to bear on resource allocation problems in the Internet. An example of a topic is the use of price mechanisms to allocate bandwidth and reduce congestion on the Internet.[10]
- *Psychology and organizational theories*—psychologists and other social scientists have also begun to study the interactions of IT with organizational and social contexts, including collaborative work, email usage, and the design of new virtual environments. Computer supported collaborative work (CSCW) is a common topic in computer science/psychology that has many applications (in this volume, Gloria Mark's chapter on collaboration discusses some aspects of CSCW).
- *Sociology*—Despite the large presence of virtual communities, sociological courses on the Internet are still rare. One example is UCLA's course on *The Sociology of Cyberspace* that examines how sociological concepts such as trust, social order, collective action, and markets, can be used to describe social interactions and community formation in cyberspace.

Traditional disciplines and their rigorous methodologies are also being applied to the study of Internet phenomena. For example, the "HomeNet" project studied how the Internet was used in home environments, including the effects on peoples' psychology (Kraut et al., 1997, and 1998). Innovative approaches for characterizing social interactions on the Internet are also being developed, such as the Netscan tool, which seeks to create a way of visualizing the "social" phenomena of the Internet (Smith, 1997).

As far as career opportunities, one group of technologically inclined social scientists and social commentators have gone so far as to speculate that a new discipline that studies virtual communities could emerge. Such a discipline could involve sociology, psychology, anthropology, and even urban planning. On the other hand, it is increasingly common to find anthropologists and other social scientists working in research institutes and computer companies in Silicon Valley and elsewhere.

Arts and design curricula

In the arts, new programs are also fusing traditional disciplines such as arts and communications with modern ones such as electronic media and computing technologies. One such program is the Renssalaer Polytechnic Institute's Electronic Media, Arts, and Communication (EMAC) degree program that combines courses and projects on electronic arts and technical communications.

Carnegie Mellon's Master of Design in Interaction Design is a two-year professional program for students who want to explore the new arena of human–computer interaction design and its many practical applications. The program exposes students to trends such as:

> "the emergence of smaller and mobile computing devices, and their contexts for interaction. Ubiquitous computing, "out of the box" solutions and web-based commerce are recent topics of exploration. They join our traditional areas of focus: effective human–computer communication, visualization of, and navigation through information spaces, time-based information design, and collaborative design among various disciplines and across distances."[11]

In the field of architecture, many programs are also taking leaps into virtual space. At Columbia University, every graduate architecture student is required to have his or her own website. Some critiques of design work are done by reviewers online and at a distance. Further, many advanced-level design studios are paperless, and some classes are held over the Internet.[12]

General observations

From our examination of the public policy, business, and library and information science schools, it appears that the Internet has brought several fields closer together. This points to a convergence in disciplines, and a constructivist way of thinking as a dominant paradigm. Examples of this are:

- *Similar types of courses being offered across a variety of departments and programs*—Where the Internet is concerned, there are no clear boundaries for where a given topic or area belongs. For instance, many types of schools and programs examined in our sample provided courses on emerging technologies, and many also carry courses on current regulations or policies regarding the IT and the Internet. In general, the technology-driven nature of the Internet requires that studies of social phenomena have a good understanding of the underlying technologies.
- *Some issues, concepts and paradigms are central to many fields of study*—For instance, courses ranging from electronic commerce to information infrastructure and intellectual property on the Internet cover technologies like agents and collaboratories; concepts like virtual communities and the nature of information; and paradigms like law, economics, and psychology. In short, in order to understand how the Internet interacts with just about any sphere of socio-economic activity, a grasp of various concepts, such as the technologies themselves, the economics, and the ideas of intellectual property, are essential.

Fortunately, the interdisciplinary nature of many courses makes them fairly accessible to students from many backgrounds, and often, the different backgrounds of students provide a richer learning experience.

Many of the professional schools surveyed, such as public policy and business programs, appear to have a comprehensive program of study on Internet, including such topics as telecommunications, the Internet's impacts on that field, e.g. electronic commerce, and policy issues. Some general developments include the joint offering of courses and degrees by more than one department, or the team teaching of a particular course by faculty from different fields of specialization. For example, the *Internet Business Strategy: Law & Policy* course at Harvard is offered jointly by the Law School, the Business School, and the School of Government, and is taught by five faculties from these departments. An example of multiple departments jointly offering a degree is the *Master of Public Policy and Information Systems* at the University of Michigan that is offered jointly by Public Policy School and the School of Information (where the latter is mainly responsible for the curriculum). Instruction for the Master of Information Systems Management Program at Carnegie Mellon University is provided by faculties from the Schools of Public Policy and Management, the Computer Science Department, the Software Engineering Institute, and other departments. Some programs may have very focused concentrations taught by experts from outside of the department.

It should be noted that even as the Internet directly transforms a specific activity within a field (e.g. business or engineering), its impact ripples through the entire field and the activities of competitors. Thus, the lower costs and more rapid (or concurrent) changes influenced by IT and the Internet the faster the propagation of the trends to many fields as well as sectors of economy/society. In light of these sea of changes, we may eventually see a trend for all types of courses to have some

Web-based components, so as to provide some recognition of these possible scenarios (in this book, Brusilovsky and Miller's chapter on delivery systems covers many Web-based learning technologies).

Learning and thinking abilities

The second component of the knowledge base consists of the learning and thinking skills needed for the emerging knowledge economy and workplaces, and the related issue of pedagogy. We will focus more generally on these than on the curricula content, hence this section should be taken as more exploratory in nature.

Learning and thinking abilities for knowledge environments

The characteristics of the knowledge economy leads to complex and rapidly changing, if not unpredictable, environments for users and providers of information. This has fueled an unprecedented demand for professionals who possess the new abilities to create knowledge and new skills to manipulate technology, data, information and knowledge. Further, these new professionals need to possess new ways of thinking, including[13]:

- *Creative ability*—the rapid diffusion and shorter life spans of knowledge makes it necessary for individuals to be able to create innovative ideas for virtual environments in fast and efficient ways.
- *Constructivist thinking*—involving the ability to integrate or fuse different types of knowledge in order to solve problems.[14] Constructivist approaches often emphasize problem-solving, collaborative and other activities. Some of the earliest constructivist approaches were developed in the area of computer education.[15] The constructivist concept is also regularly applied in higher education (though not necessarily under the same label), e.g. in educational projects that emphasize problem-solving approaches that can be adapted to people with different backgrounds and geographical contexts (Rigault et al., 1996). More generally, constructivist approaches may allow multiple values and perspectives to be represented.
- *Problem-solving ability*—for manipulating information to solve real-world problems, as is done with constructivist approaches. Problem-solving approaches are quite widespread in many existing practical environments like engineering. Given the Internet's diverse and interdisciplinary knowledge base with its multiple perspectives and value systems, problem-solving ability can be a valuable heuristic approach. We can extract value from "information-rich" environments like the Web by using open-ended approaches which emphasize new means of learning, as opposed to simply treating the Web as a delivery mechanism for content (Duchastel & Turcotte, 1996).
- *General information manipulation abilities*—These are abilities needed to manage information, e.g. searching for, organizing, summarizing, and

displaying (or visualizing) information. A familiarity with various tools is necessary for these abilities.

At the same time, individuals need to be able to *learn how to learn*, which is a direct reflection of the need to learn over a lifetime and in a constantly changing environment. This requires the fostering of open learning environments, i.e. ones that involve self-directed learning, support learning at any time and place, and are not limited to institutional boundaries.[16]

Learning and pedagogy

In general, traditional methods of learning tended to rely on classroom-based instruction, and students' participation tended to be passive. Further, what was learnt was segmented by disciplines, and research and teaching activities were also separated. All this is changing. Revised ideas of learning are coming from cognitive and social psychology, from educational studies of teachers and students, and from workplace studies (Marchese, 1997). Some are radical shifts, such as the concept of *situated learning*, typically applied to workplaces, which conceives of learning as a community-based practice rather than a teacher-led process. Other fields like neuroscience and anthropology are also influencing learning processes. As much as they alter our traditional conceptions of learning, these new perspectives also raise important questions about where learning should be headed.

These new or emerging learning pedagogies can support the teaching of the thinking abilities discussed earlier. These may also include the methods suggested for fostering "deep" learning: collaborative learning, cooperative learning, problem-based learning, service-based learning, case-method teaching, peer-based methods, undergraduate research, and multicultural learning (Marchese, 1997). Almost all of these were teachers' responses to real world needs, and not derived from theory and research. The common assumptions that these all share include learner independence and choice, intrinsic motivators and natural curiosity, rich, timely and usable feedback, occasions for reflection and active involvement in the real world, emphasis on higher-order abilities, interaction with others, high-challenge and low-threat environments, and provision for practice and reinforcement.

These insights can also be useful in designing new virtual and information-rich environments for knowledge. That is, learning environments should take into account the need for deep learning as well as the conditions for fostering them. However, it has also been noted that the information-rich nature of the Internet may not necessarily lead to more effective learning, since learning processes are still limited by technology's capabilities (Duchastel & Turcotte, 1996). We need to think of alternative modes of learning and teaching, and the skills sought to be created, which are suited to the Web (i.e. "information-rich") environment.

Thinking and learning modes in course curricula

Some of the learning and thinking abilities that have been discussed here may be found in the courses discussed previously. However, the type of information collected for our course survey was not sufficient for determining whether specific thinking skills were actually taught or conveyed in the surveyed courses. More often than not, the course design itself indicates some constructionist thinking, such as in the interdisciplinary formulation of course curricula using different concepts, theories and paradigms. The reliance on Web-based information is also increasing in course syllabi, as are the use of devices such as homepages, chat groups and in selected cases, simulation tools. The reliance on the Web logically implies that the class is becoming more open, less teacher-centered and more student-centered, but this still depends to a large degree on how central the teacher is in the class, e.g. a class oriented around classroom lectures is still a teacher-centered class.

The role of IT tools in the educational knowledge base

Technology is a critical element in the new educational knowledge base. In general, new and emerging technologies have the potential to create learning opportunities while making learning more effective and efficient. Internet technologies can also support the acquisition of the thinking and learning skills discussed earlier. Furthermore, students need specific technological capabilities for the workplace, such as the ability to manipulate information using Internet and other IT tools.

Given these needs, it is now becoming more necessary to enhance traditional programs of study with IT. The most obvious use of technology is in the offering of a variety of traditional courses through an Internet-based distance learning mode. However, many traditional universities are also making use of IT tools in their curricula. For instance, simulation tools are used to help students visualize information and scientific principles at work, while collaborative technologies such as chat software are used to help students solve problems in groups.

In this section, we will focus on the role of IT technologies or tools *in learning*, rather than on the specific skills needed for the workplace, or the uses of specific technologies. The other chapters in this book provide more detailed discussions of individual technologies, so we will not discuss specific technologies here. The level of discussion on the skills needed for curricula appear to be insufficient, particularly for core skills, possibly because of the high rate of change in the IT professions (Rada, 1999). Quite possibly, the level of discussion of the skills needed for various non-IT professions that use IT is even more immature.

In the following review of IT tools in learning, we will highlight the more positive aspects and potential of Web-based technologies for helping students learn content and skills.

Internet technologies and learning

Internet and information technologies can facilitate the development of learning and thinking abilities in a number of ways.

The Internet and IT can *change the instructional design paradigm.* The coupling of the Internet's information-rich environment with tools and ways of catering the learning experience to individual students can provide learning breakthroughs, through problem-based environments, self-guided teaching environments, and the customizing of learning for individual students (Schank & Cleary, 1994; Duchastel & Turcotte, 1996). Some examples include goal-based models (Schank, 1997) and the *CyberProf* intelligent tutor (Hubler & Assad, 1995). In addition, software can facilitate learning by doing (e.g. through repeated drills), exploration through guided tours and other means, and help students by learning from a variety of situations, including mistakes. The involvement of the learner in active and exploratory activities makes class assignments directly relevant to her. Technology can support more traditional educational objectives, such as helping students master basic concepts (e.g. through drilling) and seeing interconnections amongst concepts (Frayer, 1997).

The Internet and IT can *change the roles of learners and teachers*—In the new learning environment, instructors need to become facilitators, guides and co-investigators. Instructors will have to learn how to explicitly model and use new unstructured learning environments like the Internet, therefore, they need to be experts in both technical matters as well as applications. At the same time, as we noted above, learners become explorers, apprentices and producers (Stities, 1998).

At the same time, the Internet can *promote interaction in learning communities.* This includes interaction between students and their peers (e.g. collaborative learning), between students and teachers, and between students and information sources. One major issue that has not been considered much is the need for teachers to revise their self-image. Rather than the image of the lone teacher independently formulating course content and dwelling on research, increasingly, teachers are required to participate in broader learning communities of their own, learning information from widespread sources, as well as traditional sources put online, interacting in professional mailing lists and so on.

One of the fundamental (and perhaps overlooked) advantages of the Internet is that it can *improve knowledge representation, which promotes problem-solving and customization.* It can do this in at least three ways. The first is by enhancing learning experiences by multimedia and other Web technologies (e.g. interactive Java applets). Secondly, the Web can make information "depth-dimensional," that is, it can create another dimension for knowledge, based on the degree of depth of information. This is one way by which the web customizes the information environment. Thus, students reading a particular document can follow links to more detailed documents, depending on the extent of their interest. Finally, the Web can break up the chain of information, making content and curricula independent of levels of learning (Duderstadt, 1997). This can help different types of students to access the same information base, and bridges the gaps between informal and formal education.

The Web also breaks down boundaries in at least three ways. One is by allowing the creation of interdisciplinary environments—needed for developing the creative and problem-solving abilities. While traditional teaching methods emphasize disciplinary material and specialization, an interdisciplinary approach is necessary to integrate concepts and provide the "big picture" of the new Internet environment.

The second is by allowing the integration of a variety of academic activities—such as the combination of applied research with teaching. In problem-solving approaches, the educational experience involves applying theories learnt in the classroom to real world problems (Rigault et al., 1996). Under a slightly different emphasis, new approaches are being tested that introduce more substantive engineering research experiences into the classroom.[17] In many ways, the real world is too complex to be dealt with by conventional tools and paradigms, and complex phenomena have to be dealt with by methods that can take into account this complexity.

The third is by extending learning to informal settings and allowing the replication of real world settings. For instance, constructivist learning brings the real world into the classroom or makes school learning more like the real world. Real world experiences, practical projects, analytical tools or research methodologies can be embedded in curricula.

Despite these advantages, it should be noted that the evidence on whether or not technology actually enhances learning is still mixed. This is partly because the technological implementation often fails to consider the broader teaching/learning context. Beyond the social and individual learner's considerations, technology must be implemented with several characteristics in mind, such as flexibility, reliability, and robustness. Furthermore, functionality may have to be provided for a variety of learning contexts.

If the Web is simply used to repeat the delivery of traditional curricula in traditional ways, these efforts may fail much in the same way as the original attempts to use PCs in schools. As discussed in other chapters of this book, current constraints in the Internet (e.g. limited bandwidth for transmitting video or conferencing), costs (e.g. of codifying tacit knowledge), the experimental nature of software delivery systems, and other factors, may limit more active interaction, such as the guided tutoring of students, which reduces its effectiveness for education.

Industrial Catching-up in Developing Countries Through Curricula Reform: The Case of India

In developing countries, or indeed in any other country, curricula revision may not necessarily follow the same shape of the reforms as in the U.S. However, assuming that developing countries are following the same information sector trends as the U.S.'s, a certain amount of convergence will be unavoidable, and there will have to be a focus on reforming educational systems. These reforms will need to propagate the new learning and thinking modes, curricula subjects and IT tools while taking into consideration circumstances specific for the country or region.

In this section, we will examine a number of general issues concerning developing countries, as well as the specific case of India, to help clarify some of the issues and policy suggestions. India, one of the world's largest countries in terms of population, also has one of the fastest growing software industries in the world. The requirements of translating the latent potential within this population base into the large needs of this software industry, requires immense energies and imagination, some possibilities for which are sketched out in this section.

India's IT industry has become a world class force within a relatively short time. It currently employs 200,000 professionals in producing both domestic and export software, of which 70% are in development and operations, 11% are in marketing, 14% in support and 5% in other activities. The IT workforce is projected to grow to 300,000 in 2000 (NASSCOM, 1999). The software industry was worth $4 billion in 1997–98, and its contribution to the GDP was 2.6 percent. Despite the growing Indian software industry, both its infrastructure and manpower lag far behind the existing needs or international averages. For example, for its population of about one billion, India has about 535,000 Internet users (NASSCOM, 1999). (In comparison, the US was said to have exceeded 100 million users in 1999[18]).

The Government of India has taken steps to address the infrastructure and manpower needs of India's developing IT-based economy.[19] One of the government's first actions was the development of an *Information Technology Action Plan* to address a broad range of critical national needs in the area of information infrastructure, Internet access, software development and exports, hardware manufacture, electronic commerce, research and development in IT, manpower training and education. The Plan aims at promoting IT-based education at all educational levels. Other recommendations of the Plan includes the addition of new Indian Institutes of Information Technology networking of all universities and institutions of higher learning. This will facilitate supplementary distance education, to help achieve the goal of tripling IT graduates in the seven national level technical institutions (i.e. the six Indian Institutes of Technology and one Indian Institute of Science).[20]

Developing context-specific problems

Generally speaking, in examining the use of information technology and curricula reform, developing countries have to take into account specific characteristics that differentiate them from developed countries, namely:

- *A lack of infrastructure*—in developing countries there is a serious deficiency in information infrastructure, including everything from a severe lack of PCs for schools to poor telecommunications links.
- *Financial constraints*—capital shortages in both public and private sectors make it difficult to purchase equipment and upgrade them.
- *Equipment shortage and desire to opt for the latest technology*—while it is necessary to implement newer generations of information technology to avoid "falling behind", the latest technology is expensive. This is compounded by financial constraints, an aging IT base (consisting of legacy systems, or older generations of technology), shrinking budgets, and the growing need for IT education. To deal with such constrains some countries, such as India and Mexico, face a gradually growing trend (including K-12 in India) towards the use of free or nominally-priced public domain software, such as the LINUX operating system and LINUX-based software.
- *A lack of teachers and trainers*—a shortage of qualified trained teachers/ instructors at all levels makes it difficult to revamp national educational systems, so it is vital to create a system for training teachers within the educational system.
- *Poor learner computer skills*—unlike many of the computer-savvy young, many students in developing countries simply do not have the computer skills to be able to access Web-based environments. This requires access to computers and at least a period of hands-on exposure with teachers, which is compounded by the shortages of equipment and skilled teachers.
- *Different stages of development*—industries in developing countries are at different levels of development from developed countries, and therefore, have different needs for human resources. In these countries, vocational training has traditionally been an important bridge between traditional and emerging professionally-based sectors, thus the reform of vocational education can be as important as that of tertiary education, and many issues can be of concern. For instance, domain-specific skills are often emphasized, and the ability of students to adapt to training methods is a concern. Furthermore, many countries also rely on government planning, so government planning systems for education and technology will have to be reformed to better adapt to the continuing and rapid changes caused by technology.
- *Socio-cultural factors*—Socio-cultural factors include cultural, linguistic, and experiential issues, socio-linguistic development, and forms of cognitive learning. These backgrounds are often different from those of the developed countries, which necessitate either the adaptation of imported IT tools and courseware, or the local development of the same. The systems developed should be aware of and sensitive to local cultures and norms. For instance, cultural sensitivities may require screening out information deemed offensive (e.g. pornography or other offensive topics).

Context-independent problems

India, as well as other developed and developing countries, faces many problems in meeting the demands of a knowledge economy. In addition to the constraints discussed earlier, other major problems and issues include:

- *Inertia and resistance to change*—developing countries also face inertia in changing their curricula, because of a slow rate of technological innovation and absorption and low level of high-tech industries. As in developed countries, university faculties often resist change.
- *Needs for mass customization rather than mass production,* e.g. there is promise for virtual environments to assist in on-line evaluation, but at the massive scales involved, it is important to avoid the easiest solution of mass produced evaluations, but instead to make them sensitive to the needs of individual students.
- *Needs for innovation and a desire to take pride in traditional cultural and scientific knowledge*—in India, this has prompted the inclusion of optional courses such as "Indian System of Perception and Computing," "Cognitive models based on Sanskrit Shastras," "Vedic Mathematics," and so on. IT tools in Hindi and in a variety of other local languages are also becoming available.
- *IT use in associated industries*—the value of the IT industry and IT skills will increase as the broader society starts to use IT. In India, the advent of multimedia technology has led to the creation of a rapidly growing local content industry (e.g. services and entertainment).

These and other factors can have a major impact on the quality of IT education and training for both students and teachers. In order to address a developing countries IT industry and manpower needs, reform has to take place at many levels, including:

- Reforming institutions and their teaching (learning) models, with a focus on distance education.
- Improving the thinking skills taught to students to better provide for the new needs of society and industry.
- Reforming curricula content.

We will examine each of these in the following sections.

Reforming institutions and their teaching models

As noted earlier, in most developing countries that have emerging information economies, there is an urgent need for increasing the quantity of manpower in IT-related fields, and for newer forms of learning. Curricular reforms have to meet these needs while overcoming the barriers inherent in traditional educational systems that put emphasis on teacher-centered learning, memorization, and separating streams of academic and vocational education. However, there are limited resources available to meet these growing needs. Furthermore, historically, in many countries, the highest quality of education is reserved for a very few selected elite students. This has been the case in India, where rapid growth in needs for education at all levels (as seen in Table 4.2) is constrained by inadequate resources.

TABLE 4.2
Current Annual Enrollment in Approved Institutions at Various Levels of Education in India (all disciplines)[21]

Level	Institutes	Annual Intake	Growth rate
Diploma	1171	201,214	7%
B. Tech	662	156,343	16%
M. Tech	242	20,156	15%
Masters in Computer Applications	310	10,020	15%

In India, educational reform to broaden educational opportunities has included the use of distance education where there are eight open universities and 58 distance education programs in various universities, with a network of 90 centers. There are over 500,000 students in all disciplines currently enrolled, and this is estimated to double by 2002.

In India, SMART schools are to be set up in each state. This follows other countries, like Austria, Japan, Malaysia, Pakistan, Singapore, and the USA, which have also opened SMART schools that make use of information and communications technology. One of the key strategies for these schools involves developing IT-enabled teaching and learning environments that focuses on problem solving and collaborative learning abilities.

At the same time, despite its limited resources, the government plans to upgrade IT infrastructure in as many existing institutions as possible (including schools, polytechnics, colleges, universities) to enable Web-based and distance learning courses and programs. One such ambitious project called "Sankhya Vahini" aims to establish a data network connecting at least ten research centers and over a hundred universities. This aims to enhance the research, teaching, and learning environments of these educational institutions. It would also allow access to educational, training, and digital libraries providing content from universities in the U.S. (National Task Force on Information Technology and Software Development, 1999). Because rural and remote regions do not have Internet connectivity, efforts are being made to provide substitutes, such as CD-ROMs, and visits to Internet centers for hands-on experience.

One of the unique characteristics of the Indian IT training industry is the large contribution of the private sector in providing valuable supplemental opportunities for the workforce. Examples of this are the privately-run National Institute for Information Technology, which trains large numbers of students, and the *IT Clinics* that are to be set up in software technology parks and IT institutions of excellence. Support for the IT Clinics comes from a consortium that consists of members from the IT industry, researchers from academia, consultants from industry and trainees from IT institutions.

New models for new curricula

As with developed countries, in developing new distance education or open learning models for developing countries, particularly with the advent of Web-based courses, there is a need to develop new learning and teaching models. Constructivist and multidisciplinary approaches are potentially useful in collaborative learning environments, as they allow students to share projects and homework, exchange ideas, publish newsletters and share socio-cultural experiences. This collaborative learning has to be supported by new technologies. At the same time, the multi-cultural scenario demands other types of technology to overcome language translation issues, such as allowing text in the various users' languages.

Various forms of knowledge and thinking also need to be central to the learning process. At the most basic level of need, students in both academic and vocational institutions often require practical training. However, limited resources including equipment and facilities make it difficult to provide this. Furthermore, there is a great deal of subjectivity in certifying students' proficiency in practicals. Virtual forms of education that provide practicals can have the potential to overcome all of these problems and at lower (mass-produced) cost. However, educational systems will need to be completely reorganized if the appropriate use of virtual education is to be made.

Curricula content

In India, a national determination to become an "IT superpower" has led to the incorporation of IT skills and philosophies into a variety of social and economic activities. Curricula in not only in universities, but also in secondary and vocational schools are being transformed to include courses on IT skills and applications, where applications to a variety of disciplines and real-world domains are taught.

To address the vast new demands of the content industry, courses on new forms of knowledge and IT skills are being considered for introduction. It is also considered important to provide IT training to graduates and postgraduates of other subject disciplines such as languages, humanities, social sciences, physical sciences, etc.

Conclusions

The issue of curricula is important because employers expect graduates to have substantial knowledge (i.e. knowledge of content), as well as skills for dealing with an increasingly knowledge-intensive world. In this chapter, we have examined the changing educational knowledge base, consisting of the curricula content, learning and thinking skills, and abilities to use IT tools and Internet technologies.

In the U.S., the Internet is changing the content taught in academic programs of different disciplines. These new courses teach students how to manipulate the Internet environment, as well as to understand the impacts of the Internet on traditional fields. Many of these new programs and courses are interdisciplinary, or focus on similar issues of concern, such as public policy towards the Internet, technology and global trends. The merging of fields of study and other academic activities (e.g. the introduction of research into classrooms) can improve student's abilities to conceptualize and solve problems in the new work environments.

A second topic that we examined was the reform of curricula in India (as a representative trend in developing countries) to embrace the Internet and IT. In a way, IT and Internet-based curricula provide an opportunity for many developing countries with dysfunctional educational systems to fix them cost-effectively, and even to leapfrog stages of development. This, however, has many caveats and pitfalls. In general, we can say that even as curricula changes and reforms become more and more wholesale, crucial issues such as the following must be taken into account, in *both* developed and developing countries:

- *Accommodating emerging and continuing IT trends (including the convergence of technology)*—Any anticipated changes in educational systems have to maintain the system's flexibility for accommodating new technologies (with their rampant development). For instance, one trend is towards the merging of computers and telecommunications, which facilitates easier access to tools and other resources. Another trend is towards multiple channels or media for information access, requiring interoperability across those channels (e.g. video, broadcast, online Internet "push" and "pull" technologies, CDROM). Developing countries have more serious concerns relating to financial resources, and needs to train teachers and learners in computer skills.
- *Handling social transformations*—as with technologies, educational systems also have to accommodate wrenching social transformations, and to prepare people for facing them. While the information economy has growing needs for technological and information related skills, an even more critical question is of whether societies and cultures are positioned to handle the resulting social transformations. IT has the potential to transform economically disadvantaged societies, but at the same time, the availability of the Web and ease of access to it may allow dominant cultures on the Web to displace many pre-Web traditions, or locally held cultural systems. Thus, we have to find ways to create learning environments that preserve culture, distinctiveness, language, value systems and customs. The case of India suggests that there are some ways in which traditional knowledge could be preserved or even transmuted to provide new value in information settings.
- *Preventing inequity, i.e. the information-rich getting richer as the information-poor get poorer.* This may happen in any context, be it individual, institutional, countries or regions (developed and developing). We highlighted some instances by which developing countries could provide greater access to education through

virtual education. However, individuals still require resources in order to afford a virtual education, and inequality between the haves and have-nots could be exacerbated. We are a long way away from universal virtual education. Furthermore, it indicates that both governments and institutions have major roles to play, including the channeling of resources and teaching. At the classroom level, it may be as important to teach students skills for independent work, such as learning how to learn, as it is to teach them IT skills and content.

- *Ensuring that changes in technology, learning, and curricula are made together in an integrated way*—schools and educational systems have to balance these otherwise learning needs may not be adequately met. For example, an organization may buy a lot of computers, but without teacher and student training or a broad enough view of how learning occurs or of curricula development, it will not be able to take advantage of the equipment. This is also equally applicable in both developed and developing countries.

Notes

1 In the Information Technology Association of America (ITAA) study, *Information Technology* is defined as the study, design, development, implementation, support, or management of computer-based information systems, particularly software applications and computer hardware (ITAA 1997). Further, definitions of three core IT professions—programmers, systems analysts, and computer scientists/engineers—are provided in a more recent ITAA study (ITAA, 1998).
2 *Business Week*, March 10, 1997.
3 In the past, the creation of knowledge was mainly in the domain of universities, but the entry of corporate laboratories caused many fields of applied knowledge to migrate to private hands, including chemistry, biotechnology, and engineering product innovation. Partly as a result, universities started to specialize in fundamental research, in the hard as well as the soft (i.e. social) sciences.
4 The results for the first three fields of study—public policy, business and library science, were based on a survey of class Web sites of the highest ranked schools and programs in information technology (based on the listing from the 1998 U.S. News and World Report, Graduate Schools ranking), while the remaining fields were based on a partial study of selected schools. The survey did not include standard information systems courses such as management information systems, databases, and programming languages. In addition, other references on curricula changes were analyzed to provide a broader context. The methodology used for selecting courses involved choosing courses that were deemed "new," that is, directly reflecting the Internet's impact on that field, or information-related and technology-based courses whose scope extended beyond that of traditional IT courses. Telecommunications courses were included because of their importance to understanding the Internet as a network phenomenon.
5 The following are the identifiers for universities used in Table 1: University of Arizona (Arizona), University of California, Berkeley (Berkeley), Carnegie Mellon University (CMU), Georgia Institute of Technology (Georgia Tech), University of Pennsylvania (Penn), University of Pittsburgh (Pitt), State University of New York (SUNY), Syracuse University (Syracuse), University of North Carolina (UNC); University of Texas at Austin (UT).
6 Firefly was subsequently bought out by Microsoft, made defunct, and finally, its technology incorporated into other Microsoft products.
7 Also, http://www.ecom.cmu.edu/.
8 From the SIMS Program Overview.
9 This is based on informal and formal (published) information on agents at websites such as http://www.cdr.stanford.edu/RVPP/ and http://www.memagazine.org/contents/current/features/newfront/newfront.html. Agents are described in more detail in Chapter 2 by Ng Chong.

10 Some links on the information economy are at: http://www.sims.berkeley.edu/resources/infoecon/. For reference, see a recent book on the subject by Shapiro and Varian (1998).

11 This program stems from the belief that the human–computer interface (HCI) design paradigm was shifting such that the "boundaries between hardware and software, device and interaction, 2–d/3–d/4–d have blurred considerably and will only continue to blur and blend. The new HCI development model is one of collaborative design, with individuals who represent various fields of knowledge working together on the design and implementation of user interfaces. This new model—multidisciplinary, collaborative, and human-centered, with a concern for psychological, social, and cultural factors—is gaining wide acceptance in the HCI community. Not only is it the model for human–computer interface design, but for the entire scope of product development and human–machine interaction design; in short, interaction design."

12 Conversation with Danielle Smoller, School of Architecture, Columbia University.

13 Other types of classifications have been proposed for various purposes. The characteristics of learning, proposed by Stities (1998), can be mapped into our classification scheme. Another useful "map" of learning-related issues by Apple Computer Corporation educational researchers focused on a broader set of learning landmarks: cognitive theories (what is learning), skills and competencies (what to learn), tasks and activities (what to do to learn), pedagogy (how to foster learning), tools and materials (what artifacts foster learning), automation and amplification (what to use computers for), and diffusion (how to have an impact on practice) (Spohrer, 1992).

14 One of the important rationales behind adopting a constructivist approach is that curricula defined in a bottom-up fashion is more effective at handling information-rich environments than curricula defined in top-down fashion. The unanswered question is whether learning in environments as the Internet is any substitute for learning in disciplinary environments with somewhat more immutable standards for knowledge. One might argue that the whole edifice of human science and knowledge is built upon these standards, built up over time through the rigorous practice of the scientific method. Thus, at its extreme, constructivist learning has all the potential to yield ill-trained technicians. We will not pursue this question beyond this point since at its root, this is a philosophy of science question, and is not likely to be answered in the short space of this chapter.

15 See Resnick (1996) for a discussion of constructivism.

16 This particular open learning concept is discussed in more detail by Jain (1997).

17 Discussion with Indira Nair, Vice-Provost for Education, Carnegie Mellon University.

18 *New York Times*, November 12, 1999

19 The Office of the Prime Minister of India issued a Notification on May 22, 1998, forming a National Task Force on Information Technology and Software Development, whose goal was to work out guidelines and recommendations for making India an "IT Superpower". The first report, *The Information Technology Action Plan,* was submitted in July, 1998. The interim report is at http://it-taskforce.nic.in/

20 Source: All India Council for Technical Education (AICTE), 1998.

21 All India Council for Technical Education, 1998. Also at http://www.aicte.com/approved_inst.html.

References

Bakos Y. (1998) The Emerging Role of Electronic Marketplaces on the Internet. *Communications of the ACM,* **41**(8): 35–42.

Bollentin W. R. (1998) Can Information Technology Improve Education? *Educom Review*, **33**(1): 50–55.

Bourne J. R., Brodersen A. J., Campbell J. O., Dawant M. M., & Shiavi R. G. (1997) A Model for On-Line Learning Networks in Engineering Education. *Journal of Engineering Education,* **85**(3). Also at: http://www.aln.org/alnweb/journal/issue1/bourne.htm.

Cox R. J. & Rasmussen E. (1997) Reinventing the Information Professions and the Argument for Specialization in LIS Education: Case Studies in Archives and Information Technology. *Journal of Education for Library and Information Science,* **38**(4): 255–267.

Dalrymple P. W. (1997) The State of the Schools. *American Libraries,* **28**(1): 31–34.

Davis C. & De Matteis D. (1998) Management Skill Requirements for Electronic Commerce. http://business.unbsj.ca/users/cdavis/papers/Ecomm_mgt_skills_IC_report.pdf.

Dowling S. A. (1996) Internet Education: Reform or False Panacea? *Proceedings, INET '96 Conference, The Internet Society*. http://www.isoc.org/inet96/proceedings/c1/c1_1.htm.

Duchastel P. & Turcotte S. (1996) Online Learning and Teaching in an Information-Rich Context. *Proceedings of INET '96 International Conference, The Internet Society*.

Duderstadt J. J. (1997) The Future of the University in an Age of Knowledge. *Journal of Asynchronous Learning Networks* (on-line publication, Vanderbilt University), **1**(2). Also at: http://www.aln.org/alnweb/journal/issue2/duderstadt.htm.

Evans P. E. & Wurster T. S. (1997) Strategy and the New Economics of Information. *Harvard Business Review,* **75**(5): 71–82, Sep/Oct 1997.

Fielding R. T., Whitehead E. J., Anderson K. M. & Bolcher G. A. (1998) Web-based Development of Complex Information Products. *Communications of the ACM,* **41**(8): 84–92.

Finger S., Stirovic J., Amon C. H., Gursoz L., Prinz F. B., Siewiorek D. P., Smailagic A. & Weiss L. E. (1996) Reflections on a Concurrent Design Methodology: A Case Study in Wearable Computer Design. *Computer Aided Design,* 393–404.

Frayer D. (1997) Creating a New World of Learning Possibilities through Instructional Technology. *AAHE TLTR Information Technology Conference, Colleges of Worcester Consortium.*

Gehl J. (1996) The Curriculum Has Run Its Course. *Educom Review,* **31**(6).

Grover R., Achleitner H., Thomas N., Wyatt R. & Vowell F. N. (1997) The Wind Beneath Our Wings: Chaos Theory and the Butterfly Effect in Curriculum Design. *Journal of Education for Library and Information Science,* **38**(4): 268–282.

Hubler A. W. & Assad A. M. (1995) CyberProf: An Intelligent Human-Computer Interface for Asynchronous Wide Area Training and Teaching. In *Proceedings of The Fourth International World Wide Web Conference,* Boston.

Iansiti M. & MacCormack A. (1997) Developing Products on Internet Time. *Harvard Business Review,* **75**(5): 108–117, Sep/Oct 1997.

Information Technology Association of America (1997) Help Wanted: The IT Workforce Gap at the Dawn of a New Century, Arlington, VA.

Information Technology Association of America (1998) Help Wanted 1998: A Call for Collaborative Action for the New Millennium, Arlington, VA.

Jain M. (1997) Toward Open Learning Communities: One Vision Under Construction. *CIES '97 Conference on Education, Democracy and Development at the Turn of the Century*. Also at: http://www.education.unesco.org/educprog/lwf/lwf_docs.html.Kraut R , Scherlis W, Mukhopadhyay T., Manning J. & Kiesler S. (1996) The HomeNet Field Trial of Residential Internet Services. *Communications of the ACM*, **39**(12): 55–63.

Kraut R., Mukhopadhyay T., Szczypula J., Kiesler S. & Scherlis B. (1997) Communication and Information: Alternative Uses of the Internet in Households. At: http://homenet.andrew.cmu.edu/progress/comminfo.html.

Kraut R., Patterson M., Lundmark V., Kiesler S., Mukophadhyay T. & Scherlis W. (1998) Internet Paradox: A Social Technology That Reduces Social Involvement and Psychological Well-being? *American Psychologist,* **53**(9): 1017–1031. Also at: http://www.apa.org/journals/amp/amp5391017.html.

Kiernan V. (1997) All the World's a Lab. *New Scientist,* 12 April 1997: 24–27.

Krieger J. H. (1996) Doing Chemistry in a Virtual World. *Chemical and Engineering News*, December 9.

Marchese T. J. (1997) The New Conversations About Learning: Insights from Neuroscience and Anthropology, Cognitive Science and Workplace Studies. In *Assessing Impact: Evidence and Action*, Washington, DC: American Association for Higher Education. Also at: http://www.aahe.org/pubs/TM-essay.htm.

Marcum D. B. (1997) Transforming the Curriculum; Transforming the Profession. *American Libraries,* **28**(1): 35–37.

Mosquera M. (1998) High-tech Industry Wants More Foreign Workers. *InformationWeek*, February 1998. Also at: http://www.techweb.cmp.com/iw.

NASSCOM (1999) The Software Industry in India – A Strategic Review, 130.

National Task Force on Information Technology and Software Development (1999) Basic Background Report (BR-3), March 1999: Appendix VII.

Petrie C. J. (1996) Agent-Based Engineering, the Web, and Intelligence. *IEEE Expert*, December 1996.

Rada R. (1999) IT Skills Standards. *Communications of the ACM,* **42**(4): 21–26.

Resnick M. (1996) Distributed Constructionism. *Proceedings of the International Conference on the Learning Sciences, 1996,* Northwestern University. Also at: http://el.www.media.mit.edu/groups/el/Papers/mres/Distrib-Construc/Distrib-Construc.html.

Rigault C. R., Andre L. & Guidon J. (1996) Design, Realization and Delivery of Pedagogical Events for Telelearning in a French Context. *Proceedings of the INET '96 International Conference.*

Schank, R. (1997) *Virtual Learning: A Revolutionary Approach to Building a Highly Skilled Workforce.* New York, NY: McGraw Hill.

Schank R. & Cleary C. (1998) *Engines for Education.* Hillsdale, NJ: Erlbaum. Also at: http://www.ils.nwu.edu/~e_for_e/index.html.

Shapiro C. & Varian H. R. (1998) *Information Rules: A Strategic Guide to the Network Economy.* Harvard Business School Press.

Smith M. A. (1997) Netscan: Measuring and Mapping the Social Structure of Usenet. Presentation, 17th Annual International Sunbelt Social Network Conference, San Diego, California, February 13–16, http://netscan.sscnet.ucla.edu/csoc/papers/sunbelt97/.

Spohrer J. (1992) Mapping Learning, Research Report, Apple Computer Corporation Education Research and Technology Group.

Stanovich K. E. (1986) Mathew Effects in Reading: Some Consequences in Individual Differences in the Acquisition of Literacy. *Reading Research Quarterly,* **21**: 360–406

Stities R., Hopey C. E. & Ginsburg L. (1998) Assessing Lifelong Learning Technology (ALL-Tech): A Guide for Choosing and Using Technology for Adult Learning. *NCAL Report—Practice Guide PG 98–01* Philadelphia: University of Pennsylvania, National Center on Adult Literacy.

Wired News (1998) Virtual Communities: The Next Hot Major. *Wired News, http://www.wired.com/news/news/culture/story/8363.html.*

USA Today (1997) Hi-Tech Sector is Israel's Silver Lining. September 1997.

Part II

Transforming Learning Environments with New Technologies

5

Internet Technologies: Towards Advanced Infrastructure and Learning Applications

NG S. T. CHONG

Introduction

Lifelong learning is becoming a social and economic imperative in a networked and knowledge-intensive society. The next millennium will be a digital learning age, as computing and telecommunications technologies converge to enable human interaction and collaboration, stretching beyond geographical and cultural boundaries. The Internet now spans over 200 nations, and the World Wide Web, in particular, has been identified as the breakthrough medium for distributing resources and services to an ever-broader international audience, and facilitating interaction among groups and individuals with fewer constraints of time and space. Its potential to radically transform the future of education is both enormous and real. Universities must closely monitor Internet developments as they rethink approaches to serve the growing demand for education in a learning society and enable learners to participate in a knowledge-based economy.

In this chapter, we will examine the past, current and future developments in Internet technologies, particularly in terms of the building block technologies that form the basis for systems. Most of the information in the first few sections derives from trends in the U.S.

The second section examines the evolution and future trends of the most visible and transformational Internet technologies. Recent developments in information technology provide the building blocks for improving learning support via the Internet. Enhanced technical connectivity, faster computers, emergence of digital libraries, software agents, and computer-mediated collaborative environments, are examples of technological advances that suggest various dimensions of the future in education and research.

The third section gives an overview of the challenges, progress, and trends in information technologies that will have an impact on the learning infrastructure. The issues covered in this section are a sampler of the diversity of the ongoing research and development efforts concerning connectivity, software, and silicon. It then completes the technological landscape needed for our next discussion on the base virtual learning modes. This coverage is not intended to be exhaustive.

Traditionally, educators think of distance education as distributing study material via satellite-based systems, dedicated educational radio and television channels, or through the mail. Today the same can be done with more efficiency and flexibility through electronic mail or the World Wide Web. However, if the application of information technology is to solely achieve just-in-time distribution of learning materials, over a mass electronic medium, then we miss the chance of empowering new possibilities in the process of learning.

A fourth section places the new technologies in the context of virtual learning modes and discusses the novel approaches of application for these technologies, to enable a shift of focus in learning.

Fifthly, a number of emerging technologies show promise to enable rural and remote areas in the world to participate in the "Internet Revolution". These technologies and the barriers to their adoption and implementation in some developing countries are described.

It is not apparent as yet, whether virtual education will improve the situation: however, it is a timely response to the new learning requirements, in a borderless knowledge-based society; to the budget constraints facing universities today; and to the continuous efforts to advance the education infrastructure.

The Evolution of the Internet

Although the Internet made its public debut in the mid-1970s, it did not reach critical mass until the introduction of the World Wide Web in the 1990s. It has been evolving and expanding across several dimensions ever since, including those of scale, performance, content, and functionality.

The following survey results attest to the accelerated expansion of the Internet. According to the latest estimate by *Network Wizards*, the Internet had almost 40 million hosts in January 1999, up from 30 million hosts in the same period one year before. A related survey conducted by *Netcraft* reports a total of over 6.5 million web servers in July 1999, up from 130 in June 1993. Under the current exponential growth pattern, the number of attached Internet hosts is projected to reach *100 million* by the year 2001 (NUA Internet Surveys, 1999)

Parallel to this remarkable growth in number of connections is the phenomenal increase in the number of users accessing the Internet. According to Netcraft's most recent NUA Internet Survey, there were *179 million* users world-wide as of June 1999, or nearly 2.5% of the world's population (Rutkowski, 1998). The data in Table 5.1 indicates that the Internet infrastructure is unevenly distributed in the world. It also shows that one its strengths lies in its emergence as a universal platform which is continually expanding to interconnect more sites and more people across the U.S. and the world.

TABLE 5.1

Distribution in the number of online users in June 1999 (Sources: NUA Internet Surveys; Department of Economic and Social Affairs Population Division, UN 97)[1]

Region	Estimated number of users in millions	Percent Regional Population	Percent World population
Africa	1.14	0.15	0.02
Asia/Pacific	26.97	0.78	0.46
Europe	42.69	5.86	0.72
Middle East	0.88	0.49	0.01
Canada and USA	102.03	33.50	1.73
Latin America	5.29	1.05	0.09

At the macro level, the sheer size of the Internet user base and the underlying Internet technologies are mutually reinforcing each other, to sustain further growth as a whole. As new technologies enable faster information flow, information becomes more valuable to users, which in turn increases the demand for technologies, products, services, as well as more information. Along the human dimension, we begin to see the Internet fostering the digital extension of society. It is already replicating real-life activities. Instances include online conferencing, shopping, and access to libraries and scientific instrumentation. The Internet provides a medium for people to interact, communicate, collaborate, and socialize. Along the technology dimension, we see the Internet as a multi-disciplinary melting pot bringing together diverse fields such as software engineering, mathematics, laws, economics, sociology, artificial intelligence, and network engineering. The Internet, as a creature of advances in computers, moves at the pace of its constituent technologies and demands for new applications.

What makes the Internet so ubiquitous and popular? What is its next evolutionary step? What are the most critical Internet-related problems we face today? This section addresses these questions by tracing the changing uses and trends of the Internet from its origin to the present and projecting the current trends into the future.

The past—the bounded information medium

In its early stages the Internet was mainly intended for information exchange and sharing of computational resources among selected communities of users from government, education, and non-profit research institutions. In the beginning, the Internet mainly involved text and was intended for *skilled computer users*. Commercial use was not involved, and only a few sites were connected.

Many important applications were added over the years, including remote login procedures, file transfer capability, newsgroups, and electronic mail/e-mail.

E-mail gained widespread use and made an impact on society later. It enabled communication among individuals, and led to the development of Internet communities. The Internet expanded from only a handful of experimental networks in the 1970s to over 300 operational networks by 1985.

The present—the global information and applications platform

The Internet continues to grow by building on previous experiences. By 1985 it had become an established open, internetworking technology, interconnecting different computing platforms. The Internet was beginning to grow beyond its research roots into more practical applications.

The transition of the Internet to reach the *general user* started with the launch of the World Wide Web (i.e. "the Web") in the early 1990s. The Web gives users a unified graphical interface through which they can easily access Internet resources by following links embedded in Web pages. The Web also provides an easy page description language based on a public standard known as the HyperText Markup Language (HTML). This protocol can be used to control the appearance, and to some extent the structure, of the parts of a document. The twofold simplicity in access and publishing, sets the stage for its growing acceptance and influence on a global scale.

How big is the Web? The dynamic characteristics of the Web defeat any attempt to achieve a reliable measurement of its size. Web pages are very volatile, to say the least. Pages are frequently being added to and updated by an indeterminate number of machines. There is no guarantee that a given page can always be found in the same location in cyberspace, nor any assurance that the page is not duplicated elsewhere, or not deleted after being accessed. Results of a 1995 statistical survey conducted indicate the average lifespan of a version of a web page was only 75 days (Brake, 1997). It is currently estimated that over 600 gigabytes of the Web changes every month (Kahle,1997).

Even if we could track the number of pages existing on the Web at a given moment, it would be impossible to include those pages that are created "on the fly" in response to such events as a payment or request form. According to a recent study at Digital Systems Research Center, the lower bound of the size of the Web in March 98 was about 275 million distinct pages, up from 200 million by November '97 and 125 million by mid-'97 (Bharat & Broder, 1998). The size of the Web doubled in less than nine months, and it is currently growing at the rate of about 20 million pages per month.

The abundance of all kinds of information on the Internet sets the trend for its main use today—*access to information and electronic publishing.* Recent innovations in cross-platform Internet programming languages and telecommunications technologies are beginning to trigger new patterns of use. These new extensions of the Web give content developers the means to design more dynamic and responsive pages. Specifically, the ability to run any imaginable form of executable content on the Web, and to exchange real-time multimedia data over low bandwidth networks, are enabling innovative forms of *interactivity* among people and in the way information can be delivered.[2]

The overall result is that not only does the Web serve as the *common gateway* to the digital world, but also to *people,* and it is emerging as the *common platform* for running every possible application conceivable in the software industry. The new uses of the Internet can be thought of as developing in two directions: services and collaboration. The next section outlines the ongoing developing characteristics of the Web.

From access to services

It is undeniable that the commercialization of the Internet is leading to an expansion into the society at large. Internet services have increased steadily in both number and diversity, especially for industries where products and services can be digitized. The volume of online sales, in hard dollars, is growing at a staggering pace. Transactions on the Internet reached $500 million in 1996. However, a recent survey by the University of Texas at Austin showed that the value of e-commerce in 1998 had rocketed to $102 billion.

As the Internet begins to deliver more complex media, a wider range of services are capable of going online, and compete directly with their counterpart industries, such as marketing and advertising, radio and television broadcasting, telephony and education and training. "With just graphics, the entire book, newspaper, magazine and printing industries of $150 billion will be affected." (Bell G. website listing).

Internet commerce is attractive to both large and small businesses, owing to low set-up costs of building a Web presence; and economies of scale, which permit products and services to be marketed to a larger global customer base. The value of the Internet increases with the number of people and computers it interconnects. Electronic commerce systems can serve customers and take orders 24 hours a day. Potentially many items can be delivered directly to customers over the network faster and more cheaply.

Just as businesses can connect to a broader base of customers, consumers can enjoy wider product selection and other conveniences, such as access to customized product catalogues and shorter transaction cycles. The "global marketplace," as some might call the Internet, is too big to be ignored by merchants, but consumer confidence and trust will remain low until a truly secure transaction infrastructure is established. According to independent analysts, over 85% of Web users are hesitant about submitting their credit card numbers online.[3]

From access to collaboration

Expanding the Web with group interactivity has given enormous momentum to a global software movement which allows people from many countries of the world to work together simultaneously. The OntoLingua project at Stanford University (Farquhar et al., 1995),which aims to build a large ontology, provides evidence of this collaboration trend.[4] The huge number of specifications required by the project must necessarily rely on the co-operation of a large number of users to create and maintain the ontology data store. Because of the large Internet user base, we hope to find a sufficiently large number of people interested in being part of a collaborative project, since each participant will only need to do a little bit of work.

This observation is consistent with Maes' characterization of the Web as an intelligent machine (Maes, 1997). There always seems to be someone willing to give us an answer either in real time or after some delay as if we could ask random questions to a computer attached to the Web and get back intelligent answers from the computer.

Building a machine with human-like mental capacity has proven very difficult but by using the synergy of large-scale human collaboration and computer interaction we can combine the strengths of people and computers. We don't have to build intelligent systems, that rely solely on machine intelligence, to achieve intelligence. Instead, when the software is unable to deal with a complex situation, we could refer it to human expertise, as a last resort.

For instance, in information filtering applications, opinions left by thousands of Web site visitors can be used as a collective pool of wisdom, to alleviate the limitations of machines in correlating and ranking information—especially regarding data types such as video that cannot be easily indexed. It is also possible to connect the user to other users who share the same interests. The network is expanding both our individual problem-solving capacity and the information-processing capacity of the computer. In the foreseeable future more applications are likely to arise, which will take advantage of the capabilities and potential provided by this massive electronic infrastructure of humans and computers, which has the capacity to become the most powerful and overarching computer architecture of our time.

The future—the universal knowledge workspace

Many developments are guiding the Internet into the future. With the advent of Internet programming languages such as Java, the functional extensibility of the Internet becomes basically unlimited.[5] The Web's linkage to the Internet, can be similarly analogized with that of *C or FORTRAN* in the world of programming languages. Just as many existing applications during the 70s were re-written into C, many traditional applications that once ran on stand alone computers, such as databases, e-mail, spreadsheets, etc., are now being converted to run on the Web, from any computer connected to it.

The Internet has become a melting pot of technologies, apt to draw upon advances from all areas of, but not limited to, computer science research. Techniques and methods from Artificial Intelligence, for instance, could be applied to learning over the Internet. We could expect online course instructions which could be programmed with knowledge and intelligence, that approximate the capacity of a human tutor. This and other new applications will appear as new technologies hit the Internet.

More importantly, solving the basic problems facing the Internet makes the next transition likely to happen. In particular, concerns about the future of the Internet are centered on delays and information access, and manipulation limitations. Various national and global projects are well underway to address these concerns. The next step in the evolution is to go from *searchable* to *usable*. In the meantime the growth of the Internet is not showing any slowdown, and getting the Internet to do more for us seems to be the general interest of everybody who uses it.

The basic problems

Today's Internet infrastructure encounters two basic limitations with respect to connectivity and information management. While there are other concerns, these two problems have the most impact on our perception of the Internet and its future.

Connectivity

With multimedia becoming the lingua franca (or common data type) of the Web, network applications and user expectations have changed. Increasingly, more bandwidth-intensive traffic consisting of graphics, video, and audio streams is sent through the Internet. The typical size of a text-only Web document is about six kilobytes, but a page with images can average at least one order of magnitude bigger. Network analysts estimate that the Internet traffic is consistently doubling every six months (Flanagan, 1998), but the installed capacity is only doubling every two years (Bell G. website listing).

Users' concerns are centered on delays, bandwidth limits, and reliability. The average user connects to the Internet at modem speeds, such as 14.4 and 28.8 kilobits per second.[6] The lack of infrastructure on the user-side to meet the demand, means that videoconferencing and video on demand traffic would not be large enough to rival e-mail traffic for quite sometime (Lisa, 1997).

Even when the Internet can handle huge amounts of data transfer, the increased number of users has begun to overload it. At any given instant millions of users can be simultaneously connected to the Internet, and the aggregate network traffic could be in excess of one terabit per second. On a typical day the average Internet response time[7] is 170–270 milliseconds (The Internet Traffic Report, 1997). This delay window may already be too long to accommodate users' growing expectations for fast response times, when they browse video archives and interact with real-time three–dimensional applications. The requirements for large data and fast transfer rates demand for increasing interactive speeds in the range of 100–200 milliseconds, so the above average Internet response times are sufficient only for a portion of the time (Bailey, 1997). This also depends on the region of the world, since regions such as Asia and Africa which attempt to access sites in the U.S. or Europe, can have much longer response times.

The network topology of the Internet has grown very complex since the demise of the original university–government-based backbone—the NSFNet—in April of 1995. The Internet is a co-operative medium which is not controlled by a single entity. Packet loss and throughput sometimes changes from one second to the next. Traffic on the Internet is sent from router to router through the network to the destination, using topology and routing policy information stored on routers. These routing tables are distributed across different regions of administrative control commonly called autonomous systems (AS).

Network instability in one region can spread over to a large portion of the Internet. Fluctuation in the network topology can lead to a large number of routing updates that need to be propagated to many routers throughout the Internet. In the process a network that has not reached stability (or convergence) may result in data packets being dropped and out-of-order delivery problems.

Transmission of data packets through the network that are dropped before reaching their ultimate destination wastes bandwidth, which in turn causes congestion. Instability has a number of possible origins, including router failures and configuration errors, transient physical and data link problems, traffic congestion, and software bugs (Labovitz et al., 1997). Small problems at one part of the Internet can be spread across much larger portions of it, such as the two-hour brownout that occurred globally on the Internet in 1997 (Gareiss, 1997). In mid-April of 1998, an outage at AT&T's network disrupted service across the U.S. for several hours (Wired News, 1998). These are some examples of the incidents resulting in network performance degradations and interruptions that have led some experts to speculate on the collapse of the Internet (Gilder, 1997).

Information complexity

The continuous flow of information onto the Internet is leading to many interesting features but also creating many challenges for mining the wealth of information it offers. The potential benefits of this large amount of available data will remain unfulfilled, if the user is unable to find and retrieve the information he or she needs, with ease and efficiency—a phenomenon known as *information overload*.

The search engine is the basic tool used to search for information on the Web, and is in wide use today. Searching the Internet without an effective search engine is like looking for a needle in a haystack. But just how good are they? Many users are discovering that their sites are not indexed, or only partially indexed, by public search engines. One perspective on this is the reaction of the webmaster of the Federation of American Scientists (Brake, 1997). Upon discovering a popular search engine indexed only 10% of the 6000 pages of his Web site he observed that, "This is like buying a phone book that only has even-numbered phone numbers." A recent article estimated that combined, all of the major search engines covered just 42 percent of the Web.[8]

A few years ago a single server could index the entire Web, but this is not possible today. The problem now goes beyond the coverage limitation of search engines. The design of a good search engine for the Web presents many challenges. Resources on the Web must be discovered and continually monitored for updates and changes. Search engines need to organize huge volumes of data with higher precision[9] retrieval mechanisms. Index completeness is not the only factor affecting the quality of search results. Two other problems limit the effectiveness of search engines. They are related to the interface between users and search engines, and those between search engines and documents residing in diverse locations.

User interface barriers

Most Web search engines provide retrieval capabilities only by keyword or phrase and criteria combinations using Boolean expressions such as "and" or "or" functions. Thus, the interface for making queries may be too rigid or difficult for the user to formulate his or her search intention and expectation fully. Most users are not very fluent with Boolean logic, the dominant native "language" used in search engines. The mismatch between the user's intention and mapping it to a meaningful expression for the search engine often results in a large number of documents that the user is not interested in. For example, query words that have multiple meanings. Likewise, human spelling errors and typos can return a huge number of unrelated documents. Even when the user's query is specific enough, it is still very difficult to make sense of the obscure ranking[10] of the large number of returned documents.

Barriers to accessing content

Traditional information retrieval techniques generally apply to closed and controlled collections of homogeneous documents from a given topic or domain. In contrast, the Web is a free flowing environment, which can be thought of as the electronic memory of society. The information on the Web is highly heterogeneous—the information can potentially come from any computer connected to the Internet, and it is unstructured—the information contains very few hints about the structure and contents of the information itself (i.e., metadata).

Hypertext Markup Language (HTML) gives the structure to the Web primarily. However, HTML is essentially a presentation language, so it is not a very good tool for including *metadata,* and provides limited support of hierarchically organized information. Because of the size of the Web, automatic indexing becomes necessary and retrieval mechanisms based on keywords alone are not good enough for improving search results. Enhancements such as automatic suggestions of new and related terms, support for refinement of search results, and indexing only authoritative sources can only provide an interim solution. A long-term solution will require search engines to be able to extract structural and content information from documents. However, the lack of sufficient classification information on Web documents makes this task extremely difficult.

Web documents are not made up of just text, but also multimedia. While automatic indexing and classification of text-only documents has been studied extensively in the information retrieval community, the problem of indexing and classification of non-text media is much more difficult. For example, without automatic methods that can understand audio-visual contents, video sequences need to be interpreted manually, frame by frame, to locate the beginning and ending points of a scene.

Finally, the Web is international and information is available in many languages. However, text-searching technologies that try to take into account effectively the variations in language are still developing.

The next step—what should the Internet be like?

Information is lost if it cannot be accessed. It becomes noise if it cannot be searched, and it is effectively garbage if it cannot be used and applied in a context. Moving on to the next evolutionary step will depend to a large extent on the groundwork done to address the manifold basic problems concerning connectivity and information management. This section outlines the various efforts and plans that are likely to cause the Internet to evolve into the next stage—the global knowledge workspace.

Technologies necessary to accommodate the fast-growing Internet traffic volume, to support high-performance networked applications, to guarantee secure communications, and to provide affordable Internet access to the public are being developed under several complementary initiatives around the world. The development of the next and future generation of networks and applications, falls under several national and regional research network projects (Chon, 1998).[11] Countries in Latin America and Africa are also reporting progress in building their own high-performance network infrastructures.

Parallel to the national and regional connectivity efforts, a number of international movements such as G7 GII/GIBN (Globally Interoperable Broadband Networks) and CCIRN (Co-ordination Committee for Intercontinental Research Networks) are being formed to co-ordinate and collaborate in the interconnection of these networks within each continent, as well as between continents. Using advanced networking equipment and new protocols, the global network of the future could well operate thousands of times faster than today's Internet.

The technological issues and socio-economic implications stemming from the rapid accumulation of digital information on the Internet, have inspired many research and applications projects around the world. We find the Digital Libraries Initiative (DLI), launched in the early 90s in the U.S., among the most intensive research developments on information infrastructure. It is organized into six projects, with each project focusing on specific instances or aspects of digital libraries. The common goal is to bring search capabilities to many and varied information repositories on the Internet. The scope includes such topics as digital document preservation, information retrieval on multiple media and in multiple languages, copyright and intellectual property rights, standards, *metadata*, resource discovery, the impact of digital libraries on society, the role of digital libraries in managing information, and so on.

DLI is now moving to the start of its second phase. As local and national projects are emerging from different countries illustrates, the quest for a global digital library is now widespread. The United Kingdom, Finland, Japan, Korea, Germany, New Zealand, Brazil and China are some of the countries that have embarked on developing digital library technologies (Fox & Marchionini, 1998).

The groundwork laid by the DLI will allow unstructured and distributed information to be integrated into a searchable global information space, as well as simplify query formulation and improve accuracy in search results. In the future, new extensions of DLI may help make sense of the retrieved information, bringing more precise usability to the Internet.

Information retrieval and analysis is a non-linear process, involving a variety of feedback loops. Users often develop a better idea of what they are looking for after a sequence of steps interleaving search with analysis and data processing. For example, one might decide to start a new search or refine the search criteria after processing some search results, or even refocus interests elsewhere. The global knowledge workspace will integrate both search and analysis in a unified environment to better support users' interactive exploration and use of the information. The basic objectives will be:

1 to reduce the user's cognitive workload while in the process of discovery, selection, analysis, and assimilation of information, and
2 to enable reusability so that information and software components can be shared easily across heterogeneous software systems which can talk and collaborate with each other to produce a desired result for the user.

The last objective is reducing the "cognitive workload" of the software system which is needed to process the information, i.e., it will allow software components or applications to be more easily linked together in the information processing chain. The co-operation requires interoperability among objects (software and information) in the workspace. Interoperability will not only allow the output (information or software component) of one application to interface directly with another application, but also software components to be reusable by many applications.

A number of experimental building blocks for creating the global knowledge workspace are already beginning to appear. One promising approach consists of using 3-D representation and animation to provide the user with the ability to explore, visualize, and analyze large volumes of data and results from queries in a single view. Examples of these efforts can be found in both research and commercial arenas. For example, XEROX has developed a "hyperbolic" browser for displaying hierarchies of hundreds or thousands of pages and their links simultaneously. See Figure 5.1 for an example of this.

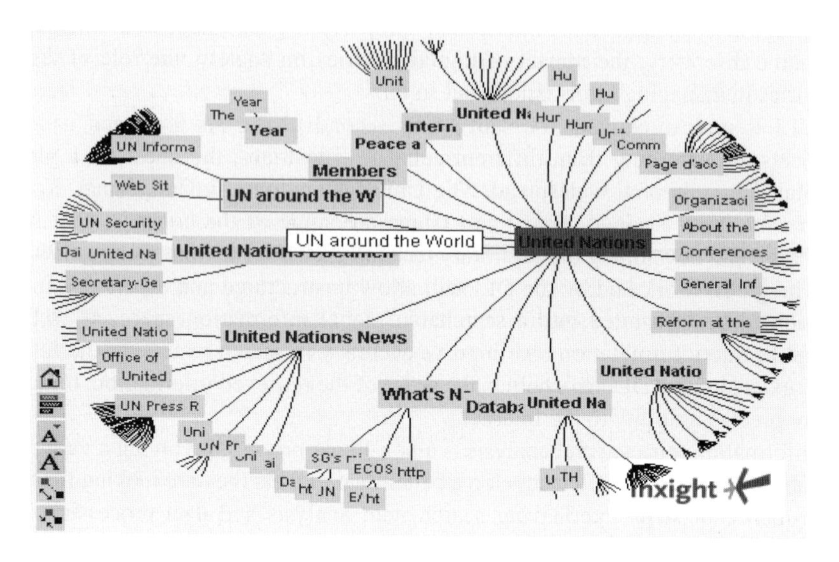

FIGURE 5.1
Example of a 3–D Interactive Visualization Environment

A listing of information retrieval interfaces can be found at the University College, London website.[12] These all try to depart from the traditional model of the page by adding depth, colors, animation, and special visual effects to enhance the information display. The result is a navigable and interactive 3-D interface that can steer users toward the information they seek, as well as help clarify relationships, patterns, and interdependencies in the data which would not otherwise be obvious using text, 2-D grids and tables alone. Users can navigate in the 3-D space without losing the context, and can zoom in for finer detail as necessary.

It is desirable to visualize the information from different perspectives to maximize opportunities for extracting meaningful and relevant knowledge from large amounts of complex information. Users, therefore, need new tools to dynamically organize information in the way that they want to see it, e.g. the ability to switch smoothly from a high-level summary view to a specific paragraph of a document or scene from a video. Thus, a major challenge is to develop a flexible and interactive environment, which can support a high degree of personalization, and customization in the way information can be displayed.

In order to increase the usability of online information, the Web will need to be organized into some form of knowledge structure, such that information processing components can reason on the information retrieved, using such automated inference methods as those researched in the field of Artificial Intelligence (AI). Or, to attain even more flexibility, we can imagine that the heterogeneous information sources are organized into, and managed by, a network of distributed interoperable intelligent components. Each intelligent component contained therein would be responsible for some information scope which could be based on network or geographic domains, or on human knowledge domains. This implies that the intelligent components have the ability to interact with and exchange knowledge from each other, working together to answer users' queries.

In the meantime, computer languages and protocols such as KQML (Knowledge Query and Manipulation Language) and KIF (Knowledge Interchange Format) have been developed to reach the goal of sharing and reusing knowledge across disparate software entities.

While many techniques exist to mitigate the problems related to information overload, many believe that the ultimate breakthrough will come from the fields of natural language understanding and image understanding.

Advances in Internet Technologies and Their Potential

The convergence of computing and telecommunications technologies, and the proliferation of high-speed, Internet-driven networks are setting new technological trends and creating new architectural opportunities in connectivity, software, and computer hardware. We will examine each in turn, focusing mainly on the software trends. Hardware and connectivity can be seen as elements that will either constrain software and its uses, or enable these uses.

While some of these trends have much broader impacts at the societal level, including at work and home, they will all have some undeniable, major, impacts on education, either because some forms of education will increasingly take place at home or at work, or because traditional educational environments will be altered. In particular, software technologies will have great impacts on education, including the facilitation of learning environments that incorporate collaborative computing, digital libraries and other advanced information functions.

Connectivity trends

The Internet is adding the freedom of time and space to virtually everything we do. This freedom has encouraged us to think of new applications such as collaborative medical diagnosis which demand speed, bandwidth, reliability, quality of service, and network scalability far beyond what today's Internet can provide. The explosive growth of Internet traffic has strained the capacity of backbones[13] built with traditional networking technologies and underestimated traffic assumptions.

The bottleneck on backbones could create traffic congestion and latencies affecting a large portion of the Internet, but the weakest link in the chain is the standard telephone service (POTS) line that connects the subscriber's home or office with the local telephone exchange. This problem is also known as the "local loop" or "last mile copper" bottleneck.

While content providers such as universities can take advantage of the latest information technologies to package information, the extent to which that is permitted depends on the capacity of the local loops and backbones through which the information is delivered or accessed. Some of the technologies that have been developed to deal with these bottlenecks include:

- Innovative "last-mile" or "local loop" technologies, which are technologies that improve the connectivity between the subscriber and the local telephone exchange—these include the Integrated Services Digital Network (ISDN), the Asymmetric Digital Subscriber Line (ADSL), wireless technologies, and cable modems. ISDN and ADSL try to squeeze as much bandwidth out of existing telephone lines, with the latter running much faster than the former. Cable modems use fast coaxial cable TV lines for data transmission to homes, while wireless technologies allow for mobile computer use. These will make real-time interactive applications (e.g. videoconferencing, a multi-user virtual reality system) and multimedia data more accessible to the average user.
- Backbone technologies—such as all-optical networks and satellite communications systems. These will enable more users to connect to the network at the same time while alleviating network congestion.

Computer hardware trends

Moore's curve, a well-known formula, predicts that the number of transistors on a silicon chip will double about every 18 months until fundamental physical limits are reached. This trend was consistent over the past 25 years and the chip industry still shows no sign of slowing down. The doublings in density mean more functional units and memory capacities can be packed onto the chip, and shorter interconnecting distances between components; which allow signals to propagate faster from component to component. Because the cost of each generation of chips has not increased thus far, users can buy more computing power for the same amount of money.

While we do not know how much longer the geometric growth will last, and technological challenges remain, research organizations and the semiconductor industry are optimistic that they can achieve further miniaturization and higher levels of integration for at least a few more generations (Bohr, 1998). Unfortunately, increasing technical barriers also come with financial and economic pressures in the form of higher costs in equipment and manufacturing, which may eventually slow down the evolution of silicon technology.

The following trends in silicon-based devices and systems will enhance computing power on both the server and client end-user sides, providing computing capability for the masses.

Computers for the masses

The computer industry has long realized that a one-size-fits-all computer platform is a stillborn vision. With the Internet cutting across all walks of life, there are more first-time computer owners and diversities of users than ever before. Inevitably, computer platforms must serve a much broader range of users while maintaining software compatibility across the fragmented standards. This movement has spurred the evolution of new hardware alternatives, ranging from TV Internet terminals and palmtop Windows PCs to 64-bit high-performance computers.

Cheap computers

Despite the popularity of PCs, the computer industry has much to catch up with home appliances (e.g. TV sets) in terms of their penetration in the consumer market. In the U.S., for instance, 39–45% of the population own PCs while 99% own television sets (Goldman, 1998). The Internet may very well be the way to narrow this gap. The crusade is on to bring the Internet to the consumer community, to every home. In this context two trends can be identified.

One trend is headed by the introduction of cheap PCs into the marketplace. This year we are starting to see an increasing number of PCs costing less than $1000. According to a recent market research report 30–40% of PCs sold through the U.S. retail stores, cost less than $1,000, up from about 5% last year (Halfhill, 1998).

Other ways to connect to the Internet

Some vendors are taking the idea of affordable PCs a step further, and producing a new breed of simple computers known as "network computers" (or NCs). Operationally speaking, the typical PC is essentially a so-called "fat-client" desktop computer; which is one that may be on a network, but where most of the application (software) and data still reside on the client, and applications are executed almost entirely locally. In contrast, the NC is a type of "thin-client" computer, which downloads both applications and data from network servers as needed, for local processing (Shah, 1998).[14] It is the network—including both the Internet and local networks—that gives life to the NC.

The trend towards thinner client computers addresses the following user concerns:

- The total cost associated with owning personal computers is very high. An often-cited research statistic by the Gartner Group in Stamford, CT indicates that the total cost of ownership (TCO) for a PC is more than $40,000 over a period of three to five years (Comerford, 1997a). The largest PC-related cost is not the initial purchase, but the hidden costs due to administration overhead (e.g., software upgrade and security set-up), technical support (because computers are often hard to use), and end-user tinkering with the computer (e.g., adding new software and/or hardware, and using it for games and other recreational purposes).
- Shorter innovation times create fear of quick obsolescence on high-priced PCs. Users face the imminent risk of not being unable to amortize a high-end system quickly enough before it becomes obsolete.
- Users in general only want to pay for what they need. For instance, if the user wants a computer for email and Internet browsing only, he/she does not need a computer with a fast CD drive, accelerated graphics or other advanced features.

NCs have some interesting attributes. They are significantly easier to use, have a lower initial cost and a lower TCO than a typical PC by simplifying administration and maintenance at the desktop. For some people, NCs have a certain psychological appeal because they refashion the computer as a true consumer appliance—or even as a disposable computer. Since they aim to give users access to networked resources, especially on the Internet and corporate intranets, a great deal of the intelligence and functionality, including such overhead as administration and configuration can be shifted to the distributed or centralized server machines. The result is a simpler architecture with fewer software and hardware components, which in turn increases system reliability.

Internet devices—embedded network computers

Given the dominant presence of PCs in the computer industry today, it is unclear whether NCs will ever grow as popular as PCs. Yet a new development direction that aims at embedding NCs into everyday items and secure a niche for NCs in the mainstream is well underway. For example, a set-top box,[15] from WebTV Networks (Palo Alto, CA), allows TV viewing, Web browsing, e-mail access and common applications such as entertainment-on-demand, gaming and electronic commerce from a TV set using a remote control or a wireless keyboard as an input device.

The increasing ratio of computers-to-users

In the era of mainframe computers, the ratio of computers to users was one-to-many—i.e., many users shared each mainframe computer. With each generation in computer evolution, the machine/user ratio has changed. When PCs first appeared the one-to-many relationship gradually shifted to one user per computer. With the advent of portable computers and the ability of computers to connect to other computers, the ratio is becoming many-to-one, that is, many computers may support one user.

The emergence of embedded NCs adds density to the relationship by integrating cheap, low-power computers into everyday appliances. The density creates the illusion that computers are everywhere, but to transform this illusion into reality we need small, low-power, computers that are in effect, wearable or mobile.[16] Mobile computer systems could allow users to literally work from anywhere. A number of working prototypes and commercial systems are already available.

Hardware trends such as embedded NCs and wearable computers support the so-called ubiquitous computing vision, which foresees a world in which computing access will be everywhere and small computational devices will be embedded seamlessly and transparently into our everyday environment, being quite "invisible" to us. Their "vanishing" into our surroundings will make our interaction with computers effortless—as whenever people learn something very well they cease to be aware of it (Weiser, 1991). These devices are aware of their surroundings and are able to react and emit information as necessary.

One implementation of ubiquitous computing technology is the active badges that can trigger automatic doors and give information about the location of a person (Want & Hopper, 1992). The ubiquitous nature of the Internet will be even more so in the vision of this technology, which can be succinctly described as follows: "The Internet revolution has barely started. It won't be done until *everything* is on the Web. Light switches, pagers, copiers, printers, as well as PCs, benefit from Web connections (Weiser, 1996)".

Trends in other devices also have direct impacts on the affordability, computing power and capabilities of computer and network systems. These include trends in integrated circuit technology, processors, and memory and mass storage.

Software trends

The Internet has achieved the widespread connectivity of networks, information and people. In the process, computing has been transformed from a stand-alone model of computing to a network model of computing that not only transcends geographic boundaries, but in the future, will provide ubiquitous access to users, including those on the move. This migration to a rapidly- expanding networked environment has created new opportunities and challenges in the software arena. We look at a snapshot of three key areas in software and systems research that are fuelling the evolution of an information infrastructure on a truly global scale:

- Ubiquitous computing environments for mobile users
- Digital libraries
- Collaborative computing

The outputs from these fields will bring important innovations to the entire life cycle of information—from production, distribution, access and use, to storage and preservation, in many cases, with benefits for users in educational (particularly university) environments.

Emergence of ubiquitous computing environments for mobile users

Today's model of how computing is done requires the user to connect to a network for work or leisure computing, but the information available through his desktop becomes inaccessible when the user changes location and/or computer. However, if users take their computers with them wherever they move, applications already in execution can be disrupted during movement as the system disconnects from the current point of attachment to the network, and reconnects to the network through another point. This is particularly true with the increasingly global nature of the Internet, since a computer connected to the Internet now becomes a portal to an unlimited number of resources, including peoples and contents around the world, but with mobile computing, users' computing environments become accessible from anywhere in the world.

Recent advances are enabling users to continue his/her computing activities with network applications irrespective of the user's physical location and the computer being used for the task, even when both the user and the computer are moving. This represents a shift to a new computing model known as *mobile computing*. Under this model, data and/or applications follow wherever users go such that users' computing environments[17] are retained or can be recreated regardless of their physical location.

Mobile computing aims to provide ubiquitous computing environments for mobile users. In particular, it attempts to give users the illusion that their computing environments (e.g. access to their usual mail folders and interfaces to running applications) are available anywhere, at anytime—right down to the final mouse click as when last accessed. There are many technical barriers to be overcome before mobile computing can become widespread, such as the ability for users to connect and disconnect with the network at will and without loss of continuity in their work, better power management for longer operations, new network protocols and security mechanisms to deal with mobility, and better displays and input interfaces which may influence the usability of mobile devices, etc. (Richardson et al., 1998).[18]

Digital libraries

The constant influx of information on the Internet is pushing the limits of our ability to make sense of the sea of data. Locating a particular piece of information on the Internet is still a major problem, and retrieving quality information is even more difficult. The variety of technologies, tools, and services that strives to bring order to this chaotic information environment can be most comprehensively delineated under the framework of *digital libraries*. As will be described later by Borgman in this book , digital libraries are a fundamental subsystem of any virtual learning environment, be it a virtual university or a stand-alone library.

The concept of a digital library is broader than its physical counterpart. In particular, it goes beyond automation to include a wider range of information sources, media types, and services, and enhances the way we deal and interact with information for different purposes—from information creation, storage and retrieval, to problem solving. Currently, many countries are developing their own digital libraries. Most of these efforts are not co-ordinated, with components developed independently and often using different technologies. But they all share the basic goal of information matchmaking, i.e., finding the information that is of interest to the user (Lync & Garcia-Molina, 1995).

Towards a truly global library

While building a digital library has its own challenges,[19] a longer-term goal centers on the integration of the individual digital libraries around the world into a single digital library. Informally, the problem of linking all the disparate sources and services from autonomously managed digital libraries in a useful and efficient way has come to be known as the *interoperability challenge* (Paepcke et al., 1998). A simple example can illustrate the importance of interoperability. Consider the case of fee-based services for searching diverse information sources from multiple digital libraries. If the differences in payment policy and mechanisms were not integrated in one consistent framework that provides some level of homogeneity, a unified front-end search service would be very awkward to implement. Depending on the degree of interoperability needed, the mediation of site-by-site variations may require human interpretation.[20]

The spectrum of interoperability issues can be broadly organized into three levels (Fox & Marchionini, 1998):

- Technical interoperability, which involves protocol level and compatibility issues in data types, applications, services, hardware, and networks
- Informational interoperability, which deals with content scope, language,[21] metadata, naming conventions, semantics,[22] and user interfaces.
- Social interoperability, which addresses personal and organizational agreements, such as interactions in policies and enforcement mechanisms concerning information security

Advances in digital library technologies

The set of concepts and technologies below provides a representative variety of building blocks that *support* the construction of digital libraries, but they are by no means an exhaustive list of all relevant technologies in the field. All of these also have some relevance to education, either directly in terms of being used in the design of educational tools, or through their use in digital libraries. Each of the technologies is an enabling technology, meaning that it provides a building block for building more complex systems and providing more functions. The first three technologies—the *architecture* of the information (i.e. its design for searchability), the use of *metadata*, and the use of *intelligent software agents* to aid users in their searches—are involved in *search* processes, a basic activity in any virtual environment. However, any of the technologies may be used to enable other more complex functions in digital libraries and other systems, e.g. software agents can be used to coordinate or allocate computing or information resources within a digital library.

Search architectures

Search architecture refers to the means of organizing information, which also involves how the search engine (or software used to search that information) does its job. The search model used by many of the search engines (e.g. Lycos, Alta Vista, Excite, HotBot, etc) today is based on a centralized architecture. In particular, they all try to maintain index references to most of the resources on the Internet in a database based on a single schema,[23] although various copies of the database can be placed in multiple locations to improve response time. These systems generally rely on software robots that download and process every web page they find to build an index database, which generally offer keyword and expression-based search services. This mode of organization offers simplicity of control because the index is all in one place. But it is also its weakness because it is increasingly more difficult for a single centralized index system to keep pace with the size and complexity of data on the Internet. As a result, the coverage is limited, indexed information tends to be outdated, searches are exhaustive, searches are based on an index that has a global scope (i.e. users cannot easily limit their search to a geographic location), and queries may become expensive as the index database grows.

The proliferation of uncoordinated and isolated search services and the use of robots to discover resources on the Internet can cause the overloading of local and global networks, and bandwidth shortages. This happens because search services duplicate efforts by sending their robots to visit the same site, sometimes simultaneously, and revisit all the sites periodically to check for updates.

Metasearch engines (e.g., the Profusion system[24]) have been introduced to address the problem of limited coverage by transparently dispatching users' queries to several search engines.[25] However, they all inherit the same fundamental problem of the centralized architecture, which is that it does not scale up to the dimensions of the Web. Other solutions handle the scaling problem by distributing the indexes across multiple servers. For instance, the distributed architecture used by search services such as *Whois++* and *Harvest* handles the scaling problem by logically partitioning the indexing information for the Web into smaller indexes that are autonomously managed by a group of index servers. Each index covers a particular scope, which represents a limited portion of the Web. This results in lower demands for computational resources on servers and implies that index servers must co-ordinate and collaborate to avoid redundant indexing. The partitioning of an index's coverage can be itself a powerful mechanism for achieving computational efficiency, as well as for restricting and focusing queries. For example, the scope of a particular index can be defined using some dimension such as the geographic or network location of Web servers. Limiting the scope in this way avoids the need to use long-distance backbones to support indexing (Goncalves et al., 1997) and reduces the number of documents that have to be indexed. Likewise, distributed indexes can be specialized into subject areas to achieve better information filtering.

The future of Internet searching will most likely be implemented on a distributed architecture because of its potential to provide a scaleable environment for indexing and searching, but the success of this architecture will depend to a large extent on how well the components in the infrastructure cooperate with each other to share indexing references and optimize performance.

Metadata

The problem of poor search engine performance in finding and ranking documents relevant to users' queries is related to the lack of sufficient machine-understandable information (i.e. semantics) in Web documents, as explained in the section discussing information complexity. What is needed is some kind of standardized metadata (i.e. a higher level indexing scheme for data) that provides machine-understandable information about the structure and content of Web documents in an interoperable manner such that it can be used by different software packages. This need has been widely acknowledged as critical to the future of the Internet search function (Guha, R V, website listing & Chang et al., website listing).

The main purpose of metadata is to facilitate the automated processing of Web documents. Metadata could describe a wide variety of information content: such as catalogue information (e.g. the parts of a document and their relationships); content ratings (e.g. quality assertion information); language, users' preferences regarding privacy, intellectual property rights; trust information (e.g. digital signatures); electronic commerce information (e.g. prices, terms of payment, etc.), (Lassila, 1998). Search engines could take advantage of metadata to improve search efficiency and accuracy.

One important development towards metadata at the international level is the World Wide Web Consortium's (W3C)[26] recent publication of the Resource Description Framework (RDF)—a new standard for expressing and exchanging all kind of metadata information (Lassila, 1998). RDF is a data model of metadata instance that uses the Extensible Markup Language (XML) as a means for encoding and transporting its metadata.[27]

Intelligent software agents

Computers are still difficult to learn and use. The software industry, with each new release of programs, tends to overload the user interface with an increasing number of features than previous versions.

At the same time it is harder for users to find the information they need, even though the access to huge amounts of information has become easier. The emergence of intelligent software agents promises to relieve users from the burden of information overload and to reduce their work on mundane or complicated tasks.

The definition of "intelligent software agents" is the subject of much controversy given the many possible roles agents can play[28] and we do not intend to settle it here. However, many researchers agree that they differ from conventional software in that they are: (Hermans, 1996 & MIT Media Lab., website listing)

- Autonomous: agents are continuously running and operate with minimal supervision.
- Adaptive: agents adjust themselves dynamically to users' changing habits, working methods, preferences, and interests.
- Reactive: agents act in response to changes in their environment.
- Proactive: agents can exhibit goal-directed behavior by taking the initiative to help the user, e.g. making suggestions.
- Social: agents interact with other agents, applications, and possibly humans via some kind of agent communication language.
- Mobile: agents have the ability to move from one computer to another and continue execution at the remote computer, e.g. personalized agents that roam the Internet autonomously for appropriate information and services, and return with answers.

This list is far from complete for some researchers, especially those in the AI community who are concerned with achieving new behaviors with human qualities such as beliefs and intentions, and even emotions[29] (e.g. facial expressions and gestures represented on an animated face or a cartoon-like graphical icon).

To give just a few examples of agent applications:

- Agents can monitor users' actions in the interface, suggest courses of action to the user and provide context-sensitive help (e.g. "wizards" that help users to step through a sequence of operations in the performance of a particular function).
- Agents can assist Web browsing, e.g. Letizia (MIT Media Lab. website listing), an agent that tracks user browsing behavior and attempts to anticipate possible items of interest by doing in parallel autonomous (i.e. without user intervention)

exploration of links from the user's current position in the Web. The recommendations can be shown in a different window.

- Agents can act as surrogates for users, e.g. Kasbah (MIT Media Lab, website listing), a Web-based multi-agent classified ad system where users create agents for selling/buying goods on the user's behalf. The agents proactively look for potential buyers or sellers and negotiate with them to make the "best deal" possible, based on the user's constraints. Xpect (Andreoli, Pacull et al., 1997) provides a framework that allows for a variety of commercial transactions
- Agents can perform a collaborative search on the Internet, e.g., Amalthaea (MIT Media Lab, website listing), an artificial ecosystem of information filtering and discovery agents that cooperate and compete in a market-like environment, evolving all the time.
- Agents can help augment peoples' memory, e.g. the Remembrance agent (MIT Media Lab, website listing), a program that continuously tracks the physical conditions of a wearable computer to provide notes that might be relevant to the wearer at the moment.

In the context of information retrieval, intelligent software agents can assist the user in many ways that are not possible using conventional search engines, such as:

- Building a user model/profile for the search process, i.e. identifying his interests and preferences, for example by learning from his actions and reactions in previous searches. The user centric model will enable agents to search using related terms or concepts and even contexts.
- Creating a personalized knowledge base of information sources that are likely to meet his information demands.
- Helping to scan resources on the Internet. Agents can adaptively plan their excursions on the basis of network conditions, user input, and the information gathered along the way, e.g. search during outside peak hours and at intervals that spread the load more evenly on the Internet.
- Notifying the user of new topics and information changes that are relevant to him.

The current Internet structure can be roughly approximated to a two-tier model, with information seekers in one-layer and information providers in another layer. But this structure is proving inadequate for dealing with the rapid accumulation of information and changes on the Internet. There has been a wide interest to contend with this problem by adding one level of abstraction and intelligence to the current structure. In particular, a framework for a system of distributed agents that act as surrogates for information seekers and providers, as well as middle agents in the role of information brokers or facilitators. In such a framework, the middle agents could support seeker agents' search of the providers' catalogues (federated and presented though a uniform interface), whereas provider agents could advertise through the middle agents about their (possibly pay-per-view) documents. This way the providers could screen out unwanted seeker agents or limit the amount of access to their sites, and hence their sites would have one less security concern and be able to better market their resources. Agent technology is under active research and development. Standardization on agent communications and interfaces is essential if agents are to be useful in a global scope.

Information security—digital watermarking

Successfully addressing the requirements for security and copyright protection in digital media is a key step toward a sustainable information infrastructure for digital libraries and electronic commerce. Unless content owners are assured that their works are properly protected against piracy, few would be willing to distribute their works online. In general, information in a private environment must offer three kinds of security (Adam & Yesah, 1996):

- confidentiality—concerns with the privacy of contents and their protection from unauthorized use
- authenticity—concerns with establishing ownership over some content
- integrity—concerns with the protection of digital objects from unauthorized alteration

Technologies such as data encryption and digital signatures address most of the critical aspects of confidentiality and authenticity. Unfortunately, beyond copyright laws there is very limited protection for decrypted digital content (graphics, images, video, and audio) against unauthorized duplication, alteration, and more importantly unauthorized commercial use.

Digital watermarking is an important emerging technology for enforcing copyright laws on digital content. Watermarking involves embedding signals and labels in digital content, often imperceptible (invisible[30] or inaudible), but always unobtrusive, that can be detected or extracted later to make an assertion about the content (e.g. verification that the content has not changed). Watermarks can be used to communicate literally any digital information in a hidden or partially hidden way, but they are usually used for the purpose of conveying ownership (e.g. an identification of the owner or intended recipient) and copyright (e.g. number of copies allowed) information. They become the "audit trails" in the data that allow the detection of copyright violation or alteration and tracing of the recipient should there be evidence of piracy.

The basic technical requirement/challenge is that a watermark must be robust enough to survive a variety of manipulations (intentional or unintentional), manipulations and conversions[31] without degrading the quality of data beyond its utility value. This is a very difficult problem, even when the set of operations performed on the digital object is specified (Memon & Wong, 1998). Making matters worse is the fact that public domain programs (e.g. UnZign[32]) have successfully removed watermarks embedded by a few commercial software products. A good discussion on watermark attacks can be found in (Memon & Wong, 1998).

Robust watermarks are resistant to common digital signal manipulations. For that reason they are very suitable for establishing ownership in digital content. On the other hand, fragile watermarks, though they can be easily corrupted signal processing procedures, are useful for verification of content integrity.

However, since no watermarking technique is known to survive all kinds of malicious attacks (Wolt, 1998), the best we can hope for is that it will deter the misappropriation of digital content.

Collaborative computing

One of the key elements of any type of virtual learning environment is its collaborative computing capability. While the social aspects of collaboration are covered later in this book by Mark's chapter, we will provide a glimpse of some of the enabling technologies here.

In technical terms, collaborative computing or CSCW (Computer Supported Co-operative Work) systems are multi-user software applications that enable people to co-ordinate and collaborate in a common task (or goal) without being in close proximity either spatially or temporally. It draws upon such fields as artificial intelligence, Human Computer Interaction (HCI), virtual reality, and the social sciences. Although CSCW systems have existed for some time (Grudin, 1994), the synergy of CSCW and the Web has emerged only recently.

CSCW on the Web is the next breakthrough in communication for the Internet community. It is transforming the Internet from being an inter-network of computers to an infrastructure for human connectivity. This movement has led to the creation of virtual communities, where geographically separated individuals and organizations can interact to socialize and work together. The interaction may be synchronous (i.e. users communicating with one another in real time) or asynchronous (e.g. users leaving messages for each other). We distinguish two classes of applications that bring CSCW capabilities to the Web:

- Web-enabled CSCW applications, which aim at leveraging non-web oriented CSCW applications to the Web. That is, they use the Web solely as the execution platform, for example, an editor shared across a community of users connected to the Web.
- Web-aware CSCW applications, which aim more closely at integrating CSCW applications into the Web by merging the application semantics with the basic control structure of a browser, which displays a set of choices, the links to other Web pages. Hence, unlike Web-enabled CSCW applications, they are tightly coupled with the behavior of a browser. Taking into account the semantics of the Web leads us to envision new kinds of CSCW applications. The real time visual sharing and navigation of the Web is one example of such applications.

Clearly, the second class of applications supersedes the first one in terms of the scope of operations and functionality that it embodies.

Technologies that support CSCW can be broadly classified into four categories according to the means used for group interaction (Mark in this volume provides another classification of a subset of these technologies, based on their use in synchronous and asynchronous applications):

- **message-based,** which uses electronic mail
- **conference-based**, which uses mailing lists,[33] online chat using either text-only or voice, and videoconferencing tools
- **task-oriented**, which uses applications that are aware of other users and their actions (groupware applications), such as a shared editor, whiteboard,[34] tele-pointer, etc.
- **virtual presence**, which uses an interactive 3-D distributed virtual environment (DVE). DVE is an extension of virtual environment (VE)[35]. DVE supports people from geographically dispersed places to meet, discuss, and collaborate inside a multi-user and networked 3-D graphical space. They see their "selves" as avatars in the computer generated 3-D world. An avatar is a 3-D graphical representation of any live object in a virtual world. Human controlled avatars, autonomous avatars (or bots, i.e. animated characters controlled by programs), and static scenery objects inhabit the DVE. People can interact with objects in the scene and with each other by controlling the motions, gestures, and speech of the corresponding avatars.
- **tele-operation,** which moves away from simulation to duplication of reality. It allows people to operate devices remotely. In a simple setting, a human operator could manipulate a remote device via a text command interface. A video camera that monitors the operation of the remote device would transmit visual feedback to the human operator via the Internet using real time video delivery software. The remote device could be a submarine robot outfitted with sensors for ocean exploration, an electron microscope, or an infrared telescope, to name a few. Advanced forms of tele-operation using input/output devices similar to those used in VR can be foreseen.

These technologies can be applied to practical work that enhances the virtual learning experience. For instance, imagine a biology student who performs a dissection on a remote specimen. A robot coupled with cameras and sensors transmits stereoscopic images and pressure signal information as well as tactile feedback to the student. Every movement of the hands and fingers of the student are duplicated by the robot. The student sees what the robot sees on his head-mounted display and feels what the robot senses on his electronic glove.

Tele-operation introduces two innovations. First, the resources that can be shared on the Internet are no longer limited to digital information and software applications, but extend to virtually any object that we can reach and manipulate with motor control. Second, it extends sharing with remote control and manipulation, which holds the potential to duplicate the laboratory experience.

The Base Virtual Learning Modes

As software and communications technologies converge, and computers become mobile and ubiquitous, educational institutions are discovering new ways of using technologies, which have the potential to enhance and make more efficient standard educational activities, such as the lecture, the library, and the laboratory. These changes are reflected in the learning modes made possible by the new information technologies.

Virtual learning modes

Learning in universities takes place inside classrooms, as well as outside classrooms. At the macroscopic level, the emerging computer-mediated distance learning modes do not diverge from this standard perception of learning in schools and universities. The real innovations lie in how and to what extent the emphasis is distributed between inside-classroom activity and outside-classroom activity. The array of choices using technologies for restructuring higher education curricula can be seen as being organized around three base learning modes: namely, teacher-led, self-paced inquiry-based, and collaboration-based. Table 5.2 gives an overview of these learning modes, the technologies they primarily depend on, and new uses in applications. More advanced learning environments support a combination of these base modes.

TABLE 5.2
Base Learning Modes

Base Mode	At a Glance	Enabling Technologies	Instances of Innovative Practices
Teacher-led	This is the most conservative of the three. It is presentation-centered and adheres closely to what happens inside the classroom, but without the constraints of space and distance.	• Audio/video conferencing over low bandwidth connections· • Synchronized presentation tools (e.g. teacher controls what is displayed on the distance learners' screens).	Cross-university lecturing, e.g. two teachers from different schools linked by communications technologies, coordinate to deliver a joint lecture to distance learners. A global perspective can be introduced in the curriculum.

Base Mode	At a Glance	Enabling Technologies	Instances of Innovative Practices
Self-paced Inquiry-based	It extends stand-alone CD-ROM multimedia applications with network access to an unbounded number of online academic library resources and tele-operable laboratory equipment. The exploration and discovery of online resources and participation in game playing or simulation activities engages learners as researchers. Note that one popular kind of learning media is a recorded classroom lecture.	• Internet programming languages· • Interactive multimedia stimulation • Authoring tools • Agents and intelligent systems • Tele-operation • Digital libraries. (e.g. information retrieval, navigation, and visualization)	Learning by doing: from interactive simulated experimetns in science and new forms of man–machine interaction made possible by virtual reality to remote operation of virtual and real instruments to resource discovery on the Internet (e.g. searching for alternate points-of-view).
Collaboration Based	This mode overcomes the limited expert knowledge intelligence that can be encoded on the digital media by providing both a delivery vehicle for real-time or differed collaborative interactions and control to augment learning opportunities through working together in groups. In addition, the interactions are not constrained by a teacher–student hierarchy, rather they are driven by the group dynamics and strategies involved in the coordination and collaboration process. This mode probably makes the best use of the human connective potential of the Internet.	• Interpersonal communication tools (e.g. e-mail, mailing lists, chat rooms, two-way interactive video, annotation systems)· • Multi-user virtual environments and virtual places • Agents and intelligent systems • Virtual realities and shared whiteboards for learning	Virtual learning communities: i.e. groups of students with common interests possibly guided by a small group of faculty, collaborate to perform such tasks as learning a concept, solving a problem, designing and creating new knowledge. In the process, learners become both producers and consumers of information and knowledge. Topics are more amenable to be examined in interdisciplinary contexts.

The new technologies have the effect of breaking away from the standard lock-step curriculum towards an open curriculum that supports more flexible and individualized learning styles. They offer more control to learners over when, where, and how they interact with instructional resources. For instance, learners would be able to choose from different schools' course syllabi, which enables them to make a compatible choice. Innovations are only possible when IT is directly integrated into curricula, and if institutions start to create partnerships and are able to adapt to different individual and group learning strategies. At the same time, established educational institutions are facing serious problems in keeping pace, with the speed and extent to which these new technologies are being developed.

A computer-centric view

With the Internet being the catalyst, computers are the most important tools in the transition from a "bricks and mortar" setting, to a more technology-enhanced educational infrastructure. Although graphical user interfaces were a major improvement over command-line interfaces, for many, computers remain difficult to learn and use. This is especially true as computer applications grow in features and capabilities. The complexity of using computers greatly diminishes the potential educational benefits of information technologies.

Intelligent software agents can support the learning and use of computer tools. As noted earlier, they can learn and adapt to user habits and preferences, and suggest alternative courses of actions when problems arise. Intelligent tutors are an example of a more specific use of agents in education. These provide a personalized "education" to a user in a specific domain of knowledge (Shute and Psotka, 1994). They can give instant or incremental advice and feedback to the user, based on the user's performance on test problems. When combined with speech technology, they can provide users with a more human or social form of interaction with the computer (Hermans, 1996).

Technology-mediated distance learning tends to move together with advances in computing and the IT industry. For instance, when computers become mobile and ubiquitous, learners would be able to continue their learning activity everywhere, even during commuting. But like many innovations that have made inroads into colleges and universities, there is some danger that their use may be constrained by existing teaching methods, or a lack of understanding about how technologies can improve student learning. In many cases, universities are adopting the use of technologies because of their belief that technologies are needed to remain competitive as an institution, and to make the delivery of education more cost effective (The Institute for Higher Education Policy, 1999). But often the adopted technology is too difficult to use and even unreliable. This leads to higher operating costs caused by users demanding frequent technical support. In fact, not a single published study provides convincing evidence that the use of information technology results in improvements in teaching and learning (Sheppard et al., 1998). However, these criticisms can be constructively used in thinking about future technologies and their applications.

Access to Virtual Learning in Developing Countries and Rural Areas

Much has been said and demonstrated about the Internet's use in education and information diffusion, but very little consideration has been given to the feasibility of Internet development in less-developed countries or rural areas within a country. There are growing concerns that Internet technologies will further widen the gap between north and south in access to information and technologies, which are critical resources for societal development and growth in the 21st century.

By 1998, the last remaining unconnected developing countries became connected to the Internet (Kibati, 1999) However, Internet access in these countries is still mainly concentrated in the capital and a few urban centers. About 70% to 80% of populations live in rural and often isolated areas that lack basic infrastructure such as roads, telephone service or electricity. In fact, disparity in access is a global problem. The access gap persists even in developed countries like the United States.[36] According to the "digital divide" report released by the U.S. Commerce Department in July 1999, White households are much more likely to have computer and Internet access from home than Hispanics or Blacks from any location (NUA Internet Surveys, 1999). Despite these findings, the Internet seems to have a definite place in an information-based society and economy. Developed countries are starting to reap the economic and social benefits derived from the use of the Internet. At the same time, the Internet's explosive growth is having an even more profound impact on developing countries than elsewhere—it is driving the installation of the basic communications infrastructure in countries like Haiti, Senegal, India, China, Brazil, and Argentina. As Christian Huitema[37]—a distinguished figure in the Internet community—surmised in the IEEE Spectrum's third annual roundtable, it would make sense for rural areas to go straight to the Internet rather than to go back to the wired telephone system (Comerford, 1997b). Unconnected places like the city of Merida in Venezuela and Haiti, are experiencing such a transition path, thanks to wireless technologies (Peha, 1999 & Pietrosemoli, 1999).

Recognizing the importance of the Internet for development and growth, many governments worldwide are taking steps and changing their policies to accelerate the diffusion of the Internet in their countries. This section gives a bird's-eye view of the Internet development in emerging nations and explores some key technologies that can address the telecommunications needs in the disconnected areas of the world.

Internet development—obstacles

The growth of the Internet is uneven in developing countries, due in part to differences in governmental and societal attitudes toward the Internet, human and financial resources, telecommunication infrastructure, and level of support from local and international entities. Barriers to Internet growth facing developing nations can be attributed to a combination of the following circumstances:

- Under-developed or non-existing telecommunication infrastructure. The bulk of the infrastructure concentrates in large cities and does not go beyond a few urban places. Often the waiting time for a phone line is very long. The number of lines per 100 inhabitants in industrialized countries is normally well above 50 (Chaumeil, 1999), which means it could take decades for the least developed countries to reach similar telephone penetration rates. In the case of Haiti, for example, the telephone density is 0.9 per 100 inhabitants and there is a waiting list of nearly 100,000 people (Kahle, 1997).
- High telecommunications costs—often caused by monopoly control of public entities and because many Internet Service Providers are outside the countries. Internet users in developing countries pay much more than users in the U.S. and Finland for lower-quality services (NUA Internet Surveys, 1999). Market liberalization and competition can help reduce the cost of information and communication services. On the other hand, the trend towards privatization in telecommunications may not attract investment to improve and expand service in rural areas because of the dispersed markets in these regions, (Garriott, 1997). However, owing to the feasibility of E-commerce, rural populations can promote tourism and market indigenous products in a global marketplace, which could lead to more favorable conditions towards financing the infrastructure.
- Low per-capita income, amongst other things, results in low numbers of computers per person. Furthermore, computing and networking equipment is also scarce. For example, the distribution of PCs per 100 inhabitants in Asia in 1995 was 18.7 in northern countries and 0.7 in southern countries (Rao, 1996). It has improved since then, but PCs are still only within the reach of affluent people.
- Low awareness of decision-makers and developers.
- Low computer literacy.
- Language and cultural barriers.
- Fear that the Internet will undermine cultural values, national security or even the political regime—the latter may lead to restricting access to information.
- High cost of credit which hinders the growth of start-up technology companies. (Bell G.,- website listing)
- Restrictions on the export of technologies, such as U.S. restrictions on the transfers of equipment, such as routers and digital decoding equipment used in satellite-based Internet access, to developing countries like China and the former Soviet Union.

All in all, beyond technical constraints, government policies are among the principal barriers to Internet growth. Governments, universities, industries, and

NGOs all have roles to play as stakeholders in addressing the issues related to access and information content.

Digital empowerment of the rural and isolated areas

A number of technologies can narrow the information inequity gap between the north and the south. But technologies alone are not enough to address the problem of the widespread lack of access to information. For instance, technologies must be made affordable and national communications entities must be flexible in their pricing and regulatory policies. Although access is important, it can do little if the capacity for selecting and using the information is poorly developed. Hence, it is imperative to create a critical body of skilled workers who can take the responsibility for developing and enhancing the skills base and information infrastructure.

Communications infrastructure

There is not a single technology that will meet the needs of all developing countries. The choices depend on the information needs, existing infrastructure, and local circumstances. Wireless technologies have been identified as one of most viable solutions to connect the hinterlands and poorer regions in Asia, Africa, and Latin America to the public telephone network and to the Internet backbone (Brake, 1997; Rutkowski, 1998; NUA Internet Surveys, 1999). One good example of a low-cost wireless solution is based on direct sequence spread spectrum[38] (DSSS). It has a typical connectivity speed of 2 megabits per second and costs less than US$1,000 for the wireless LAN adapter, high-gain antenna, and cable (Barzins et al. 1999). In addition, the two-way satellite communication costs for accessing the Internet are falling from $25,000 (on equipment alone) a couple of years ago to as low as $5,000, and it is expected to drop to $1,000 or less within the next few years (Hegener, 1999). The price per megabyte sent/received may decline to as little as one cent.

Wireless networks are particularly useful when the geography of a region (e.g. rugged mountains) poses a big obstacle for laying cables or the infrastructure is not amenable to an upgrade or retrofit. They require less maintenance and operating costs and are quicker to install than conventional wireline solutions. Although in the interim, voice and low speed data services can meet the communication needs of developing countries, eventually they will need to upgrade their data infrastructure to accommodate high-bandwidth data services. Results based on cost models suggest that it is realistic to incorporate voice and high speed data capabilities in any new installations, while adopting technology that has an easy migration path to the next generation of wireless networks (NUA Internet Surveys, 1999).

Computers and software

Although the costs of hardware are falling, "cheap" computers may still be too expensive for developing countries. A new breed of software may prove very effective for recycling the old PCs that are around for developing countries. WebSuite from NewDeal, a cheap software solution provides pre-Pentium class PCs (e.g. those with 80286 microprocessors) with email access, Web surfing and other common applications. Analysts estimate that there are between 30 million and 60 million pre-Pentium PCs, (Toft, 1999). Other affordable hardware solutions discussed earlier, such as NCs, are also good candidates for poorer regions of the world.

As software grows more complex, their prices will also become higher. Not-for-profit movements like Linux and others suggest the need for more worldwide collaboration to build open source software, which could be useful for users and regions whose financial resources are too limited to allow them to afford commercial software.

After the connectivity infrastructure and computers are in place, a morass of problems related to language seeps to the surface. For someone from an underdeveloped country, chances are that most of the software environment on his computer is in a language with which he is not comfortable. Hence, if computer programs are already difficult enough to use and learn, the language barrier will further deter the use of IT in the developing world. Likewise, the person may find most of the information on the Internet is not in his own language. If the information could not be understood, it would negate the benefits of getting connected to the global Internet. In response to this pressing problem, the United Nations University/Institute of Advanced Studies has launched the UNL (Universal Networking Language) project, probably the largest of its kind, to provide a platform for international exchange of information in users' mother tongues. In the future, through the UNL, any text in software and documents could be readily translated into many different languages.

Conclusions

The Internet forms the next technological breakthrough in distance learning tools. Advances in the Internet concerning access and quality of information are fundamental for making the Internet efficient and effective as a learning tool. Along these two broad dimensions, this chapter provides a chart of the technological landscape describing the current state of the Internet and the critical trends that are driving the evolution of the Internet. This chapter also gives a characterization of the base learning modes and highlights some of the Internet developments in less developed countries.

One unique feature of the Internet is that it is not just a cross-platform for just running any kind of computer application and service. It is quickly becoming a social place for people to meet and work together on a global scale. However, the use of the Internet for synchronous cooperation, particularly co-participation in joint tasks using shared software tools is still rare. This follows from the uneven growth of bandwidth in different locations and the shortage of collaboration facilities and the needed tools for building such software systems. At the same time, constraints in bandwidth may restrict the type, quality, and quantity of the media used for designing the content as well as the methodologies used for remote collaboration. Most of the course materials available on the Web are still static with little if any interactivity. This is probably because of bandwidth limitations and high development cost of interactive learning.

Seekers of information often perceive the Internet to be a huge library without a catalog. Digital library projects have the potential to make the Internet more useful for learning and more. They aim to interconnect disparate information repositories through a consistent search and navigation interface. Solving this problem requires coping with the challenge of ensuring interoperability of technologies. Interoperability could lead to an environment where programs cooperate with each other, sharing interchangeable data and software to facilitate the deciphering and use of the information returned by a search operation or a user selection.

Although the concept of distance learning is not new, it is just beginning to expand its capability to fit into an increasingly networked society. Distance learning environments are moving from stand-alone CD ROM-based courses to highly flexible and personalized environments that are more social and human-based. Because the Internet does not just connect computers around the world, but also humans, learning environments for the Internet appear to be intelligent, in the sense that there is a high likelihood for someone to almost always find a human online to talk to, when in need of help. In addition, drawing on advances from mobile computing and networking, access to learning in the not-too-distant future will be ubiquitous in the true sense, even when the learner is moving from location to location.

Year after year, the literature on educational technology has documented the problems, and provided partial to complete solutions. At the same time, there are still many challenging technical problems and open issues. In addition, there is a paucity of published works on the effectiveness of IT in learning and teaching. As a result, most educational tools are ad hoc and are necessarily catered to the objectives of a curriculum. But in the case of developing countries where education is under-served and under-developed, the question of effectiveness is probably not as important as the cost of not getting an education at all.

Notes

1 http://www.undp.org/popin/wdtrends/p98/p98toc.htm
2 Bandwidth is the measure of data-carrying capability of a communication medium.
3 http://www.verisign.com/server/whitepaper/view/index.html
4 Ontology in the context of the project, is a formal specification which provides a vocabulary for representing terms about some topic and the logical statements that describe what the terms are and how they are related to each other, and how they can or cannot be related to each other.

5 A simple, general-purpose, object-oriented, application-development language, developed by Sun Microsystems, Inc. Java provides a "write once, run anywhere" architecture, that is, Java programs can run on any platform, that is, any computer that can connect to the Internet will be able to run Java programs. It is a very useful language for bringing true dynamic applications to Web browsers.

6 Compare these figures with the bandwidth that would be required to transmit a ¼ screen video (320×240) using 256 colours and a frame rate of 10 frames per second (or 1/3 of the rate for motion picture quality)—6.14 megabits per second without compression or a typical bandwidth of 610 kilobits per second after compression.

7 Response time measures the round trip delay in sending some chunk of data from one point to another and back.

8 *Los Angeles Times*, July 8, 1999.

9 Precision and recall are two measures that are widely used in the field of information retrieval. Precision is the percentage of the documents retrieved that are actually relevant and recall is the percentage of all relevant documents retrieved.

10 The algorithms that order the search results for queries are usually kept as trade secrets by companies offering search services [16]. This is not necessarily bad in all cases, for example, if content providers do not know the details of these algorithms it becomes more difficult for them to alter the ranking of their pages, a practice known as "search engine spamming".

11 In Asia—APAN (Asia-Pacific Advanced Network), Europe—TEN34 (Trans-European-Network), North America—CA*net2 (Canada's Internet 2 Initiative), vBNS (very high performance Backbone Network Service), NGI (Next Generation Internet), and Internet2.

12 http://www.geog.ucl.ac.uk/casa/martin/geography_of_cyberspace.html

13 In a hierarchical network, the backbone is the top level, employing high-speed data transmission and serving as a major access point; smaller networks connect to the backbone.

14 Some variations—Java based thin clients, for example—are split clients that run some applications and all graphical processing on the client side and leave some of the heavier-end application partially or completely running on the server. Although there are software technologies that allow Java-based access to other existing application base such as Windows, many NC implementations run Java applications-only by default.

15 A device that converts a TV broadcast signal into an input signal to the TV set.

16 Wearable computers are entire computer systems that can be "worn" as a piece of clothing. They usually consist of a small computer and a battery pack, and may be worn on a belt with a head-mounted display screen. Wearable computers provide the ability to have constant access to services of such computers without having to use both hands. Thus, they help mold the computer to the ergonomic requirements of difficult work environments, such as aircraft maintenance.

17 The term computing environment used here refers to user's data (e.g. data associated with user interfaces to applications, mail messages, etc) and applications (e.g. the code being executed such as a document editor) that may execute locally or elsewhere on the network.

18 One example of a possible advance in networking software that supports the ubiquitous computing vision is Sun Microsystem's JINI technology. JINI theoretically allows any device using it to plug into the network easily and seamlessly, and takes away the nightmare of having to configure a network to accept new devices every time. A key characteristic in JINI is "spontaneous networking", which allows devices to be immediately available as soon as they are plugged into a network like plugging in a phone today. For example, once a printer is connected to a network, it announces itself and tells the network what it is able to do and offers its services to other devices. Likewise it can also look up other information about other devices that already joined the system.

19 For example, some of the key technical challenges are multilingual information access and cross-lingual data services, evaluation of information quality, multimedia information capture, organization, indexing, and distributed search services from heterogeneous repositories, and data visualization. The information services must also be supported with security and privacy mechanisms before content providers will make their information available.

20 For example, a search function requires as input parameter a search string of no more than 40 alphanumeric characters (this is a variation of the example mentioned in [38]). This level of interface information is critical for achieving syntactic interoperability, which allows everybody to call the function without causing an invocation error. However, if the function also published the fact that the string should be in Spanish, it would provide some level of semantic interoperability.

21 Overcoming language barriers across national boundaries can be viewed as a cultural interoperability problem. According to Global Reach (http://www.euromktg.com), the fastest growing groups of Web-users are non-English-speaking: Spanish, 22.4%; Japanese, 12.3%; German, 14%; and French, 10%. An estimated 55.7 million people access the Web whose native language is not English.
22 One major problem in semantics deals with the interoperability of terminology across different domains. For example, a biologist should be able to search for information in the domain of physics using the biology terminology.
23 A description of a database to a database management system (DBMS) in the language provided by the DBMS. A schema defines aspects of the database, such as attributes (fields) and domains and parameters of the attributes [Microsoft Bookself 98].
24 Center for Research, Inc., U of Kansas, Profusion metasearch engine, at http://profusion.ittc.ukans.edu/profusion/
25 In some implementations users have direct control on which search engines to be queried. Others have automatic query dispatchers that choose the best search engines for each specific query based on some sort of knowledge.
26 W3C is the organization that controls the standardization of many WWW-related formats and protocols.
27 XML itself is an important development because of its rapid diffusion into and adoption by the Internet community as a common syntax used in standards such as OSD (Open Software Description) for automatic software distribution, and CDF (Channel Definition Format) to describe push channels. Because XML emphasises structure, not formatting or layout, it is media-neutral. The conversion of XML-based documents into a media-dependent format can be done automatically by using style sheets. There is no limitation on the number of different style sheets that can be applied on a XML encoded document. This characteristic is appealing to content providers, such as publishing industries which need to deliver information via multiple media.
28 Any attempt to define Intelligent Software Agent will inevitably open up the old AI can-of-worms about definitions of intelligence.
29 Emotional agents are potentially important in training applications. Studies suggest that training in a simulated environment is much more effective if the agents adopting the roles of their real counterparts are believable [48].
30 Sometimes it is desirable to make the watermark entirely visible, as is the case of content protection—a company that wants to distribute freely some content may want to stamp a hard to remove visible logo on the data to assert ownership of the data.
31 For example, an image is often resized, filtered, contrast/color enhanced, sharpened, and compressed before publishing.
32 http://www.altern.org/watermark
33 Mailing lists provide more support for group activities. They provide a broadcast environment that allows people to view the (old and new) discussions and decide to join them. Information posted by members of the group is usually organized into a tree of responses that are archived, sorted, and converted into documents that are accessible through the Web. Both e-mail and mailing lists share the objective of allowing participants to augment the shared information content in different times.
34 A whiteboard in its basic form is a shared electronic work area for drawing and writing that is synchronized across all the users in a what-you-see-is-what-I-see manner.
35 VE leverages simulation closer to an audio/visual experience of a reality or imagined space. But unlike virtual reality (VR), their primary goal is not to provide a total sensory or immersion experience indistinguishable form reality, using expensive input/output hardware (e.g., data gloves/ head mounted displays (HMD)) (Grudin, 1994). Rather VE aims at a navigational 3-D computer representation of a real or imagined space for a mass audience.
36 http://www.internetnews.com/ec-news/article/0,1087,4_158901,00.html, July 9, 1999
37 He is a member of the Internet Society's board of trustees and former chair of the Internet Architecture Board, as well as chief scientist in the Internet Architecture Research Laboratory at Bellcore, Livingston, NJ.
38 DSSS is a modulation technique that distributes the transmitted signal over a wide frequency spectrum while minimizing the amount of power present at any frequency. The net result is reliable communications in an environment prone to frequency jamming.

References

Adam N., Yesah Y. et al. (1996) Strategic Directions in Electronic Commerce and Digital Libraries: Towards a Digital Agora. *ACM Computing Surveys,* **28**, (4), December 1996, 819–835.

Andreoli J. M., Pacull F. & Pareshi R. (1997) XPECT: A Framework for Electronic Commerce.*IEEE Internet Computing,* July–August, 40–48.

Bailey, R. (1982) *Human Performance Engineering.* Prentice Hall, New Jersey, pp.44.

Barzins G, Tully J, & Riekstins A. (1999) Applications of High-Speed Wireless Solutions for Developing Countries: Lessons Learned in Latvia and Moldova. *INET.*

Bell, G. Mapping the Internet: Yesterday, Today & Tomorrow", http://www.uvc.com/gbell/ltranscript.htm.

Bharat, K. & Broder, A. (1998) *A Technique for Measuring the Relative Size and Overlap of Public Web Search Engines.* 7th International World Wide Conference, April 1998.

Bohr, M. (1998) Silicon Trends and Limits for Advanced Microprocessors. *CACM,* **41**, (3), 80–87.

Braham R. & Cromerford R. (1997) Sharing Virtual Worlds. *IEEE Spectrum,* March, 18–25.

Brake, D. (1997) Lost in Cyberspace. *New Scientist,* June 28.

Chang K., Garcia-molina, H., et al. STAIRS version 1.0", http://www-diglib.stanford.edu/diglib/WP/PUBLIC/DOC82.html.

Chaumeil, T. (1999) The Internet in Argentina: Study and Analysis of Government Policy. *On the Internet,* May/June , pp. 22–30, 38–39.

Chon, K. (1998) Global High Performance Research Network: An Asia-Pacific Perspective. *Lecture Notes in Computer Science 1386, Worldwide Computing and its Applications—WWCA '98, Tsukuba, Japan ,* March, pp. 6–17.

Comerford, R (1997a) The Battle for the Desktop. *IEEE Spectrum,* May, 21–28.

Comerford, R. (1997b) State of the Internet: Roundtable 3.0 *IEEE Spectrum* October, pp. 29–38.

Craver S. Yeo B.L. , & Yeung M. (1998) Technical Trials and Legal Tribulations. *CACM,* **41** (7), 45–54.

Dearle A. (1998) Toward Ubiquitous Environments for Mobile Users. *IEEE Internet Computing,* January–February, 22–32.

Farquhar, A., Fikes, R. Pratt, W., & Rice, J. (1995) *Collaborative Ontology Construction for Information Integration*, Knowledge Systems Laboratory Department of Computer Science, KSL-95-63, August 1995.

Flanagan, P. (1998) This Year's 10 Hottest Technologies in Telecom. In: *Telecommunications Online Newsletter,* May.

Filman, E. R. & P. Sangam (1998) Searching the Internet. *IEEE Internet Computing,* July–August, 21–23.

Fox, E A. & Marchionini, G. (1998) Toward a Worldwide Digital Library *CACM*, April, **41**, (4), 29–32.

Gareiss, R. (1997) Is the Internet in Trouble? *Data Communications,* September 21.

Garriott, G.L. (1997) Low Earth Orbiting Satellites and Internet-Based Messaging Services. *INET.*

Geppert, L. (1998) Solid State – Technology 1998 Analysis and Forecast. *IEEE Spectrum,* January 23–28.

Gilder G. (1994) The Bandwidth Tidal Wave. *Forbes ASAP,* December 5, http//sunsite.unc.edu/horizon/gems/gemlong.html.

Gilder G. (1997) On the Bandwidth of Plenty. *IEEE Internet Computing*, January–February, 9–18.

Goldman, P. (1998) Viewpoint: Bringing the Internet to the Consumer. *IEEE Spectrum*, January, 60

Goncalves P. F., Merina S. L., & Salgado A.C., 1997 A Distributed Mobile Code-Based Architecture for Information Indexing, Searching, and Retrieval on the World Wide Web", *INET 97.*

Grudin, J. (1994) Computer-Supported Cooperative Work: History and Focus.*Computer,* **27**, (5), May, 19–26.

Guha, R. V. Meta-Content Format. *Apple Computer Technical Draft,* http://mcf.research.apple.com/hs/mcf.html.

Gupta V., Montenegro G. & Rulifson J. (1998) *"Complete Computing", Lecture Notes in Computer Science No. 1368, Worldwide Computing and its Applications—WWCA '98,* pp.174–189.

Halfhill, T. R. (1998) Disposable PCs. *Byte magazine,* February.

Hegener M. (1999) The Internet Satellites and Human Rights. *On the Internet,* March/April, pp. 20–29, p.40.

Hermans B. (1996) *Intelligent Software Agents on the Internet.*-Thesis. Tilburg University: Tilburg, The Netherlands.

Huhns M. N. & Singh P. M. (1998) Agents on the Web – Anthropoid Agents. *Internet Computing,* January–February, 94–95.

Hutscheson, G. D. & Hutscheson, J. D. (1997) Technology and Economics in the Semiconductor Industry. *Scientific American,* October.

IEEE Internet Computing, January–February, pp. 33–38.

Institute for Higher Education Policy (1999) A Review of Contemporary Research on the Effectiveness of Distance Learning in Higher Education. *The Institute of Higher Education Policy,* April, 1999.

Kahle B. Archiving the Internet, extended version of the article "Preserving the Internet" that appeared in *Scientific American*, March 1997, http://www.sciam.com/0397/issue/0397kahle.html.

Kibati M. (1999) What is the Optimal Technological and Investment Path to "Universal" Wireless Local Loop Deployment in Developing Countries? *INET.*

Labovitz, C. M., Rober G. & Farnam J. "*Internet routing instability",* Technical report CSE-TR-332-97, U. of Michigan, department of EE &CS.

Lassila O. (1998) Web Metadata – A Matter of Semantics. *IEEE Internet Computing,* July–August, 30–37.

Lisa, B. (1997) Hello? Still There? In: *Wall Street Journal Reports,* June 16.

Lync, C., & Garcia-Molina, H., "Interoperability, Scaling, and the Digital Libraries Research Agenda", IITA Digital Libraries Workshop, August 22, 1995.

Maes, P. Towards a Truly Personal Computer, Experience ACM 97: The Next 50 Years of Computing, http://www.acm.org/events.

Martin R. L. (1997) Fancy Femtoseconds, Humble Bits, and Innovation Systems International Symposium. *Management of Technology under Global Competition and Collaboration towards 21st Century,* June, pp. 11–15.

Memon N. & Wong P. W. (1998) Protecting Digital Media Content. *CACM,* **41**, (7), July, 35–43.

NUA Internet Surveys (1999) http://www.nua.ie/surveys/how_many_online, June.

Paepcke A., Chang C. C. K., Garcia-Molina H. & Winograd T. (1998) Interoperability for Digital Libraries Worldwide. *CACM,* **41**, (4), April 33–43.

Peha J. M. (1999) Alternative Paths to Internet Infrastructure: The Case in Haiti. *INET.*

Pietrosemoli E. (1999) Data Transmission in the Andes: Networking Merida State. *INET.*

Rao M. (1996) Education and Research in the Virtual University: The Internet Challenge for Developing Nations. Virtual University Symposium, November http://www.edfac.unimelb.edu.au/virtu/rao.htm.

Richardson T., Stafford-Fraser Q., Wood K. R. & Hopper A. (1998) Virtual Network Computing.

Rutkowski, T. (1998) Dimensioning the Internet. *IEEE Internet Computing,* **2**, (2), March-April.

Shah R. (1998) All the Network-centric Systems Exposed! *NC World, June* http://www.ncworldmag.com/ncworld/ncw-06–1998/ncw-06–netlaunch.html.

Sheppard, S. D., Reamon D., Friedlander L., Kerns C., Leifer L., Marincovich M., Toye G., members of the Stanford Learning Laboratory, Assessment of Technology-Assisted Learning in Higher Education: It Requires New Thinking by Universities & Colleges, *Frontiers in Education Conference 1998.*

Shute, V. J. & Psotka J. (1994) Intelligent Tutoring Systems: Past, Present and Future. In: Jonassen D. (ed) *Handbook of Research on Educational Communications and Technology,* Scholastic Publications.

The Internet Traffic Report http://www.internettrafficreport.com, September 1997.

Toft D. (1999) New Life for Old PCs. *PC World*, July 7.

Want R. & Hopper A. (1992) Active Badges and Personal Interactive Computing Objects *IEEE Transactions on Consumer Electronics,* February, pp.10–20.

Weibel, S. et al. (1995) OCLC/NCSA Metadata Workshop Report, *The March 1995 Metadata Worshop. MIT Media Lab., Software Agents Group: Research,* http://agents.www.media.mit.edu/groups/agents/research.html.

Weiser, M. (1996) *"Xerox Names Computing Pioneer As Chief Technologist For Palo Alto Research Center"* http://www.ubiq.com/weiser/weiserannc.htm, August.

Weiser, M. (1991) The Computer for the 21st Century. *Scientific American,* September pp. 94–104.

Wired News (1998) *AT&T Restored Disrupted Service* http://www.wired.com/news/news/technology/story/11665.html, April 1998.

Wolt, S. (1998) *"On the limitations of digital watermarks: A cautionary note", Proc. Of the 4th Int. Conf. Of Info. Systems. Analysis and Synthesis,* (Orlando, Fla., July 12–16).

6

Course Delivery Systems for the Virtual University

PETER BRUSILOVSKY and PHILIP MILLER

Introduction

This chapter analyzes course delivery systems in the context of distance learning using the World Wide Web (WWW) and Internet in general. Many projects are focused on Web-based education (WBE) systems and tools, covering the spectrum of teacher, student and institutional needs for virtual education. These needs are addressed in traditional, face-to-face residential, education (we will call this human-based education or HBE) by a variety of methods: lectures, textbook, tests and exercises, labs and seminars, office hour contacts with the teachers and teaching assistants (TA), the registrar's office etc.

Systems and teams that produce systems are changing rapidly, but the underlying educational needs are not. The goal of this chapter is to provide the reader with a sense of the landscape of systems and features plus some idea of who the players have been up to this time. For the builder of on-line courses and virtual universities we hope to help shape thinking and provide a good bibliography.

Educational components

We divide the needs of virtual university courses into four main components: presentation, activities, communication, and administration.

Presentation comprises all functions related to the delivery of new material. It is one core of education. In HBE, presentation is achieved through lectures, textbooks, and video.

Activities comprise the active and interactive learning materials which involve students in doing something. In most WBE systems, activities are assessment oriented; i.e. their goal is assessment of student progress (it includes self-assessment). Assessment, provided in HBE through quizzes, tests and homework, is important in credit-bearing courses and critical in education. This is because it diagnoses the student's state of knowledge, and then prescribes ways to move the student to a higher state. A number of WBE systems can also offer activities specially developed to support learning-by-doing. These systems are more concerned with providing student support and are similar to HBE labs.

Communication comprises all ways of communicating (group or one-to-one) between teachers and students, or between student groups. The university student who has problems is able to ask questions of a teacher or of peers. Communication is an important way that teachers and students diagnose and remedy problems. This is done with problem-solving feedback, additional explanations and suggestions of additional work. Communication between students is so important it is often taught as a separate skill. In HBE, communication is achieved in contact during classroom activities, office hours and informally.

Administration comprises all record-keeping activities of registering students, payments, course cancellations, course credits and grades, auditing student progress against degree requirements etc. Administration is performed in HBE by teachers and administrative personnel. While at first it may seem that tasks such as recording payments and student withdrawals are not really a part of education, they most certainly are a part of what allows universities to operate. Without administration, university education would not take place.

Multiple tools and systems have been developed to provide all four components in virtual universities (Table 6.1). Within each component these tools and systems could be further divided into two groups—tools used in course preparation (various authoring[1] and set-up tools), and tools used to support a running course (delivery and run-time management tools).

TABLE 6.1
Groups of Tools Required to Prepare and Run a Virtual University Course. Course Delivery Tools are Shaded Gray

	Presentation	Activities	Communication	Administration
Before Start	Authoring tools	Authoring tools	Set-up tools	Set-up tools
Delivery Time	Delivery tools	Delivery tools	Support tools	Run-time tools

This review is centered on two groups of tools and systems (shaded gray on Table 6.1) which are required to support what is called *course delivery* for a virtual university course. We will divide these groups of tools into further subgroups and analyze the completeness and sophistication of existing WBE tools and systems. Although other groups of tools are outside the scope of this paper, we will be talking briefly about presentation and exercise authoring and grade viewing for the purpose of clarifying comment on delivery tools.

What kinds of tools are available

The various authoring and delivery tools used in Web-based education could technically be divided into two groups—specialized tools and integrated tools. There are many functions which could be performed by a tool during the process of design and delivery of a Web-based course. Specialized tools usually support a narrow range of these functions. It could be a single function, such as test authoring, or a small group of functions centered on a particular type of educational material, such as Web lecture tools. Integrated tools are aimed at supporting a wide range of activities. The goal of all integrated tools is to provide a comprehensive universal environment, which could support at least all the most important functions. If this goal is achieved, the integrated tool could be the one and the only educational tool required for creating and delivering a Web-based course. There are multiple benefits of using integrated tools as opposed to a set of many independent tools. Very few examples of integrated tools were available just three years ago, but due to the high demand, dozens of university-developed and commercial integrated tools are available today. There is a clear tendency now towards developing integrated tools. Existing specialized tools are becoming part of "kits" and "suites" (CourseWeb Team, 1997; Instructional Toolkit Team, 1997; Siviter, 1997; POLIS Team, 1998). Narrow-range integrated tools are being extended with more functionality. Companies that had a system which covers only a part of the needs are rapidly enhancing their tools to the level of an integrated tool by development and mergers with other companies (like SoftArch with their FirstClass, Centra with its Symposium (Centra, 1998), or Lotus with its Learning-Space (Lotus, 1999).

Who makes the tools

According to their origin, existing authoring and delivery tools could be divided into four categories, where the first two categories form a group of *university-level tools* and the second two form a group of *commercial tools*.

- University research-level systems
- University-supported products
- University-grown commercial systems
- Full-fledged commercial systems

University research-level systems are often quite advanced and demonstrate innovative ideas, but these tools are not supported so other universities and companies are unlikely to put them into the critical path of materials development or delivery. Some of these tools are available as freeware (or through technology license) and are used by teachers in other universities. The following are the most interesting examples of this kind of system: QuestWriter (Bogley et al., 1996; QuestWriter Team, 1998)—Oregon State University, ClassNet (ClassNet Team, 1998; Gorp & Boysen, 1996) from Iowa State University, CourseWeaver

(Rebelsky, 1997) and ASML (Owen & Makedon, 1997) from Dartmouth College, ONcourse (Jafari et al., 1998) from Indiana University, Online (Rehak, 1997a; Rehak, 1997b) and InterBook (Brusilovsky et al., 1998) from Carnegie Mellon University, ARIADNE (Forte et al., 1996a; Forte et al., 1996b) and MTS (Graf & Schnaider, 1997; IDEALS Project, 1998) from two European University-driven consortia, WebMapper (Freeman & Ryan, 1997) and FLAX (Routen & Graves, 1997a; Routen et al., 1997) from De Montfort University.

In America and Canada a strong market push helped many research-level tools evolve into *university-supported products*. These systems are essentially products offered at low cost or sometimes no cost from their home universities. These tools are well-developed and tested. Since installation and maintenance help is available for this group of products, they are quite popular and are used in many universities, mainly in USA and Canada. Some well-known integrated systems of this category are: CyberProf (Hubler & Assad, 1995; Hubler & CyberProf Team, 1998; Lam & Hubler, 1997; Raineri et al., 1997) and Mallard (Brown, 1997; Brown & Mallard Team, 1998; Graham et al., 1997; Graham & Trick, 1997) from University of Illinois at Urbana-Champaign, Serf™, from University of Delaware (Serf Team, 1998), WebAssign from North Carolina State University (Titus et al., 1998), and Virtual-U from Simon Fraser University (Fisher et al., 1997; Harasim et al., 1997; Virtual-U Research Team, 1997).

For some *university-grown* tools, the success in their home university leads to the establishment of a company that usually ships some version of the tool as a *commercial system* and continues the development of this tool on an industrial basis. The most well-known examples are WebCT, from University of British Columbia (Goldberg, 1997a; Goldberg, 1997b; Goldberg, 1997c; Goldberg et al., 1996; WebCT, 1999), Web Course in a Box (WCB) from Virginia Commonwealth University (madDuck, 1997), and CourseNet from SUNY (Whitehurst, 1997).

Full-fledged commercial tool systems is now the fastest growing category. Until recently, only two products in this group were available: TopClass and LearningSpace. TopClass (WBT Systems, 1999) started several years ago as a university-grown tool. For a long time it was the only commercial system on the market, and has become very strong and comprehensive. LearningSpace (Lotus, 1999) started from Lotus' strength in database and Web-base server technology. (Later on in this volume, Martin's chapter illustrates the use of LearningSpace at the Monterrey Institute of Technology.) Since that time many commercial tools have appeared on the market. Several major players of software industry such as Asymetrix, Macromedia, and Oracle has joined the list. Some often referred-to commercial systems are: CourseInfo (Blackboard, 1999), Docent 2.0 (Docent, 1998), The Learning Manager (Campus America, 1998), Symposium (Centra, 1998), WebMentor (Avilar, 1998), Enterprise Education Server (Mentorware, 1998), Web University (IMG, 1998), LearningEngine (Knowledge Navigators, 1999), and IntraKal (Anlon, 1998).

We have to note that the difference in functionality and service between categories of tools is not clear-cut. Some research level tools are stronger and more mature than some commercial tools. However, commercial tools evolve faster. The market pushes them to be more generic and user friendly. Commercialization brings better service. Naturally, a number of tools are in the process of moving from one category to a higher one. Some research-based tools are becoming products; some university-level products are becoming commercial tools; companies (like Blackboard and ULT) are acquiring some university-grown commercial tools.

Levels of sophistication

To characterize different systems, we identify three levels of advancement in each component: Base, State-of-the-art and Research.

Base level: These are components that provide minimum requirements. All usable systems and their components cannot be below this level. A system which offers less is not given credit for the component and will not be able to survive on the market. Still, systems offering no more than the Base level may survive because they are well supported, inexpensive, reliable or because customers do not yet demand more sophistication.

State-of-the-art level: The best systems provide this level of functionality. Few systems will be State-of-the-art on all components. Systems with State-of-the-art features are often presented at development-oriented conferences (such as WebNet, ED-TELECOM, or EDUCOM). As this chapter is being written, commercial products have some, but not all of the State-of-the-art features, with the rest being at the base level.

Research level: Experimental systems, often incorporating advanced elements of artificial intelligence or cognitive psychology, are referred to as research-level systems. These systems are not ready to be commercially distributed. Often these systems are primitive in some components, but very advanced in components, which are of research interest. No commercial systems exist with research-level features. This work is typically presented at research conferences.

A few words about advances in authoring and delivery tools are in order. There is a tacit implication that education is better served as systems are more heavily populated with components that are state-of-the-art, and are even better when research level advancements are folded in. While the authors generally subscribe to this view, they are not prepared to make such claims. It is an empirical question as to the educational impact of "advancements" in Web delivery. Indeed this is, in part, why educational research is ongoing. Similarly advances in content creation tools certainly seem to be a good idea, but user studies (not reviewed here and largely not yet undertaken) will have the final word on this.

The progress in WWW-related technology is fast, and many techniques, which were State-of-the-art just a year ago are now only Base level. Essentially when features are incorporated into multiple commercial systems, those features define a new Base level. The life cycle is that Research-level techniques move into the State-of-the-art and State-of-the-art features are rapidly incorporated into commercial systems. This is natural and appropriate. Software vendors have user communities that cannot be abandoned. Therefore advances must be painstakingly engineered into systems that continue to support cash-paying customers with systems built on earlier technology.

Presentation: Learning by Reading and Watching

Existing options and systems for WBE content presentation differ in five main aspects: structure, type of content material, media, authoring, and delivery. *Structure* is the most essential aspect. In the early days of WBE an unstructured presentation or *a heap* was the dominant form of content presentation for Web courses. A heap is essentially an assorted set of static HTML[2] pages connected to the course home page. While many universities still apply a heap approach, an ability to support structured presentation is now a base level requirement. We distinguish two types of structuring the learning material: a hierarchical structuring which we call an *electronic textbook*, and a sequential structuring which we call an *electronic presentation* (in the latter, we distinguish two special subclasses: the *electronic lecture* and the *guided tour*). From our point of view, electronic textbooks and electronic presentations essentially differ from each other, so we will compare existing systems within each of the identified classes.

Type of content material and *media* are closely connected. The currently used types are text, static figures, dynamic figures, audio material, video material, and, very rarely, interactive simulation (this type will be considered separately later). Type usually determines the media used: HTML is used for text, various graphical formats for figures and various canned or streamed formats for audio and video material. There are also several multimedia and animation formats like Shockwave (Plant, 1998) or HyperStudio (Wagner, 1998) that can be used to represent dynamic figures and simulation, though simple dynamic figures are more often done with animated GIFs. All kinds of dynamic behavior are brought about by Java and JavaScript technologies. They are also used to "spice" up the other types of media. Currently, a good number of commercial authoring tools are available to develop audio, video, and multimedia fragments in several formats. However, most of the university level Web courses and systems are still far from using advanced media types.

From the *authoring* point of view, we can distinguish three levels of advancement.[3] In the original authoring technology, all content was developed in plain HTML using a text editor (we call it zero-level authoring). Zero level authoring is now below the base level. Nearly all serious content developers are using either general tools (we call this *level one* authoring) or special educational tools (*level two* authoring).

General tools are not specific for educational application. Using general tools is typical for base level courses and systems. There are two kinds of general tools. In the first, authors develop content with a traditional word processor in some standard format (Microsoft Word/RTF, Emacs/LaTeX, etc.). Most word processors have an option to save as HTML. Alternatively a separate converter could be used to transfer the material into an HTML format. Since this option is especially attractive for producing electronic textbooks we will conider it in more detail later. Another kind of general tool that is numerous is the HTML WYSIWYG[4] editor, including products such as Microsoft Front Page or Adobe Pagemill which produce content directly in HTML which in turn is displayed as interpreted HTML (typically as attractive graphics) to the author. The most advanced of these tools include commercial products like the NetObjects Fusion (NetObjects, 1998), GoLive Cyberstudio (GoLive, 1998), university-developed W3Lessonware (Siviter, 1997) and Gentler (Thimbleby, 1997), which can handle not only single pages but also groups of pages and complete sites.

Education-oriented authoring tools can provide additional support for the authors of course material. A number of current university research projects are aimed at developing more useful authoring tools. These tools are classified into two groups: educational mark-up tools and structured content authoring tools. Educational markup tools assume the use of some educational mark-up language (HTML-like or LaTeX-like) and a text editor for authoring. Usually an educational mark-up language offers some extensions to HTML. Two means of providing educational mark-up extensions are: support for rendering formulae as in (Bryc & Pelikan, 1999; Hubler & Assad, 1995; Xiao, 1999) and support for exposing a multi-document structure in one document (Owen & Makedon, 1997). Structured content authoring tools provide additional support for authoring one of the structured types of course material. There are specific authoring tools for developing textbooks, lectures, and guided tours (which we will talk about later). In the near future, education-oriented authoring tools will allow course developers to index all developed material with education-specific metadata.[5]

From the *delivery* point of view, we can distinguish three levels of sophistication. The base level for Web courses still uses static HTML and media pages with static links. More advanced systems usually add some functionality, implemented as simple CGI[6] scripts, e.g. adding coherent footer, header, or navigation buttons. However, the state-of-the-art level now uses database technology in which course information and content is stored in a database, where most of the pages are generated on the fly. A system with a database core can provide a lot more functionality while being more manageable at the same time. Illustrative of the advantages are generation of tests from equivalence classes of test items, archival for system usage, and automatic handling of multiple simultaneous file accesses. A number of advanced commercial and university-level delivery systems like TopClass (WBT Systems, 1999), LearningSpace (Lotus, 1999), CourseInfo (Blackboard,

1999), ARIADNE (Forte et al., 1996a; Forte et al., 1996b) or Carnegie Mellon Online (Rehak, 1997a) are database-driven.

The electronic textbook

The electronic textbook offers a hierarchically structured representation of material. It mirrors a printed textbook with its subdivision into chapters, sections and subsections. The material is usually presented in a text form, and is extended with figures. Hierarchical structuring implies special hierarchical navigation aids.[7] A typical set of hierarchical links includes links to all subordinate sections, a link to the higher level section, and a link to the beginning of the electronic document.[8] Sequential navigation links (i.e. next-page/previous-page operations) are provided in most electronic textbooks. These links are often provided in the form of "arrow" icons. Additional navigation aids include a navigable "path" (i.e. a list of direct ancestors usually shown at the top of a page) with the possibility to move in a single click to any direct ancestor, and sometimes, to any page at the same level, a table of contents, or an index. Pages from the same textbook should have a uniform look sharing a similar design, and header and footer. If possible, a glossary of terms should be provided (Barbieri & Mehringer, 1997; Brusilovsky et al., 1998; Goldberg et al., 1996; Langenbach & Bodendorf, 1997; WebCT, 1999).

Base level: We consider the following configuration as the base level for electronic textbooks. A textbook is a tree of static HTML pages with still figures. It usually provides simple navigation buttons (previous-next-top) and table of contents. Base level textbooks are usually produced with general authoring tools. In particular, they could be very efficiently produced using a structured document preparation system and a special converter. Currently a number of advanced converters are available: LaTeX to HTML (Dracos, 1998), RTF to HTML (Hector, 1998), FrameMaker to HTML (Harlequin, 1998). These converters preserve the hierarchical structure of the content, and generate a hierarchy of HTML pages with a separate table of contents and all the basic navigation buttons. A number of early advanced Web-based electronic textbooks were produced using these converters (Antchev et al., 1995; Brusilovsky et al., 1996; Marshall & Hurley, 1996a; Marshall & Hurley, 1996b).

State-of-the-art level: The state-of-the-art electronic textbooks extends the base level in both media and authoring. State-of-the-art textbooks include some choice of more advanced media items such as audio and video fragments, and simple animations and dynamic figures developed with commercial-of-the-shelf (COTS) software tools. From the authoring side, state-of-the-art electronic textbooks are usually produced with advanced content authoring tools. It could be either some education-oriented authoring tools mentioned above or a combination of general authoring tools and special tools for uploading (and on-the-fly structuring) of course material. The main reason for application of these tools is to ensure the generation of consistent links and to provide a homogeneous navigation and look-and-feel. A number of university developed tools such as ASML (Owen & Makedon, 1997), WebMapper (Freeman & Ryan, 1997), FLAX (Routen & Graves, 1997a; Routen & Graves, 1997b; Routen et al., 1997), WebCT (Goldberg et al., 1996; WebCT, 1999), CourseWeaver (Rebelsky, 1997) and most of the advanced commercial tools for development and delivery of Web courseware such as TopClass (Schwarz et al., 1996; WBT Systems, 1999), CourseInfo (Blackboard, 1999), LearningSpace (Lotus, 1999) provide one of these authoring options. From the delivery point of view, we still can count textbooks produced from static HTML files as state-of-the-art electronic textbooks. At the same time, most state-of-the-art systems use CGI scripting extensively and a number of them use database technology. We expect that very soon a database core will be a requirement for the state-of-the-art level. Databases become increasingly important as system usage increases. While a file system and hand coded CGI is sufficient for managing a few students interacting with simple courses, more sophistication is essential in maintaining large numbers of students in entire universities.

Research level: There are several directions in which research level systems are trying to advance the state-of-the-art level. The most active research direction is now in what we call "reusability authoring":[9] examples are ARIADNE (Forte et al., 1996a;b), MTS (Graf & Schnaider, 1997; IDEALS Project, 1998), Educational Broker (Langenbach & Bodendorf, 1998), and others (Neumann & Zirvas, 1998; Rebelsky, 1997). Reusability authoring is a technology which allows authors and groups of authors to develop pools of various content objects indexed with meta-data, and to re-use previously authored objects along with new objects to build a new course. This technology implies a database core.

Another research direction is aimed at improving the structure of the educational hyperspace by identifying concepts (i.e. the knowledge atoms) behind the content and, as a result, providing more advanced navigation facilities. It includes work on concept-based hyperspace design (Abou Khaled et al., 1998; Adams & Carver, 1997; Brok, 1997; Eklund, 1995; Fröhlich & Nejdl, 1997; Neumann & Zirvas, 1998; Nykänen, 1999; Pilar da Silva et al., 1998) and work on concept-based navigation (Brusilovsky & Schwarz, 1997).

An interesting research direction is "intelligent figures." Illustrative of this are Java-based figures in Medtec (Eliot et al., 1997), which could be used for both exploration (e.g. one can point to a part of a figure and see some additional information) and quiz mode (e.g. one can get a question, then point to a part of the figure as an answer). Intelligent figures integrate presentation, exploration, and assessment and further extend the notion of "figures" in a textbook.

The most challenging research direction is in the area of adaptive presentation and adaptive navigation support. The goal of *adaptive presentation technology* is to adapt the content of a course page to the student's goals and knowledge, or other information known about the student and his or her condition of work. In a system with adaptive presentation, the pages are not static, but adaptively generated or assembled from pieces for each user. For example, with several adaptive presentation techniques, expert users receive more detailed and deep information, while novices receive additional explanation. Existing works on adaptive presentation for Web courses cover the adapting of media types selection to the learners' individual traits and equipment connection speed (Carver et al., 1996; Danielson, 1997), and the adapting of educational content to different groups of learners (Lemone, 1997) and learners individual level of knowledge (Ahanger & Little, 1997; Calvi & De Bra, 1997; Eliot et al., 1997; Kay & Kummerfeld, 1994a; Kay & Kummerfeld, 1994b; Yoon et al., 1997). Adaptation to the learner's individual needs is not limited to a page-level adaptation—a sequence of content from a small presentation to a whole course could be individually generated on demand (Ahanger & Little, 1997; André et al., 1998; Vassileva, 1997).

The goal of *adaptive navigation support technology* is to support the student in hyperspace orientation and navigation by changing the appearance of visible links. In particular, the system can adaptively sort, annotate, or partly hide the links of the current page to make easier the choice of which link is to be proceeded to next. Adaptive navigation support (ANS) for Web-based courses has been researched in a number of projects: (Brusilovsky et al., 1998; 1996; De Bra, 1997; Nakabayashi et al., 1997; Neumann & Zirvas, 1998; Pilar da Silva et al., 1998; Specht & Oppermann, 1999).

The electronic presentation

An electronic presentation is a way to represent essentially sequential material such as a lecture or a slide show. A main criterion for considering material as sequentially structured is that sequential navigation aids are present. The presentation should have a distinct "top" node with some introduction to and a link to the first node. All nodes (we call them "slides") should contain "next" and "previous" navigation links as well as a link to the top node. A table of contents with a possibility of direct navigation to any slide should be also provided. Each slide in a presentation usually has a content and a narration. In this sense a zero-level presentation is just a sequence of slides where both content and narration are authored together as pure text (or as text with figures). While this style of presentation is still very popular, we will not consider it here since it is actually equal to the one-level-deep electronic textbook analyzed above. In this section, we will consider two special types of electronic presentations used in Web-based courses: the electronic lecture and the guided tour. An electronic lecture is a sequence of slides extended with audio or audio/video narration. It is modeled after a regular lecture. Usually both the content (slides) and the narration are created by the same author who takes the role of a virtual lecturer. A guided tour is usually a multiple-author presentation. The content of a guided tour's slides could be previously developed by different authors and located anywhere on the Web. The role of a guided tour author is to sequence this content and extend it with a narration. (Here, the narration is usually in the form of text). This type of presentation is modeled after a museum's guided tour.

The electronic lecture

There are two kinds of "electronic lectures" used in distance education: synchronous lectures and asynchronous lectures. Synchronous lectures simply provide distant access to a real lecture theatre. While this technology is quite popular in distance education, we will not consider it here. A synchronous lecture is a mode of distance education but is not an element of courseware. Also, the current implementation of synchronous lectures is most often based on special video-conferencing software, rather then on a Web-based solution. Asynchronous lectures are recorded and could be viewed at any time. This makes it an eligible courseware element which can be stored, "owned", and distributed. A number of advanced suits of tools support both synchronous and asynchronous lectures: a synchronously presented lecture can be recorded, enhanced, and turned into an asynchronous lecture (Eisenstadt & Domingue, 1998; Synnes et al., 1998).

Asynchronous lectures come in "single piece" or "chunked" form. A "single piece" lecture is usually a continuous record of a classroom presentation. We will not consider single piece lectures here. First, this type of lecture is much less useful than a full-fledged Web lecture because it is almost impossible to either read a video-recorded content of presented slides (Abowd et al., 1998) or navigate within the content of the lecture. Second, this type of lecture is technically a piece of video or audio learning material instead of a special form of Web presentation. What we are considering below is a "chunked" asynchronous lecture, i.e. an asynchronously presented sequence of slides with video and/or audio narration. We will call this type of electronic lecture a "Web lecture". Though this kind of electronic lecture was originally developed for an older CD-based technology (Dannenberg, 1998), its Web implementation now dominates.

The Web lecture is becoming a more and more popular technology for presenting course material on the Web. First of all, it is still the best replacement for classroom lectures. Neither textbooks nor handouts adequately replace an up-to-date lecture done by a leading researcher or professional. Secondly, lectures provide distance students with the "feeling of the classroom". Third, it is often the easiest way to place some course content on the Web. Some research claims that Web lectures are at least as efficient as regular lectures (LaRose & Gregg, 1997).

The developers of modern Web lecture technologies are driven by three reasonably different goals. For some developers (Abowd et al., 1998), the Web lecture is the way to support regular "classroom" students by recording "what happened in the classroom". Since these students can always re-play the lecture at their own pace, they can spend less time taking notes and more time understanding the actual lecture. For other teams (Bacher & Ottmann, 1996; Harris & DiPaolo, 1996; Severance, 1998; Stanford Online Team, 1998) the goal is "on-the-fly" authoring, i.e. providing a "fast copy" of a real lecture for students (mainly distance learners) who can't be in the classroom. Yet other developers aimed at providing archival material especially for distance education (Barbieri & Mehringer, 1997; Dannenberg, 1998). While the target audience (that is, classroom students, distance students, or both) may influence different features of Web lectures, any existing Web-lecture system can serve all three kinds of audiences.

Current technologies for producing Web lectures differ significantly in two aspects—the *content structuring* level and the *media level.* A high-quality "archival" Web lecture should allow a fine-grained sequential and random access to the lecture content. First, the video/audio stream has to be divided into the smallest meaningful chunks, which usually corresponds to a line or a piece of a slide. Chunking is important for slide synchronization, random access to lecture parts and retrievability. Synchronization means that each audio or video chunk has to be associated with a corresponding portion of the slide presentation. With *slide-level* synchronization a chunk of narration is associated with the whole slide. With *line-level* synchronization a chunk is associated with each line (or a fragment) of the slide. Random access means that each chunk and slide (or slide line) can be individually addressable via a lecture table of contents. Retrievability means that the user can retrieve lecture chunks satisfying some criteria, for example, if it contains some keywords. To be retrievable by request, each chunk should be indexed with keywords or domain concepts. This is very important because it enables students having problems with later lectures or exercises to retrieve a helpful piece of a teacher's explanation. Finally, an author may be willing to provide annotation, i.e. associate comments, references, and links to additional resources with any chunk of material. Annotations could also be used for finding a relevant lecture piece with full-text search (Stanford Online Team, 1998). On the other hand, a simple Web lecture provides no line-level synchronization, and no indexing.

Currently very few systems like MANIC (Stern et al., 1997a), CALAT (Nakabayashi et al., 1996), and JITL (Dannenberg, 1998) support line-level chunking and synchronization and even fewer (Dannenberg, 1998; Smeaton & Crimmons, 1997) support indexing and search. The problem is that reliable chunking, synchronization and indexing requires several hours of manual work for one hour of lecturing. Special advanced synchronized recording software used in AOF (Bacher & Ottmann, 1996), Sync-o-Matic 3000 (Severance, 1998), WLS (Klevans, 1997b), mStar (Synnes et al., 1998), and Class 2000 (Abowd et al., 1998) can to some extent solve the problem of synchronization. This software records the time when the presenter changes each slide or slide line and uses this data to change the slide or to highlight next line at the proper time when re-playing the presentation. The same technology could also be used to build a table of contents, however, it may not be able to solve the indexing problem or support retrievability.

We distinguish three media levels in existing Web lecture systems: recorded audio, recorded video, and high-quality video. The difference between recorded audio and recorded video mainly depends on the capabilities of the client computer. A number of development teams specially restricted themselves to audio because their students used low-bandwidth Web connections, and because they saw that video provided little additional value in exchange for the long waiting time. With new streamlined video technologies and the availability of increasing bandwidth, recorded video will become more and more popular. The difference between recorded video and high-quality video depends on the required equipment and processing time. High-quality video requires special recording equipment—preferably a studio with a *blue screen*—and several hours of processing time for one hour of lecture.

There is an obvious relationship between the goals in developing Web-lectures and the level of used technology. Web-lectures made as a record for classroom students (Abowd et al., 1998) may stay with audio or low-quality video and without fine grained chunking and random access (since a teacher is available to solve any problems!) On-the-fly authoring with the goal to provide a copy of a classroom lecture for distance audience as soon as possible can afford to perform only 2–3 hours of processing for one hour of lecturing (Stanford Online Team, 1998), and can use only a minimal level of manual content processing and video processing. However, an archival Web lecture specially prepared for asynchronous distance education needs the highest level of content structuring and a high quality video. It currently requires many hours of processing for one hour of lecturing (Dannenberg, 1998).

Base level: A base level Web lecture is currently a sequence of static slides with a separate audio or video narration file for each slide (Barbieri & Mehringer, 1997; Ingebritsen et al., 1997; LaRose & Gregg, 1997; Smeaton & Crimmons, 1997) or video (Bacher & Ottmann, 1996; Harris & DiPaolo, 1996; Radhakrishnan & Bailey, 1997; Stanford Online Team, 1998). One evaluation claims that such base level Web lectures are as efficient as ordinary face-to-face lectures (LaRose & Gregg, 1997). Base level Web lectures could be developed with COTS tools such as Microsoft Power Point.

State-of-the-art level: The requirement for state-of-the-art lectures is to have in-slide synchronization with audio and video. It could be either authored synchronization as in MANIC (Stern et al., 1997a), CALAT (Nakabayashi et al., 1996), JITL (Dannenberg, 1998) or recorded time synchronization (Abowd et al., 1998; Bacher & Ottmann, 1996; Klevans, 1997a; Severance, 1998; Synnes et al., 1998). Authored synchronization is better since it allows a line-level access to the material, but it is much more expensive. Slides or even lines in state-of-the-art systems should be indexed or annotated to allow keyword or full-text search. State-of-the-art lectures are usually developed with special authoring tools (Abowd et al., 1998; Bacher & Ottmann, 1996; Klevans, 1997a; Severance, 1998; Synnes et al., 1998) or technologies (Dannenberg, 1998).

Research level: Research teams working on Web lectures are trying to extend Web lecture technologies in the following directions: developing more powerful authoring tools (Bacher & Ottmann, 1996; Dannenberg, 1998; Klevans, 1997a; Severance, 1998), adding advanced options for finding relevant audio/video chunks (Smeaton & Crimmons, 1997), adding adaptivity to user knowledge to systems with Web lectures (Nakabayashi et al., 1996; Stern et al., 1997a; Stern et al., 1997b; Stern, 1997), and using life-like intelligent agents ("virtual lectors") for presenting the material (André et al., 1997).

The guided tour

Guided tours or paths are a traditional feature of hypertext systems (Trigg, 1988). A classic guided tour is a sequential path through complex hyperspace extended with some narration. As with their museum predecessors, guided tours could be developed to introduce a user into a complex hyperspace or to express a specific aspect of the available information. Though the first work on educational Web-guided tours started as early as other trials to use Web for education (Hauswirth, 1995; Nicol et al., 1995), this technology is still not very popular. Some possible reasons are the lack of material to be reused, legal problems with the reuse of previously authored content, and the limited applicability of guided tours in university-level education. In addition, a good guided tour requires a special authoring and delivery system. Currently, a few authoring systems for Web-guided tours have been developed: FootSteps (Nicol et al., 1995), Hierarchical Guided Tours (Hauck, 1996), Walden Paths (Furuta et al., 1997; Shipman III et al., 1998; Shipman III et al., 1997), Ariadne (Jühne et al., 1998) and others (Hauswirth, 1995; Langenbach & Bodendorf, 1997; Wright & Jones, 1997). All these systems allow the author of the tour to specify a sequence of existing Web pages for the tour and to add a narration. Some systems also support hierarchical (Hauck, 1996) or other branching tours (Jühne et al., 1998). Most tour development systems provide a graphical interface for designing a tour, though some systems require a tour to be written in a special text-based language (Hauswirth, 1995; Langenbach & Bodendorf, 1997). At the delivery stage, these systems support user navigation by providing a panel with bi-directional sequential navigation buttons (backward, forward, top), a table of contents, and (for a non-linear tour) a graphical map (Jühne et al., 1998). The narration and the navigation buttons are either shown in the same window with the content of the original WWW page (Furuta et al., 1997; Hauck, 1996; Hauswirth, 1995; Nicol et al., 1995) or in a separate window (Jühne et al., 1998; Wright & Jones, 1997). The latter is probably better from a copyright point of view. An important navigation feature in a guided tour-delivery mechanism is a tool (preferably a simple button) for the user to return to the "main trail" after "side excursions" to Web pages outside the tour (Hauswirth, 1995; Langenbach & Bodendorf, 1997; Nicol et al., 1995). Until recently, a "tour server" with a CGI script was a feature of any tour delivery system. More recent systems have relied more on JavaScript (Wright & Jones, 1997) or Java (Jühne et al., 1998; Langenbach & Bodendorf, 1997) functionality.

Since we only know a handful of reported guided tour systems, we cannot provide a well-grounded level taxonomy for guided tours. It is reasonably clear, though, that state-of-the-art guided tours should provide a navigation overview (i.e. a table of contents or map) and a "back to the tour" feature, and should be developed with special authoring systems. Recent studies report that Web-guided tours are a promising educational tool (Shipman III et al., 1998). This technology may become very popular in the near future as soon as COTS authoring systems appear and if the copyright problem is solved. The reuse of existing material is already becoming a better option due to the increase in the amount of already authored material and improvements in Web search technologies. Progress in the reusability authoring approach will probably make the guided tour technology very popular, since a guided tour of a set of pre-authored resources is the fastest way to develop a customized course (Langenbach & Bodendorf, 1998).

Actively Engaging Students: Assessment and Learning-by-doing

Students learn, in part, by performing activities. To go beyond presentation-only courseware (e.g. learning-by-reading, learning-by-watching etc.), Web-based education systems have students *do* something. There are two major goals in adding various student activities to an educational system. The first goal is assessment: the only reliable way to evaluate students' understanding and progress is to let them do something which could show this understanding, and to evaluate results. The second goal is learning-by-doing itself. It is through activities, like problem solving and exploration, that students can come to grips with the subject; and intellectually make it their own. The difference between assessment and learning-by-doing activities is generally not clear-cut, for example, such typical activities as problem solving promote learning-by-doing, but the results could still be used for reliable assessment. In current academic and Web-based education practice, there is a clear distinction between two groups of activities that we will call *objective* and *applicative*. Objective activities (true/false questions, multiple choice questions, short-answer questions) are designed to check student understanding and involve little creativity. As a rule, objective activities are easily graded automatically. Applicative activities (we will call them exercises) involve students in serious problem solving, development, or exploration. In most cases, applicative activities must be evaluated and graded by a human teacher or assistant. While objective activities tend towards evaluation and applicative activities are more learning oriented, activities may emphasize each of these two major goals to a greater or lesser degree. At one end of the objective activity spectrum are those activities used solely for assessment: the results are going directly to the teacher and students are not even provided with feedback (a precondition of learning-by-doing). At the other end we have objective activity developed solely for self-assessment. The students are provided with comprehensive feedback and the teacher does not see the results at all. The same spectrum of goals could be observed for applicative activities. There are applicative activities issued solely for assessment and there are applicative activities aimed solely to promote learning-by-doing. Since current technologies for dealing with objective activities and applicative activities are quite different, we will consider these types separately in the following subsections.

Assessment: quizzes and objective activities

Objective tests and quizzes are among the most widely used and well-developed tools in higher education. A classic test is a sequence of reasonably simple questions. Each question assumes a simple answer that could be formally checked and evaluated as correct, incorrect, or partly correct (for example, incomplete). Questions are usually classified into types by the type of expected answer. Classic types of questions include yes/no questions, multiple-choice/single-answer (MC/SA) questions, multiple-choice/multiple-answer (MC/MA) questions, and fill-in questions with a string or numeric answer. More advanced types of questions include matching-pairs questions, ordering-questions, pointing-questions (the answer is one or several areas on a figure) and graphing-questions (the answer is a simple graph). Also, each subject area may have some specific types of questions.

Life cycle and anatomy of quizzes and questions

To compare existing options, we have analyzed the life cycle of a question in Web-based education (see Table 6.2). We divided the life cycle of a question into three stages: preparation (before active life), delivery (active life), and assessment (after active life). Each of these stages is further divided into smaller stages. For each of these stages, we have investigated a set of possible support technologies.

The life of a question begins at *authoring* time. The role of WBE systems at the authoring stage is to support the author by providing a technology and a tool for question authoring. All authored questions (the content and the metadata) are *stored* in the system. The active life of a stored question starts when it is *selected for presentation* as a part of a test or quiz. This selection could be done statically by a teacher at course development time or dynamically by a system at run time (by probability or according to some cognitive model).

Next, the system *delivers* a question: it presents the question, it provides an interface for the student to answer; it gets the answer for evaluation. At the *assessment* stage, the system should do the following things: evaluate the answer as correct, incorrect, or partly correct, deliver feedback to the student, grade the question and record student performance.

Existing WBE tools and systems differ significantly on the type and amount of support they provide on each of the stages mentioned above. Simple systems usually provide partial support for a subset of the stages. Cutting edge systems provide comprehensive support for all the listed stages. The power of a system and the extent of support provided is seriously influenced by the level of technology used at each of the main stages—preparation, delivery and assessment. Below, we analyze the currently explored options.

TABLE 6.2
Life Cycle Stages of a Test Question. The Choice of Technology for Storing, Interacting and Rcording Results Usually Determines the "Testing Power" of a WBE System.

Before	During	After
Preparation: • Author • *Store* • Select	Delivery: • Present • *Interact* • Get the answer	Assessment: • Evaluate • *Grade and record* • Deliver feedback

Preparation

A question is created by a human author—a teacher or a content developer. A state-of-the-art question has the following components: the question itself (or *stem*), a set of possible answers, an indication as to which answers are correct, a type of interface for presentation, question-level feedback that is presented to the student regardless of the answer, and specific feedback for each of the possible answers. Questions used primarily for self-assessment may include additional scaffolding information such as "hints" or "help" (Raineri et al., 1997). Finally, an author may provide metadata such as topics assessed, keywords, the part of the course a test belongs to, question weight or complexity, allowed time, number of attempts, etc. This metadata could be used to select a particular question for presentation as well as for grading the answer.

The options available for *authoring support* usually depends on the technology used for storing an individual question in the system. Currently, we can distinguish two different ways to store a question, which we call *presentation format* and *internal format*. In WBE context, storing a question in presentation format means storing it as a piece of HTML code (usually, in HTML form). Such questions could also be called static questions. They are "black boxes" for a WBE system: static questions can only be presented "as is". The authoring of this type of question is often not supported by a WBE system. It could be done with any of the HTML authoring tools.

Storing a question in an internal format usually means storing it in a database record where different parts of the question (stem, answers, and feedback) are stored in various fields of this record. A question as seen by a student is generated from the internal format at the delivery time. Internal format opens the way for more flexibility: the same question could be presented in different forms (for example, fill-in or multiple choice) or with different interface features (for example, radio buttons or selection list). Options in multiple choice questions could be shuffled (Carbone & Schendzielorz, 1997). This provides a higher level of individualization. It is also pedagogically useful and decreases the possibility of cheating. There are two major ways for authoring questions in internal format: a form-based *graphical user interface* (GUI) or a special *question markup language* (Brown, 1997; Campos Pimentel et al., 1998; Hubler & Assad, 1995). Each of these approaches has its benefits and drawbacks. Currently, a GUI approach is much more popular. It is used by all advanced commercial WBE systems (such as Blackboard, 1999; Question Mark, 1998; WBT Systems, 1999; WebCT, 1999). Note, however, that some WBE systems use a GUI authoring approach but do not store questions in internal format. Instead, these systems generate HTML questions "right away" and store them as static questions.

The simplest means for *question storage* is *a static test or quiz*, i.e. a static sequence of questions. The quiz itself is usually represented in plain HTML form and authored with HTML-level authoring tools. Static tests and quizzes are usually "hardwired" into some particular place in a course. One problem with this simplest technology is that all students get the same questions at the same point in the course. Another problem is that each question hardwired into a test is not reusable. A better option for question storage is a *hand-maintained pool of questions*. The pool could be developed and maintained by a group of teachers who teach the same subject. Each question in a question pool is usually static, but the quizzes are more flexible. Simple pool management tools let the teacher re-use questions; all quizzes may be assembled and added to the course pages when it is required. This is what we call *authoring time flexibility*. The same course next year, a different version of the course, or sometimes even different groups within the same course may get different quizzes without the need to develop these quizzes from scratch.

An even better option is to turn a hand-maintained pool into a database of questions. A database adds what we call *delivery time flexibility*. Unlike a hand-maintained pool, a database is formally structured and is accessible by the delivery system. With a database of questions, not only can the teacher assemble a "quiz-on-demand", the system itself can generate a quiz from a set of questions. Naturally, the questions could be randomly selected and placed into a quiz in a random order (Asymetrix, 1999; Brown, 1997; Byrnes et al., 1995; Carbone & Schendzielorz, 1997; Ni et al., 1997; Radhakrishnan & Bailey, 1997; WBT Systems, 1999; WebCT, 1999). As a result, all students may get personalized quizzes (a thing that a teacher can not realistically provide manually), significantly decreasing the possibility of cheating. Note that implementation of a database of questions does not require the use of a commercial database management system. Advanced university systems like QuestWriter (Bogley et al., 1996) or Carnegie Mellon Online (Rehak, 1997a) and many commercial systems such as TopClass (WBT Systems, 1999) or LearningSpace (Lotus, 1999) use full-fledge databases such as ORACLE or Lotus Notes for storing their pools of question in internal format. However, there are systems which successfully imitate a database with the UNIX file system by using specially structured directories and files (Byrnes et al., 1995; Gorp & Boysen, 1996; Merat & Chung, 1997).

A problem for all systems with computer-generated quizzes is how to ensure that these quizzes include a proper set of questions. The simplest way to achieve it is to organize a dedicated question database for each lesson. This approach, used for example in WebAssessor (ComputerPREP, 1998), reduces question reusability between lessons. More advanced systems like TopClass (WBT Systems, 1999) can maintain multiple pools of questions, and can use several pools for generating each quiz. With this level of support, a teacher can organize a pool for each topic or each level of question complexity, and specify the desired number of questions in a generated quiz to be taken from each pool.

A database of questions in internal format is currently a state-of-the art storage technology. Research teams are trying to advance it in three main directions. One direction is related to different kinds of *parameterized questions* as in CAPA (Kashy et al., 1997), EEAP282 (Merat & Chung, 1997), WebTester (Sapir, 1999), OES (Bryc & Pelikan, 1999), WebAssign (Titus et al., 1998) or Mallard (Brown, 1997; Graham et al., 1997). This allows one to create an unlimited number of tests from the same set of questions and can practically eliminate cheating (Kashy et al., 1997). The second direction of research is related to question metadata. If the system knows a little bit more *about* the question (for example, type, topics assessed, keywords, part of the course a test belongs to, weight or complexity) then the system can generate customized and individualized quizzes by the author's or system's request. This means that the authors could specify various parameters for the quiz their student needs at some point of the course—e.g. total number of questions, proportion of questions of specific types or for specific topics, difficulty, etc.—and the system will generate a customized quiz on demand (that is still randomized within the requirements) (Byrnes et al., 1995; Merat & Chung, 1997; Rehak, 1997a; Rios et al., 1998). This option is definitely more powerful than simple randomized quizzes. Systems that make extensive use of metadata really "know" about the questions and their functionality. The third direction of research is the adaptive sequencing of questions. This functionality is based on an overlay student model which separately represents student knowledge of different concepts and the topics of the course. Intelligent systems such as ELM-ART (Weber et al., Submitted), Medtec (Eliot et al., 1997), (Lee & Wang, 1997), SIETTE (Rios et al., 1998), Self-Learning Guide (Desmarais, 1998; Khuwaja et al., 1996) can generate challenging questions and tests adapted to the student level of knowledge as well as reduce the number of questions required to assess the students' state of knowledge.

Delivery stage

The interaction technology used to get an answer from the student is one of the most important parameters of a WBE system. It determines all delivery options and influences authoring and evaluation. Currently, we distinguish five technologies: HTML links, HTML/CGI forms, scripting language, plug-in, and Java.

HTML links is a simple interaction technology that presents a set of possible answers as list of HTLM links. Each link is connected to a particular feedback page. The problems here are that questions are hard to author (because the question logic must be hardwired into the course hypertext) and that it supports only yes/no and MC/SA questions. This technology was in use in the early days of WBE, when more advanced interaction technologies like Common Gateway Interface (CGI), JavaScript or Java were not established (Holtz, 1995).

The most well-established technology for Web testing which is used now in numerous commercial and university-grown systems is a combination of HTML forms and CGI-compliant evaluation scripts. HTML forms are very well suited for presenting the main types of questions. Yes/no and MC/SA questions are represented by radio buttons, selection lists, pop-up menus, MC/MA questions are represented by multiple selection lists or checkboxes. Fill-in questions are implemented with input fields. More advanced questions such as matching pairs or ordering can also be implemented using forms. In addition, hidden fields can be successfully used to hold additional information about the test that a CGI script may need. There are multiple benefits to using server-side technology such as form/CGI technology and a similar server-side map technology that can be used for implementing graphical pointing questions. These are that test development is relatively simple and can even be done with HTML authoring tools; and that sensitive external information which is required for test evaluation (such as question parameters, answers, feedback) may be safely stored on the client side, thus preventing students from stealing the question (i.e. the only external information which is required in a well-developed system to evaluate a test is the test ID and the student ID). Server-side evaluation makes all assessment time functions (such as recording results, grading, and providing feedback) easy to implement. In fact, the same server-side evaluation script could perform all these functions. The main problem of server-side technology is its low expressive power. It is well suited only for presenting basic types of tests. More advanced types of tests as well as more interactive types of tests (for example, tests which involve drag-and-drop activities) can not be implemented with pure server-side technology. Authoring questions with server-side evaluation is tricky because a question's functionality is spread between its HTML presentation (either manually authored or generated) and a CGI evaluation script. Another serious problem is that CGI-based questions do not work when a user's connection to the server is broken or very slow.

A newer technology for question delivery and evaluation is JavaScript (McKeever et al., 1997). The interface provided by the JavaScript interaction technology is similar to that of form/CGI technology's. At the same time, JavaScript functionality supports more advanced, interactive questions, for example, selection of a relevant fragment in a text. With pure JavaScript technology, all data for the question evaluation and feedback as well as the evaluation program should be stored as a part of the question text. It means that a JavaScript question can work in standalone mode. This means that the question is self-sufficient: everything for presentation and evaluation is located in the same file, making it a very attractive option for authoring. But this also means that students can access the source of the question and crack it. Also, with pure JavaScript evaluation technology, there is no way for recording the results and grades. With all the above features, JavaScript technology is a better choice for self-assessment tests than for assessments used in grading. We think the proper place for JavaScript in WBE is in a hybrid JavaScript/server technology. With this technology, JavaScript can be used to present more types of questions, do it more interactively and with compelling user interfaces leaving evaluation and recording to be done by traditional CGI for reasons of security (ComputerPREP, 1998; WebCT, 1999).

A higher level of interface freedom can be achieved by using a plug-in technology.[10] The classic example of serious use of this technology in education is the Shockwave player (Macromedia, 1999b) which can run multimedia presentations prepared with Macromedia authoring tools. Currently, Shockwave technology is used in WBE mainly for delivering "watch only" animations, but this technology is more powerful. In fact, a variety of very attractive Shockwave-deliverable questions could be developed using Macromedia tools with relatively low effort. Some examples could be provided by AST (Specht et al., 1997) and Medtec (Eliot et al., 1997). The negative side is the same as with JavaScript: recording assessment results requires a connection to the server. Until recently, Shockwave provided no Internet functionality, and its users had to apply special techniques (e.g. saving the evaluation results in a temporary file). Due to Shockwave communication problems, some teams that started with Shockwave migrated later to the more powerful Java technology (Eliot et al., 1997). Currently, after the release of the Authorware Web Player (Macromedia, 1999a)—a special WBE-targeted update of Shockwave player, this technology has become an attractive platform platform for delivering various self-assessment questions.

The highest level technology for question delivery is provided by Java. An important advantage of Java is that it is a complete programming language designed to be integrated with browser functionality and the Internet. Java combines the connectivity of form/CGI technology and the interactivity of Shockwave and JavaScript. Any question interface can be developed with Java, and, at the same time, Java-made questions can naturally communicate with the browser as well as with any Internet object (a server or a Java application). Examples of systems which heavily use Java-based questions are FLAX (Routen et al., 1997), NetTest (Ni et al., 1997), Mallard (Graham & Trick, 1997), and Medtec (Eliot et al., 1997). Developing question interfaces with Java is more complicated than with form/CGI technology and it is not surprising that all the examples mentioned above were produced by advanced teams of computer science professionals. However, the complexity will not stop this technology. Java is currently the way to implement a variety of question types non-implementable with form/CGI technology, such as multiple pointing questions, graphing questions, and specialized types of questions. Developing Java-based questions can become suitable for ordinary authors with the appearance of Java-based authoring systems (Ni et al., 1997; Routen et al., 1997).

Assessment stage

As we noted, the choice of interaction technology significantly influences *evaluation* options. Evaluation is the time when an answer is judged as correct, incorrect, or partially correct (for example, incomplete). Usually, correct and incorrect answers are provided at the authoring time, so evaluation is either hard-wired into the question like in MC/SA questions, or performed by simple comparison (as in fill-in questions). There are very few cases that require more advanced evaluation technology. In some domains, the correct answer may not be literally equal to a stored correct answer. Examples are a set of unordered words, a real number, and a simple math expression (Bryc & Pelikan, 1999; Holtz, 1995; Hubler & Assad, 1995; Xiao, 1999). In this situation special comparison programs are required. Some systems includes special evaluation modules for advanced answer matching (Hubler & Assad, 1995). Other systems rely on a "domain expert" such as the Lisp interpreter for Lisp programming (Weber & Specht, 1997) or a computer algebra system for algebra domain (Antchev et al., 1995; Pohjolainen et al., 1997; Sapir, 1999) to evaluate the answer. The first two evaluation options are very simple and could be implemented with any interface technology—even JavaScript could be used to write a simple comparison program. If more advanced computation is required (as in the case of intelligent answer matching) the choice is limited to full-function programming with either Java or a server side program using a CGI interface. If a "domain expert" is required for evaluation, the only option currently is to run a domain expert on the server side with a CGI-compliant gateway. In fact, a number of "domain expert" systems (for example, Mathematica computer algebra system) have a CGI gateway.

The usual options for the *feedback* include: simply telling if the answer is correct, not, or partially correct, giving the correct answer, and providing some individual feedback. Individual feedback may communicate information such as: what is right in the correct answer, what is bad in the incorrect and partially incorrect answer, as well as provide some motivational feedback, and provide information or links for remediation. All individual feedback is usually authored and stored with the question. A system that includes assessed concepts or topics as a part of question metadata can provide good remedial feedback without direct authoring since it "knows" what knowledge is missing and where it can be found. It means that the power of feedback is determined by authoring and storage technology. The amount of information presented as feedback is determined by the context. In self-assessment the student usually receives all possible feedback—the more the better. This feedback is a very important source of learning. In a strict assessment situation the student usually gets neither a correct answer, nor knowledge of whether the answer is correct. The only feedback for the whole test might be the number of correctly answered questions in a test (Rehak, 1997a). This greatly reduces the student's chances for cheating and chances to learn. To support learning, many existing WBE systems make assessment less strict and provide more feedback by trying to combat cheating by other means. The only way to combine learning and strict assessment is to use more advanced technologies such as parameterized questions (Brown, 1997; Bryc & Pelikan, 1999; Kashy et al., 1997; Merat & Chung, 1997; Titus et al., 1998) and domain-specific test generation (Eliot et al., 1997; Sapir, 1999; Weber & Specht, 1997), which can generate an unlimited number of questions. In this situation, a WBE system can provide full feedback without promoting cheating.

If a test is performed purely for self-assessment, then generating feedback could be the last duty of a WBE system in the "after-testing" stage. The student is the only one who needs to see test results. In the assessment context, the last duty of a WBE system in the process of testing is to *grade student performance on a test and to record these data* for future use. Grades and other test results are important for teachers, course administrators, and students themselves (a number of authors noted that the ability to see their grades online is the WBE system feature that students appreciate the most). Early WBE systems provided very limited support for a teacher in test evaluation. Results were either sent to the teacher by e-mail or logged into a special file. In both cases a teacher was expected to complete grading and recording manually: to process test results and grade them, to record the grades, and to ensure that all involved parties get access to data according to university policy. This lightweight support is easy to implement and does not require that teachers learn any new technology. For the latter reason, this technology is still used as an option in some more advanced systems (Carbone & Schendzielorz, 1997). However, a system that provides no other options for grading and recording is now below the base level. A state-of-the-art WBE system should be able to grade a test automatically and record the test results in a database. It also should provide properly restricted[11] access to the grades for students, teachers, and administrators. Many university-level systems (Bogley et al., 1996; Brown, 1997; Carbone & Schendzielorz, 1997; Gorp & Boysen, 1996; Hubler & Assad, 1995; MacDougall, 1997; Ni et al., 1997; Rehak, 1997a) and almost all commercial level systems (Lotus, 1999; WBT Systems, 1999; WebCT, 1999) provide this option in a more or less advanced way. Less advanced systems usually store the grades in structured files and provide limited viewing options. Advanced systems use database technology to store the grades, and provide multiple options for viewing the grades and other test performance results (such as time on a test or the number of efforts made). Database technology makes it easy to generate various test statistics involving the results of many students on many course tests. In a Web classroom, where student-to-student and student-to-teacher communication is limited, comparing statistics is very important for both teachers and students, so as to get the "feeling" of the classroom. For example, by comparing the class average with personal grades, a student can determine his or her class rank. By comparing class grades for different tests and questions, a teacher can find the questions that are too simple, too difficult, or even incorrectly authored.

Levels for quizzes

Base level: A base-level WBE should satisfy three requirements: be able to administer tests, grade tests and store the results. Any Web-based testing technology that is unable to record testing results is below this base level. Self-assessment orientation can hardly be considered as an excuse for the absence of tools for recording and viewing grades. As was noted above, even in a self-assessment context where tests are not expected to be graded, there are multiple benefits to maintaining a record of student test results.

State-of-the-art level: To qualify for State-of-the-art a WBE system should provide several advancements over a base level system (Table 6.3). The system should support at least all the basic question forms: true/false, MC/SA, MC/MA, and fill-in. An authoring system should be provided—at least for the development of these basic question types. Questions should be stored in a pool that allows at least authoring-time flexibility in selecting questions for a test. The test results should be automatically graded and recorded in a database for further access by students and teachers. Grade-viewing tools allowing teachers to view grades in several basic ways should be provided. Competition between state-of-the-art systems is providing better authoring tools, including features as: test assembly from question pools; multiple question types; options for restricting questions by time and repeated attempts; and multiple ways to view results.

Research level: Research systems investigate advanced Web-based testing options in several directions. A summary of the main research directions is as follows:

- *Metadata:* Several research teams are working on tools which can support the author in adding metadata to questions (Forte et al., 1996ab), and selection tools which either help the author in selecting a proper question from a pool at authoring time (Forte et al., 1996ab; Graf & Schnaider, 1997), or use metadata to generate a customized test on demand using criteria supplied by the author (Byrnes et al., 1995; Merat & Chung, 1997; Rehak, 1997b).
- *Variety–interactivity:* The extending of variety and interactivity is accomplished using Java and similar technologies. It includes developing more interactive types of questions (Graham & Trick, 1997; Ni et al., 1997; Raineri et al., 1997) and developing authoring tools which support a greater variety of questions (Ni et al., 1997; Routen & Graves, 1997a; Routen et al., 1997). From another direction, some teams are working on extending the variety of question types by providing more advanced evaluation technologies such as advanced comparison (Hubler & Assad, 1995; Raineri et al., 1997) and "domain experts" (Antchev et al., 1995; Bryc & Pelikan, 1999; Pohjolainen et al., 1997; Sapir, 1999; Weber & Specht, 1997; Xiao, 1999).
- *Parameterized questions:* This work (Brown, 1997; Bryc & Pelikan, 1999; Kashy et al., 1997; Merat & Chung, 1997; Sapir, 1999; Titus et al., 1998)

covers several related issues: authoring systems and languages for creating parameterized questions, generation of a parameterized question, and evaluation.

- *Artificial intelligence:* Artificial intelligence technologies are being adapted for test generation from a knowledge base (Eliot et al., 1997; Specht et al., 1997; Weber et al., Submitted) and adaptive question sequencing (Desmarais, 1998; Eliot et al., 1997; Khuwaja et al., 1996; Lee & Wang, 1997; Rios et al., 1998).

TABLE 6.3
Levels and Features for Web-based Testing

	Basic	State-of-the art	Research
Types of questions	2-3 of basic types	All basic types	Greater variety
Authoring	COTF tools	Authoring tool or language	Authoring for advanced questions
Storage	Any	Pool, database	Enriched with metadata
Selection	Static	Authoring time or delivery time flexibility	Metadata for test assembling, intelligent sequencing
Interaction	CGI	CGI, some Java, restricted questions	Interactive questions, extensive use of Java
Evaluation	Simple matching	Simple and advanced matching	Intelligent comparison and domain experts
Grading/ recording	Any form	Database	
Viewing	Simple	Advanced	

Learning-by-doing: exercises

A nature of learning-by-doing is to push students to do something which requires application of knowledge being taught: to write an essay or a program, to draw a diagram, to translate a phrase into a foreign language, to solve a problem, to run an experiment with a simulator. An exercise is aimed at achieving an expected result that could be evaluated, if required. Typically, an *artifact* is produced. It could be a program, a problem solution, an essay, a set of simulation parameters—anything which could be created and evaluated. Still, within these limits, there are many parameters which distinguish different exercise settings. The size could differ from small, several minutes-long activities to projects lasting days or weeks. An exercise could be assigned by a teacher or selected by the student. Teacher evaluation may or may not be expected.

From a technical point of view it is important to distinguish exercises where the result is a "executable" artifact (such as a program or simulation) from those where the result is static such as an essay. If an exercise result is executable the student can use the computer to perform a self-check. The usual cycle of work with an executable artifact is: develop—run—view—submit—evaluate—grade. If a student is not satisfied with the result at any of these stages, the process can be interrupted, returning to development. If the artifact is not executable, the only way to get a reliable evaluation is to submit it to the teacher. Here the cycle of work is simply develop—submit—evaluate—grade. An exercise with a "executable" result enables a student to perform many develop-run-view cycles before involving the teacher.

One of the ways to compare WBE systems is to classify them in regards to the support they could provide to the students for various stages of work with exercises. From this point of view, almost all WBE systems which support exercises fall into two groups: assessment systems and learning environments. Assessment systems assist students in submissions. It is expected that the desired artifact is developed outside the system. What the system provides is an interface to deliver the solution to a teacher—either an HTML form box to paste the artifact (or a part of it (Hitz & Kögeler, 1997)) if it could be represented as text, or an interface to upload a file with the solution. For these kinds of systems, assessment is the goal and learning-by-doing is more of a by-product.

In contrast, learning environments are more oriented towards learning-by-doing and involve little or no teacher assessment.[12] These systems are trying to assist the student on different stages of their work on an exercise. At the design stage, learning environments provide an interface to develop the solution, such as a simple editor or *grapher* (McKenna & Agogino, 1997), a structural or diagram editor (Suthers & Jones, 1997), or a domain-oriented design interface, for example, for developing a microprocessor program (Graham et al., 1997). Other kinds of design support environments include "virtual labs" where the student can solve a case or design an experiment (EDesktop, 1997a; Faulhaber & Reinhardt, 1997; Johnson & Shaw, 1997; Johnson et al., 1998; Pearce & Livett, 1997). At run-time, learning environments "execute" the designed artifact—a program, a mathematical expression (Xiao, 1999), or an "experiment"—and deliver a static result to be viewed, for example, simulation results (EDesktop, 1997b; Tardif & Zaccarin, 1997). Some learning environments are not limited to showing the final static result—they could show either continuous or interactive animation of a running artifact—a computer program (Bilska et al., 1997; Brown & Najork, 1996; 1997; Brusilovsky et al., 1996; Campbell et al., 1995; Ibrahim, 1994; Jehng et al., 1997; Naps, 1997; Procopiuc et al., 1996; Rodger, 1997), a mathematical object (Xiao, 1999), or a simulation (Hampel et al., 1998; Kirkpatrick et al., 1997; Marshall & Hurley, 1997; McKenna & Agogino, 1997; Milton et al., 1997; Warendorf & Tan, 1997). At the evaluation stage, some learning environments could evaluate developed artifacts and either simply judge it as correct/incorrect (Brusilovsky et al., 1996; Graham & Trick, 1997), provide canned feedback, or provide an intelligent diagnosis with explanations and analysis of student's misconceptions (Brusilovsky et al., 1996; Okazaki et al., 1997; Yang & Akahori,

1997). Assessment-oriented environments can also calculate the grade (Foxley et al., 1998). Most advanced intelligent learning environments could provide evaluation and assistance for any step of problem solving. Examples are ADELE (Johnson & Shaw, 1997; Johnson et al., 1998), D3WWW-Trainer (Faulhaber & Reinhardt, 1997), Belvedere (Suthers & Jones, 1997), and PAT Online (Ritter, 1997).

Currently, assessment systems and learning environments constitute two different approaches to exercise implementation in WBE systems, so we will separately consider them in more detail. Potentially, any WBE system could combine assessment and learning-by-doing support. However, very few existing systems combine features of assessment systems and learning environments. Some assessment systems provide more support for development, for example, letting a student edit a program, run it, and see the results in a browser window (Barbieri & Mehringer, 1997; Hitz & Kögeler, 1997). From another side, some learning environments can formally evaluate the developed artifact. Also, all systems with automatic evaluation such as WebCeilidh (Foxley et al., 1998) or ELM-ART (Brusilovsky et al., 1996) and all systems with intelligent diagnosis (Brusilovsky et al., 1996; Okazaki et al., 1997; Suthers & Jones, 1997; Yang & Akahori, 1997) are bridging the gap. Any of the above systems provides evaluation support without a teacher's involvement. These systems promote learning-by-doing by increasing the number of design-evaluate cycles even for the domains without executable artifacts (Okazaki et al., 1997; Suthers & Jones, 1997; Yang & Akahori, 1997).

Assessment systems

The functionality of learning-by-doing assessment systems is similar to that of test systems. It includes authoring, selection, delivery, submission, evaluation, feedback and grading. For most of these steps, there is no difference between assessment and test systems, so whatever was discussed about tests in the previous chapter is also relevant for assessment systems. The authoring and delivery of exercises is usually simple, with authoring of HTML text enhanced with figures (and probably some media items), and delivery with a form box or another interface used for submitting the result. Existing assessment systems differ from test systems and from each other in respect to the type and level of support provided on submitting and evaluation stages. Some systems still use non-Web options for submitting the answer—it is either sent by regular or custom e-mail to the teacher, or uploaded to a special site by FTP. The regular options are either a form/CGI interface with a box for entering the solution or a simple browser interface for uploading the solution. The first option is better, but it supports only text solutions. If the solution is, for example, an MS Word file, uploading is the only option. More advanced systems are trying to combine some development support with the interface for submissions, for example, the ability to compile the program to be submitted, run it, and view the results (Barbieri & Mehringer, 1997; Hitz & Kögeler, 1997; Mehringer & Lifka, 1998). Some other systems make special provisions to integrate a Web system with a stand-alone problem-solving environment, for example, a programming environment. The student is expected to develop the solution in this stand-alone system, copy it, and paste into the browser window (Lawrence-Fowler & Fowler, 1997). To simplify the work with two applications, a WBE system could provide special buttons on an exercise page to launch all required stand-alone systems, such as an editor or a compiler. (Westhead, 1996).

The regular evaluation options for both base level and state-of-the-art level systems are "manual" evaluation and feedback generation written by a teacher. More advanced systems provide better interfaces for the teacher to evaluate submitted work. On the same screen, the teacher can view the result and enter both the grades and the feedback. Less advanced systems expect the teacher to maintain a paper log with grades and enter them all in once.

The main research efforts here are centered on automatic evaluation. So far, the most visible progress has been achieved for mathematical exercises, where an evaluation module could rely on modern computer algebra systems available via a Web interface. A number of existing Web-based systems for such areas as algebra or calculus are able to automatically evaluate student solutions to a wide range of exercises—from to simple questions to relatively big problems (Bryc & Pelikan, 1999; Pohjolainen et al., 1997; Sapir, 1999; Xiao, 1999). Automatic evaluation for computer programming is another well investigated area. There are some pre-Web systems like Ceilidh (Benford et al., 1993) or TRY (Reer, 1988), and Web-based systems like WebCeilidh (Foxley et al., 1998; Marshall & Hurley, 1996b; 1997) and ELM-ART (Brusilovsky et al., 1996) which evaluate student programs by running them against test data. Similar technology could be used for the automatic evaluation of other executable artifacts. Web-based systems in areas with non-executable artifacts have to rely on artificial intelligence techniques for automatic evaluation and diagnosis (Okazaki et al., 1997; Suthers & Jones, 1997; Yang & Akahori, 1997).

Base level: A base-level WBE should satisfy very modest requirements of: being able to deliver exercise results to the teacher for evaluation (e-mail is a legitimate option) and providing an interface for entering feedback and grades. As in the case of tests, there should be an interface for viewing grades.

State-of-the-art level: A state-of-the-art WBE should provide a Web-based student interface for submitting results and an integrated teacher interface for evaluation and entering grades. Special provisions (at the least, providing hints and help) should be given so as to integrate a WBE and a stand-alone problem-solving environment (such as a programming environment). Alternatively, for executable artifacts, some interface for running a solution and viewing results in a browser should be provided. Similar to state-of-the-art test systems, state-of-the art learning-by-doing assignment systems should also provide authoring time or run-time flexibility for selecting exercises from a pool of exercises as well as various grade viewing options. Finally, advanced systems should integrate communication facilities with exercise interfaces to let a student cooperate with other students in working on the exercise or to discuss it with a teacher.

Research level: As we have already noted, the main research directions are in automatic evaluation, which includes auto-checking and auto-grading systems in mathematics (Bryc & Pelikan, 1999; Pohjolainen et al., 1997; Sapir, 1999; Xiao, 1999) and programming (Brusilovsky et al., 1996; Foxley et al., 1998; Marshall & Hurley, 1996b; 1997), in systems which could check various properties of a developed artifact, such as plagiarism or compliance to standards (Foxley et al., 1998), and in systems which can do intelligent analysis of solutions (Brusilovsky et al., 1996; Okazaki et al., 1997; Yang & Akahori, 1997).

Learning environments

Unlike learning-by-doing assessment systems, existing Web-based learning environments are rarely oriented toward assessment of user performance. The goal here is to promote learning-by-doing. Moreover, every learning environment should allow students not only to work on exercises which require some kind of result, but also to support free exploration, for example, by playing with existing examples. In the latter case, no assessment is possible because there is no result to assess. Learning environments are a well-investigated domain in pre-Web education. The main problem for developers of Web-based learning environments lies not with the level of ideas but with the level of technology. The Web is still a less than perfect platform for implementing highly interactive learning environments. Here we just provide a review of various technologies which could be used for implementing Web-based learning environments.

To start with, the oldest and the easiest to implement technology is that of a "helper". A helper is an application which is called by the browser to work with a special type of file. The key here is to declare all design artifacts in a particular area as files of a special kind, and to assign a helper—such as a CASE design tool (Lawrence-Fowler & Fowler, 1997), LogicWorks (McCartney et al., 1997), or simply a text editor (Westhead, 1996)—for this type of stand-alone design environment. After that, clicking on a link could launch the desired environment with an example to play with or a partial solution to work on. Alternatively, clicking on a link could just launch the required helper, while an example (Tilbury & Messner, 1997; Lawrence-Fowler & Fowler, 1997) has to be copied from the Web page and pasted into the helper environment. This technology is simple and efficient. The only technical problem here is that all client computers must be properly configured.

The first "pure Web" learning environments were developed with form/CGI technology for algorithm animation (Campbell et al., 1995; Ibrahim, 1994) and program visualization (Brusilovsky et al., 1996). Currently, CGI technology can not be considered relevant for implementation of interactive animations because each step of interaction requires communication with the server. However, if the environment is not providing animated interactive simulation, but only shows the result of executing a program (Brusilovsky et al., 1996) or an experiment (EDesktop, 1997a; EDesktop, 1997b), then this server-based technology could be very relevant.

Currently, with development and maturation, Java is becoming a dominant technology for developing Web-based learning environments, mainly for supporting interactive simulations and animations. There are multiple examples of using Java for interactive and one-shot simulations in such fields as data structures (Marshall & Hurley, 1997; Warendorf & Tan, 1997), microprocessor simulation (Graham et al., 1997; Graham & Trick, 1997), algorithm animation (Ben-Ari, 1997; Bilska et al., 1997; Brown & Najork, 1996; 1997; Procopiuc et al., 1996; Rodger, 1997), simulations for chemistry (Milton et al., 1997), engineering (Kirkpatrick et al., 1997), physics (McKenna & Agogino, 1997) and electronics (Tardif & Zaccarin, 1997). Java could also be used to provide full-scale Web labs for solving a case (Faulhaber & Reinhardt, 1997; Johnson & Shaw, 1997), or running an experiment (Pearce & Livett, 1997). In some cases when a simulation engine already exists in a stand-alone form or is just too big to be completely implemented in Java, the best solution is a combination of Java and CGI technologies where Java is used to support interaction (Faulhaber & Reinhardt, 1997) or to render animation (Naps, 1997) while cycle intensive processes are done on the server side. A very unique example of connecting a Java interface with an external "simulation engine" is described in (Kirkpatrick et al., 1997). Here the simulation engine is a real device, and Java provides a safe interface to play with it over the Web.

With the maturity of Macromedia's Shockwave technology, traditional interactive multimedia authoring tools developed by this company are becoming a very attractive alternative for developing Web labs. While Java stays as the only tool for developing complex Web-based learning environments, Shockwave technology supported by excellent authoring tools could often provide a better choice for developing small and medium size interactive simulations and exercises.

Web-based learning environments are new and labor-consuming. Most of the developments in this area should be considered as research. The only systems that could be placed on the state-of-the-art level are simple Java-based or Shockwave-based animations and simulations. With the current level of authoring support, any reasonable development team could develop it. We have found no examples of Base level Web-based learning environments.

Conclusions

This paper provides an analysis of a group of tools required for building a virtual university, which we call course delivery tools. We have tried to analyze various aspects of currently existing tools and technologies. We have also tried to provide some kind of grading system by categorizing the various tools and systems into three levels of advancement. We expect that this review could help Virtual University practitioners to judge various tools that are available for course delivery. We also hope that the paper will be useful for researchers on Web-based education by helping them to find the place of their work in the overall picture, preventing some teams from "re-inventing the wheel", and showing most attractive and demanding directions of research for new teams.

Notes

1 By *authoring tool* we mean the tools that help put lectures, tests and so forth onto the Web.
2 HTML, HyperText Markup Language, a page definition language, is a defacto standard for Web publishing. Using HTML allows one to publish a Web page and not worry about the computer, operating system or application software sued by its readers.
3 As we stated, authoring issues are left outside of the scope of this paper. Here we provide a very simple classification and a brief review which we need to classify existing Web courses and systems.
4 What You See Is What You Get (WYSIWYG) editors became popular first in text editors when graphics oriented displays emerged (pioneered by Xerox with its Alto system, later commercialized by Apple with the Macintosh). The key feature is that complex editing commands are performed directly and their results (not a marked-up and yet to be compiled text stream) instantaneously appear on the computer's display.
5 Metadata means "data about data". In WBE metadata are used to provide educational information about fragments of learning material such as media files, tests, or lessons. Educational metadata usually includes such fields as author of the fragment, language, difficulty, related topics. Several research groups are working on developing a standard set of educational metadata fields.
6 Common Gateway Interface defines a standard way to transfer data entered by a user into a HTML form from browser to the client, process this data with a special program (called CGI script) on the server, and return the results back to the browser.
7 Navigation, in this context, means the traversal of the electronic text. Just as the reader of a book can read one page at a time, check out the table of contents, use footnotes or rely on an index or glossary the reader of the electronic text can jump from place to place in the electronic document. Good navigation aids guide the reader in this process, making it easy to get to related portions of the text and easy to get back into the main flow of the document.
8 The reader may recognize this hierarchy as the a-cyclical directed graph or "tree"
9 This direction of research is only partly relevant to electronic textbooks
10 Plug-in technology enables independent vendors to extend the browser functionality by developing specially structured programs called *plug-ins*. At start-up time, a browser loads all plug-ins located in a special directory and they become parts of the browser code.
11 Restrictions are usually determined by university policies. For example, a student may not be allowed to see grades of other students or a teacher could be allowed to change the automatically assigned grades.
12 This distinction is empirical. It is possible to have learning environments that are also rich in assessment. But such instances are just now coming onto the scene.

References

Abou Khaled O., Pettenati M. C., Vanoirbeek C. & Coray G. (1998) MEDIT: a Distance education prototype for teaching and learning. *WebNet'98, World Conference of the WWW, Internet, and Intranet*, Orlando, FL, November 7–12, 1998, pp. 503–508.

Abowd G. D., Atkeson C. G., Brotherton J., Enqvist T., Gulley P. & LeMon J. (1998) Investigating the capture, integration and access problem of ubiquitous computing in an educational settings. *CHI'98*, Los Angeles, CA, 18–23 April, 1998, pp. 440–447.

Adams W. J. & Carver C. A., Jr. (1997) The effects of structure on hypertext design. *ED-MEDIA/ED-TELECOM'97 – World Conference on Educational Multimedia/Hypermedia and World Conference on Educational Telecommunications*, Calgary, Canada, June 14–19, 1997, pp. 1–6.

Ahanger G. & Little T. D. C. (1997) Easy Ed: An integration of technologies for multimedia education. *WebNet'97, World Conference of the WWW, Internet and Intranet*, Toronto, Canada, November 1–5, 1997, pp. 15–20.

André E., Rist T. & Müller J. (1997) WebPersona: A Life-Like Presentation Agent for Educational Applications on the World-Wide Web. *Workshop "Intelligent Educational Systems on the World Wide Web" at AI-ED'97, 8th World Conference on Artificial Intelligence in Education*, Kobe, Japan, 18 August 1997, pp. 78–85, http://www.contrib.andrew.cmu.edu/~plb/AIED97_workshop/Andre/Andre.html.

André E., Rist T. & Müller J. (1998) Guiding the user through dynamically generated hypermedia presentations with a life-like character. *International Conference on Intelligent User Interfaces, IUI'98*, San Francisco, CA, January 6–9, 1998, pp. 21–28.

Anlon (1998) IntraKal, Anlon Systems, Inc., http://www.anlon.com/aboutset.html.

Antchev K., Luhtalahti M., Multisilta J., Pohjolainen S. & Suomela K. (1995) A WWW Learning Environment for Mathematics. *4th International World Wide Web Conference*, Boston, USA, December 11–14, 1995, http://www.w3.org/pub/Conferences/WWW4/Papers/89/.

Asymetrix (1999) Librarian, Asymetrix Learning Systems, Inc., Bellevue, WA, http://www.ASYMETRIX.com/products/librarian/.

Avilar (1998) WebMentor, Avilar Technologies, Inc., Laurel, MA, http://avilar.adasoft.com/avilar/msubfrm.html.

Bacher C. & Ottmann T. (1996) Tools and services for authoring on the fly. *ED-MEDIA'96 – World conference on educational multimedia and hypermedia*, Boston, MA, June 17–22, 1996, pp. 7–12.

Barbieri K. & Mehringer S. (1997) Techniques for enhancing Web-based education. *WebNet'97, World Conference of the WWW, Internet and Intranet*, Toronto, Canada, November 1–5, 1997, pp. 27–32.

Ben-Ari M. (1997) Distributed algorithms in Java. *ACM SIGCSE bulletin* **29**(3): 62–64.

Benford S., Burke E. & Foxley E. (1993) Learning to construct quality software with the Ceilidh system. *Software Quality Journal* **2**: 177–197.

Bilska A. O., Leider K. H., Procopiuc M., Procopiuc O., Rodger S. H., Salemme J. R. & Tsang E. (1997) A collection of tools for making automata theory and formal languages come alive. *SIGCSE Bulletin* **29** (1): 15–19.

Blackboard (1999) CourseInfo, Blackboard Inc., http://company.blackboard.com/CourseInfo/index.html.

Bogley W. A., Dorbolo J., Robson R. O. & Sechrest J. A. (1996) New pedagogies and tools for Web based calculus. *WebNet'96, World Conference of the Web Society*, San Francisco, CA, October 15–19, 1996, pp. 33–39.

Brok E. (1997) Hypertext and semantic nets: Experiences with modular authoring. *ED-MEDIA/ED-TELECOM'97 – World Conference on Educational Multimedia/Hypermedia and World Conference on Educational Telecommunications*, Calgary, Canada, June 14–19, 1997, pp. 106–111.

Brown D. J. (1997) Writing Web-based questions with Mallard. *FIE'97, Frontiers in Education Conference*, Pittsburgh, PA, November 5–8, 1997, pp. 1502.

Brown D. J. & Mallard Team (1998) Mallard, University of Illinois at Urbana-Champaign, Urbana, IL, http://www.cen.uiuc.edu/Mallard/.

Brown M. H. & Najork M. A. (1996) Collaborative active textbooks: A Web-based algorithm animation system for an electronic classroom. *IEEE Symposium on Visual Languages (VL'96)*, Boulder, CO, Sept. 3–6, 1996, pp. 266–275, http://www.research.digital.com/SRC/JCAT/vl96.

Brown M. H. & Najork M. A. (1997) Collaborative active textbooks. *Journal of Visual Languages and Computing* **8**(4): 453–486.

Brusilovsky P., Eklund J. & Schwarz E. (1998) Web-based education for all: A tool for developing adaptive courseware. *Computer Networks and ISDN Systems* **30**(1–7): 291–300.

Brusilovsky P. & Schwarz E. (1997) Concept-based navigation in educational hypermedia and its implementation on WWW. *ED-MEDIA/ED-TELECOM'97—World Conference on Educational Multimedia/Hypermedia and World Conference on Educational Telecommunications*, Calgary, Canada, June 14–19, 1997, pp. 112–117.

Brusilovsky P., Schwarz E. & Weber G. (1996) ELM-ART: An intelligent tutoring system on World Wide Web. In: Frasson C., Gauthier G. & Lesgold A. (eds.) *Intelligent Tutoring Systems. Lecture Notes in Computer Science,* Vol. **1086** pp. 261–269.

Bryc W. & Pelikan S. (1999) Online Exercises System. University of Cincinnati, Cincinnati, OH, http://math.uc.edu/onex/demo.html.

Byrnes R., Debreceny R. & Gilmour P. (1995) The development of a multiple-choice and true-false testing environment on the Web. *Ausweb95: The First Australian World-Wide Web Conference*, Ballina, Australia, April 30–May 2, 1995, http://elmo.scu.edu.au/sponsored/ausweb/ausweb95/papers/education3/byrnes/.

Calvi L. & De Bra P. (1997) Using dynamic hypertext to create multi-purpose textbooks. *ED-MEDIA/ED-TELECOM'97 – World Conference on Educational Multimedia/Hypermedia and World Conference on Educational Telecommunications*, Calgary, Canada, June 14–19, 1997, pp. 130–135.

Campbell J. K., Hurley S., Jones S. B. & Stephens N. M. (1995) Constructing educational courseware using NCSA Mosaic and the World-Wide Web. *Computer Networks and ISDN Systems* **27**(6): 887–895.

Campos Pimentel M. d G., dos Santos Junior J. B. & de Mattos Fortes R. P. (1998) Tools for authoring and presenting structured teaching material in the WWW. *WebNet'98, World Conference of the WWW, Internet, and Intranet*, Orlando, FL, November 7–12, 1998, pp. 194–199.

Campus America (1998) The Learning Manager, Campus America, Calgary, Canada, http://www.campuscan.com/.

Carbone A. & Schendzielorz P. (1997) Developing and integrating a Web-based quiz generator into the curriculum. *WebNet'97, World Conference of the WWW, Internet and Intranet*, Toronto, Canada, November 1–5, 1997, pp. 90–95.

Carver C. A., Howard R. A. & Lavelle E. (1996) Enhancing student learning by incorporating student learning styles into adaptive hypermedia. *ED-MEDIA'96 – World Conference on Educational Multimedia and Hypermedia*, Boston, MA, June 17–22, 1996, pp. 118–123.

Centra (1998) Symposium, Centra Software, Inc., Lexington, MA, http://www.centra.com/product/index.html.

ClassNet Team (1998) ClassNet, Iowa State University, http://classnet.cc.iastate.edu/.

ComputerPREP (1998) WebAssessor, ComputerPREP, Inc, Phoenix, AZ, http://www.webassessor.com.

CourseWeb Team (1997) CourseWeb, Pennsylvania State University, State College, PA, http://projects.cac.psu.edu/CourseWeb/.

Danielson R. (1997) Learning styles, media preferences, and adaptive education. *Workshop "Adaptive Systems and User Modeling on the World Wide Web" at 6th International Conference on User Modeling, UM97*, Chia Laguna, Sardinia, Italy, June 2, 1997, pp. 31–35, http://www.contrib.andrew.cmu.edu/~plb/UM97_workshop/Danielson.html.

Dannenberg R. (1998) Just-in-Time Lectures, Carnegie Mellon University, Pittsburgh.

De Bra P. (1997) Teaching through adaptive hypertext on the WWW. *International Journal of Educational Telecommunications* **3**(2/3): 163–180.

Desmarais M. C. (1998) Self-learning guide, CRIM, Montreal, http://www.crim.ca/hci/demof/gaa/introduction.html.

Docent (1998) Docent, Docent, Inc., Mountain View, CA, http://www.docent.com/contact/index.html.

Dracos N. (1998) LaTeX2HTML, University of Leeds, http://www.cbl.leeds.ac.uk/nicos/tex2html/doc/latex2html/latex2html.html.

EDesktop (1997a) Virtual EarthQuake, California State University, Los Angelos, http://vearthquake.calstatela.edu/edesktop/VirtApps/VirtualEarthQuake/VQuakeIntro.html.

EDesktop (1997b) Virtual Flylab, California State University, Los Angelos, http://vearthquake.calstatela.edu/edesktop/VirtApps/VflyLab/IntroVflyLab.html.

Eisenstadt M. & Domingue J. (1998) KMi Studium, Open University, Milton Keynes.

Eklund J (1995) Cognitive models for structuring hypermedia and implications for learning from the world-wide web. *Ausweb95: The First Australian World-Wide Web Conference*, Ballina, Australia, April 30–May 2, 1995, pp. 111–116, http://elmo.scu.edu.au/sponsored/ausweb/ausweb95/papers/hypertext/eklund/index.html.

Eliot C., Neiman D. & Lamar M. (1997) Medtec: A Web-based intelligent tutor for basic anatomy. *WebNet'97, World Conference of the WWW, Internet and Intranet*, Toronto, Canada, November 1–5, 1997, pp. 161–165.

Faulhaber S. & Reinhardt B. (1997) *D3–WWW-Trainer: Entwiklung einer Oberfläche für die Netzanwendung*. München: Technische Universität München, 31–40.

Fisher B., Conway K. & Groeneboer C. (1997) Virtual-U developmental plan: Issues and process. *ED-MEDIA/ED-TELECOM'97 – World Conference on Educational Multimedia/Hypermedia and World Conference on Educational Telecommunications*, Calgary, Canada, June 14–19, 1997, pp. 352–357.

Forte E., Forte M. W. & Duval E. (1996a) ARIADNE: A framework for technology-based open and distance lifelong education. *7th EAEEIE International Conference "Telematics for Future Education and Training"*, Oulu, FInland, June 12–14, 1996, pp. 69–72.

Forte E., Forte M. W. & Duval E. (1996b) ARIADNE: A supporting framework for technology-based open and distance lifelong education. *Educating the engineer for lifelong learning. SEFI Annual Conference '96*, Vienna, Austria, September 11–13, 1996, pp. 137–142.

Foxley E., Benford S., Burke E. & CEILLIDH Development Team (1998) WebCeilidh, University of Nottingham, http://www.cs.nott.ac.uk/~ceilidh.

Freeman H. & Ryan S. (1997) Webmapper, a tool for planning, structuring and delivering courseware on the Internet. *ED-MEDIA/ED-TELECOM'97 – World Conference on Educational Multimedia/Hypermedia and World Conference on Educational Telecommunications*, Calgary, Canada, June 14–19, 1997, pp. 372–377.

Fröhlich P. & Nejdl W. (1997) A database-oriented approach to the design of educational hyperbooks. *Workshop "Intelligent Educational Systems on the World Wide Web" at AI-ED'97, 8th World Conference on Artificial Intelligence in Education*, Kobe, Japan, 18 August 1997, pp. 31–39, http://www.contrib.andrew.cmu.edu/~plb/AIED97_workshop/Froehlich/Froehlich.html.

Furuta R., Shipman III F. M., Marshall C. C., Brenner D. & Hsieh H.-w (1997) Hypertext paths and the World-Wide Web: Experience with Walden's paths. *Eight ACM International Hypertext Conference (Hypertext'97)*, Southampton, UK, April 6–11, 1997, pp. 167–176.

Goldberg M. W. (1997a) CALOS: First results from an experiment in computer-aided learning of operating systems. *SIGCSE Bulletin* **29**(1): 48–52.

Goldberg M. W. (1997b) Using a Web-based course authoring tool to develop sophisticated Web-based courses. In: Khan B H (ed.) *Web Based Instruction.* Educational Technology Publications, Englewood Cliffs, New Jersey, pp. 307–312.

Goldberg M. W. (1997c) WebCT and first year: student reaction to and use of a Web-based resource in first year computer science. *ACM SIGCSE bulletin* **29**(3): 127–127.

Goldberg M. W., Salari S. & Swoboda P. (1996) World Wide Web – course tool: An environment for building WWW-based courses. *Computer Networks and ISDN Systems* **28**: 1219–1231.

GoLive (1998) CyberStudio, GoLive Inc., http://www.golive.com/.

Gorp M. J. V. & Boysen P. (1996)ClassNet: Managing the virtual classroom. *WebNet'96, World Conference of the Web Society*, San Francisco, CA, October 15–19, 1996, pp. 457–461.

Graf F. & Schnaider M. (1997) *IDEALS MTS – EIN modilares Training System für die Zukunft.* München: Technische Universität München, 1–12.

Graham C. R., Swafford M. L. & Brown D. J. (1997) Mallard: A Java enhanced learning environment. *WebNet'97, World Conference of the WWW, Internet and Intranet*, Toronto, Canada, November 1–5, 1997, pp. 634–636.

Graham C. R. & Trick T. N. (1997) An innovative approach to asynchronous learning using Mallard: Application of Java applets in a freshman course. *FIE'97, Frontiers in Education Conference*, Pittsburgh, PA, November 5–8, 1997, pp. 238–244.

Hampel T., Keil-Slawik T., Ferber F. & Müller W. H. (1998) How to bring cooperative structures and hypermedia into the field of technical mechanics? – Our experiences. *WebNet'98, World Conference of the WWW, Internet, and Intranet*, Orlando, FL, November 7–12, 1998, pp. 380–386.

Harasim L., Calvert T. & Groenboer C. (1997) Virtual-U: A Web-based system to support collaborative learning. In: Khan B H (ed.) *Web Based Instruction.* Educational Technology Publications, Englewood Cliffs, New Jersey, pp. 149–158.

Harlequin (1998) WebMaker™, Harlequin Inc, Cambridge, MA, http://www.harlequin.com/products/dpp/webmaker/.

Harris D. & DiPaolo A. (1996) Advancing asynchronous distance education using high-speed networks. *IEEE Transactions on Education* (August).

Hauck F. J. (1996) Supporting hierarchical guided tours in the World Wide Web. *5th International World Wide Web Conference*, Paris, France, May 6–10, 1996, http://www5conf.inria.fr/fich_html/papers/P30/Overview.html.

Hauswirth M. (1995) Pooh's guided tour service Wien, http://www.infosys.tuwien.ac.at/GuidedTour/GuidedTour.html.

Hector C. (1998) RTFtoHTML 4.10, Sunrise Packaging Inc., Blaine, MN, http://www.sunpack.com/RTF/.

Hitz M. & Kögeler S. (1997) Teaching C++ on the WWW. *ACM SIGCSE bulletin* **29**(3): 11–13.

Holtz N. M. (1995) The tutorial gateway, Carleton University, Ottawa, CA, http://www.civeng.carleton.ca/~nholtz/tut/doc/doc.html.

Hubler A. W. & Assad A. M. (1995) CyberProf: An intelligent human–computer interface for asynchronous wide area training and teaching. *4th International World Wide Web Conference*, Boston, USA, December 11–14, 1995, http://www.w3.org/pub/Conferences/WWW4/Papers/247/.

Hubler A. W. & CyberProf Team (1998) CyberProf, University of Illinois at Urbana-Champaign, Urbana, IL, http://prof.ccsr.uiuc.edu/.

Ibrahim B. (1994) World-Wide algorithm animation. *Computer Networks and ISDN Systems* **27**(2): 255–265.

IDEALS Project (1998) MTS, Zentrum für Graphische Datenverarbeitung e. V., Darmstadt, Germany, http://ideals.zgdv.de/.

IMG (1998) Web University, The Information Management Group, Chicago, IL, http://www.imginc.com/webuniversity/.

Ingebritsen T., Brown G. & Pleasants J. (1997) Teaching biology on the Internet. *WebNet'97, World Conference of the WWW, Internet and Intranet*, Toronto, Canada, November 1–5, 1997, pp. 235–240.

Instructional Toolkit Team (1997) Instructional Toolkit, University of Virginia, Charlottesville, VA, http://cti.itc.Virginia.EDU/toolkit2/.

Jafari A., Mills D. & ONcourse R&D Team (1998) ONcourse, Indiana University, http://www.weblab.iupui.edu/projects/ONcourse.html.

Jehng J.-C. J., Tung S.-H. S. & Chang C.-T. (1997) Visualization strategies for learning recursion. *ED-MEDIA/ED-TELECOM'97 – World Conference on Educational Multimedia/Hypermedia and World Conference on Educational Telecommunications*, Calgary, Canada, June 14–19, 1997, pp. 532–538.

Johnson W. L. & Shaw E. (1997) Using agents to overcome deficiencies in Web-based courseware. *Workshop "Intelligent Educational Systems on the World Wide Web" at AI-ED'97, 8th World Conference on Artificial Intelligence in Education*, Kobe, Japan, 18 August 1997, pp. 70–77, http://www.contrib.andrew.cmu.edu/~plb/AIED97_workshop/Johnson/Johnson.html.

Johnson W. L., Shaw E. & Ganeshan R. (1998) Pedagogical agents on the Web. *Workshop "WWW-Based Tutoring" at 4th International Conference on Intelligent Tutoring Systems (ITS'98)*, San Antonio, TX, August 16–19, 1998, http://www.isi.edu:80/isd/ADE/ITS98–WW.htm.

Jühne J., Jensen A. T. & Grønbæk K. (1998) Ariande: a Java-based guided tour system for the World Wide Web. *Computer Networks and ISDN Systems* **30**(1–7): 131–139.

Kashy E., Thoennessen M., Tsai Y., Davis N. E. & Wolfe S. L. (1997) Using networked tools to enhanse student success rates in large classes. *FIE'97, Frontiers in Education Conference*, Pittsburgh, PA, November 5–8, 1997, pp. 233–237.

Kay J. & Kummerfeld R. (1994a) Adaptive hypertext for individualised instruction. *Workshop on Adaptive Hypertext and Hypermedia at Fourth International Conference on User Modeling*, Hyannis, MA, Aug 17, 1994, http://www.cs.bgsu.edu/hypertext/adaptive/Kay.html.

Kay J. & Kummerfeld R. J. (1994b) An individualised course for the C programming language. *Second International WWW Conference*, Chicago, IL, 17–20 October, 1994, http://www.ncsa.uiuc.edu/SDG/IT94/Proceedings/Educ/kummerfeld/kummerfeld.html.

Khuwaja R., Desmarais M. & Cheng R. (1996) Intelligent Guide: Combining user knowledge assessment with pedagogical guidance. In: Frasson C, Gauthier G & Lesgold A (eds.) *Intelligent Tutoring Systems. Lecture Notes in Computer Science,* Vol. **1086** pp. 225–233.

Kirkpatrick A., Lee A. & Willson B. (1997) The engine in engineering – development of thermal/fluids Web based education. *FIE'97, Frontiers in Education Conference*, Pittsburgh, PA, November 5–8, 1997, pp. 744–747.

Klevans R. L. (1997a) Automatically generated Web-based multimedia presentations: The Web Lecture System (WLS) North Carolina State University, Raleigh, NC, http://renoir.csc.ncsu.edu/WLS/white_paper/wls.htm.

Klevans R. L. (1997b) The Web lecture system (WLS). *WebNet'97, World Conference of the WWW, Internet and Intranet*, Toronto, Canada, November 1–5, 1997, pp. 669–670.

Knowledge Navigators (1999) LearningEngine™, Knowledge Navigators International, Inc, Halifax, Canada, http://knav.com/content.htm.

Lam M. M. & Hubler A. W. (1997) NetworkTA: The CyberProf conferencing system, a complex systematic approach to creating a collaborative learning environment. *ED-MEDIA/ED-TELECOM'97 – World Conference on Educational Multimedia/Hypermedia and World Conference on Educational Telecommunications*, Calgary, Canada, June 14–19, 1997, pp. 613–618.

Langenbach C. & Bodendorf F. (1997) A framework for WWW-based learning with flexible navigation guidance. *WebNet'97, World Conference of the WWW, Internet and Intranet*, Toronto, Canada, November 1–5, 1997, pp. 293–298.

Langenbach C. & Bodendorf F. (1998) An education broker toolset for Web course customization. *Journal of Universal Computer Science* **4**(10): 780–791.

LaRose R. & Gregg J. (1997) An evaluation of a Web-based distributed learning environment for higher education. *ED-MEDIA/ED-TELECOM'97 – World Conference on Educational Multimedia/Hypermedia and World Conference on Educational Telecommunications*, Calgary, Canada, June 14–19, 1997, pp. 1286–1287.

Lawrence-Fowler W. A. & Fowler R. H. (1997) Heterogeneous information resources and asynchronous workgroups: Creating a focus on information discovery and intergation in computer science. *FIE'97, Frontiers in Education Conference*, Pittsburgh, PA, November 5–8, 1997, pp. 833–837.

Lee S. H. & Wang C. J. (1997) Intelligent hypermedia learning system on the distributed environment. *ED-MEDIA/ED-TELECOM'97 – World Conference on Educational Multimedia/Hypermedia and World Conference on Educational Telecommunications*, Calgary, Canada, June 14–19, 1997, pp. 625–630.

Lemone K. (1997) Experiences in virtual teaching. *WebNet'97, World Conference of the WWW, Internet and Intranet*, Toronto, Canada, November 1–5, 1997, pp. 306–311.

Lotus (1999) LearningSpace, Lotus, Cambridge, MA, http://www.lotus.com/products/learningspace.nsf.

MacDougall G. (1997) The Acadia advantage adademic development centre and the authomatic courseware management systems. *ED-MEDIA/ED-TELECOM'97 – World Conference on Educational Multimedia/Hypermedia and World Conference on Educational Telecommunications*, Calgary, Canada, June 14–19, 1997, pp. 647–652.

Macromedia (1999a) Authorware Attain Web Player, Macromedia, Inc., http://www.macromedia.com/software/authorware/productinfo/webplayer/.

Macromedia (1999b) Shockwave player, Macromedia, Inc., http://www.macromedia.com/shockwave/.

madDuck (1997) Web course in a box (WCB), madDuck Technologies, Richmond, VA, http://www.madduck.com/wcbinfo/wcb.html.

Marshall A. D. & Hurley S. (1996a) Delivering hypertext-based courseware on the World-Wide-Web. *Journal of Universal Computer Science* **2**(12): 805–828.

Marshall A. D. & Hurley S. (*1996b)* The design, development and evaluation of hypermedia courseware for the World Wide Web. *Multimedia Tools and Applications* **3**(1): 5–31.

Marshall A. D. & Hurley S. (1997) Courseware development for parallel programming and C programming. *ED-MEDIA/ED-TELECOM'97 – World Conference on Educational Multimedia/Hypermedia and World Conference on Educational Telecommunications*, Calgary, Canada, June 14–19, 1997, pp. 689–697.

McCartney R., Weiner B. & Wurst K. R. (1997) Delivering a lab course in a Web-based learning environment. *FIE'97, Frontiers in Education Conference*, Pittsburgh, PA, November 5–8, 1997, pp. 849–855.

McKeever S., McKeever D. & Elder J. (1997) An authoring tool for constructing interactive exercises. *WebNet'97, World Conference of the WWW, Internet and Intranet*, Toronto, Canada, November 1–5, 1997, pp. 695–696.

McKenna A. & Agogino A. (1997) Engineering for middle schools: A Web-based module for learning and designing with simple machines. *FIE'97, Frontiers in Education Conference*, Pittsburgh, PA, November 5–8, 1997, pp. 1496–1501.

Mehringer S. & Lifka D. (1998) The virtual workshop companion: a Web interface for online labs. *WebNet'98, World Conference of the WWW, Internet, and Intranet*, Orlando, FL, November 7–12, 1998, pp. 647–652.

Mentorware (1998) Enterprise Education Server, Mentorware, Inc., Sunnyvale, CA, http://www.mentorware.com/.

Merat F. L. & Chung D. (1997) World Wide Web approach to teaching microprocessors. *FIE'97, Frontiers in Education Conference*, Pittsburgh, PA, November 5–8, 1997, pp. 838–841.

Milton J., Thomas B. & Yaron D. (1997) Java programs for physical chemistry, Carnegie Mellon University, Pittsburgh, PA, http://www.chem.cmu.edu/milton/Gamelan/.

Nakabayashi K., Maruyama M., Kato Y., Touhei H. & Fukuhara Y. (1997) Architecture of an intelligent tutoring system on the WWW. In: Boulay B d & Mizoguchi R (eds.) *Artificial Intelligence in Education: Knowledge and Media in Learning Systems*. IOS, Amsterdam, pp. 39–46.

Nakabayashi K., Maruyama M., Koike Y., Fukuhara Y. & Nakamura Y. (1996) An intelligent tutoring system on the WWW supporting interactive simulation environments with a multimedia viewer control mechanism. *WebNet'96, World Conference of the Web Society*, San Francisco, CA, October 15–19, 1996, pp. 366–371, http://www.contrib.andrew.cmu.edu/~plb/WebNet96.html.

Naps T L (1997) Algorithm visualization on the World Wide Web – the difference Java makes. *ACM SIGCSE bulletin* **29**(3): 59–61.

NetObjects (1998) NetObjects Fusion, NetObjects, Inc., http://www.netobjects.com/products/html/nof.html.

Neumann G. & Zirvas J. (1998) SKILL – A scallable internet-based teaching and learning system. *WebNet'98, World Conference of the WWW, Internet, and Intranet*, Orlando, FL, November 7–12, 1998, *pp. 688–693*, http://nestroy.wi-inf.uni-essen.de/Forschung/Publikationen/skill-webnet98.ps.

Ni Y., Zhang J. & Cooley D. H. (1997) NetTest: An integrated Web-based test tool. *WebNet'97, World Conference of the WWW, Internet and Intranet*, Toronto, Canada, November 1–5, 1997, pp. 710–711.

Nicol D., Smeaton C. & Slater A. F. (1995) Footsteps: Trail-blazing the Web. *Third International World-Wide Web Conference*, Darmstadt, 10–14 April, 1995, pp. 879–885.

Nykänen O. (1999) Visualizing navigation in educational WWW hypertext by introducing partial order and hierarchy. *ED-MEDIA/ED-TELECOM'99 – World Conference on Educational Multimedia/Hypermedia and World Conference on Educational Telecommunications*, Seattle, WA, June 19–24., http://matriisi.ee.tut.fi/~onykane/papers/navtutor/2444.html.

Okazaki Y., Watanabe K. & Kondo H. (1997) An implementation of the WWW based ITS for guiding differential calculations. *Workshop "Intelligent Educational Systems on the World Wide Web" at AI-ED'97, 8th World Conference on Artificial Intelligence in Education*, Kobe, Japan, 18 August 1997, pp. 18–25.

Owen C. B. & Makedon F. (1997) ASML: Web authoring by site, not by hidnsight. *ED-MEDIA/ED-TELECOM'97 – World Conference on Educational Multimedia/Hypermedia and World Conference on Educational Telecommunications*, Calgary, Canada, June 14–19, 1997, pp. 677–682.

Pearce J. M. & Livett M. K. (1997) Real-world physics: a Java-based Web environment for the study of physics. *AUSWEB'97, The Third Australian World Wide Web Conference*, Queensland, Australia, July 5–9, 1997, pp. 211–215, http://ausweb.scu.edu.au/proceedings/eklund/paper.html.

Pilar da Silva D., Durm R. V., Duval E. & Olivié H. (1998) *Concepts and documents for adaptive educational hypermedia: a model and a prototype*. Computing Science Reports, Eindhoven: Eindhoven University of Technology, 35–43.

Plant D. (1998) *Shockwave! Breathe new life into your web pages*. Ventana Communications Group.

Pohjolainen S., Multisilta J. & Antchev K. (1997) Matrix algebra with hypermedia. *Education and Information Technologies* **1**(2): 123–141.

POLIS Team (1998) POLIS, University of Arizona, Lexington, KY, http://www.u.arizona.edu/ic/polis.

Procopiuc M., Procopiuc O. & Rodger S. H. (1996) Visualization and Interaction in the Computer Science Formal Languages Course with JFLAP. In: Vol. 29 IEEE, Salt Lake City, Utah, pp. 121–125.

Question Mark (1998) Perception, Question Mark Corporation, Stamford, CT, http://www.question-mark.com/.

QuestWriter Team (1998) QuestWriter, Oregon State University, http://iq.orst.edu/doc/final/QWHome.html.

Radhakrishnan S. & Bailey J. E. (1997) Web-based educational media: Issues and empirical test of learning. *WebNet'97, World Conference of the WWW, Internet and Intranet*, Toronto, Canada, November 1–5, 1997, pp. 400–405.

Raineri D. M., Mehrtens B. G. & Hubler A. W. (1997) CyberProf™ – An intelligent human–computer interface for interactive instruction on the World Wide Web. *Journal of Asynchronous Learning Networks* **1**(2): 20–36.

Rebelsky S. A. (1997) CourseWeaver: A tool for building course-based Webs. *ED-MEDIA/ED-TELECOM'97 – World Conference on Educational Multimedia/Hypermedia and World Conference on Educational Telecommunications*, Calgary, Canada, June 14–19, 1997, pp. 881–886.

Reer K. A. (1988) The TRY system or How to avoid testing student's programs. *ACM SIGCSE bulletin* **20**(1): 112–116.

Rehak D. (1997a) A database architecture for Web-based distance education. *WebNet'97, World Conference of the WWW, Internet and Intranet*, Toronto, Canada, November 1–5, 1997, pp. 418–425.

Rehak D. R. (1997b) Carnegie Mellon Online: Web-mediated education. *FIE'97, Frontiers in Education Conference*, Pittsburgh, PA, November 5–8, 1997, pp. 1510–1516.

Rios A., Pérez de la Cruz J. L. & Conejo R. (1998) SIETTE: Intelligent evaluation system using tests for TeleEducation. *Workshop "WWW-Based Tutoring" at 4th International Conference on Intelligent Tutoring Systems (ITS'98)*, San Antonio, TX, August 16–19, 1998, http://www-aml.cs.umass.edu/~stern/webits/itsworkshop/rios.html.

Ritter S. (1997) Pat Online: A model-tracing tutor on the World-wide Web. *Workshop "Intelligent Educational Systems on the World Wide Web" at AI-ED'97, 8th World Conference on Artificial Intelligence in Education*, Kobe, Japan, 18 August 1997, pp. 11–17, http://www.contrib.andrew.cmu.edu/~plb/AIED97_workshop/Ritter/Ritter.html.

Rodger S. H. (1997) Visual and interactive tools: JFlap, visual and interactive Tools JFLAP, Pâté, JAVAA. PA, Duke University, Durham, NC, http://www.cs.duke.edu/~rodger/tools/tools.html.

Routen T. & Graves A. (1997a) FLAX: Interactive courseware on the Web, De Montfort University, http://www.cms.dmu.ac.uk/coursebook/flax/.

Routen T. W. & Graves A. (1997b) Activating the Student: Authoring Interactive Exercises in Java with Flax http://www.cms.dmu.ac.uk/~twr/flaxp/activating.html.

Routen T. W., Graves A. & Ryan S. A. (1997) Flax: Provision of interactive courseware on the Web. *Cognition and the Web '97*, pp. 149–157, http://www.cms.dmu.ac.uk/coursebook/flax/.

Sapir M. (1999) The WebTester and the linear algebra WebNotes. *ALN Magazine* **3**(1).

Schwarz E., Brusilovsky P. & Weber G. (1996) World-wide intelligent textbooks. *ED-TELECOM'96 – World Conference on Educational Telecommunications*, Boston, MA, June 17–22, 1996, pp. 302–307, http://www.contrib.andrew.cmu.edu/~plb/ED-MEDIA-96.html.

Serf Team (1998) Serf, University of Delaware, Newark, DE, http://www.udel.edu/serf/.

Severance C. (1998) Sync-O-Matic 3000, Michigan State University, East Lansing, http://www.egr.msu.edu/~crs/projects/syncomat/.

Shipman III F. M., Furuta R., Brenner D., Chung C.-C. & Hsieh H.-w. (1998) Using paths in the classroom: Experience and adaptations. *Ninth ACM International Hypertext Conference (Hypertext'98)*, Pittsburgh, USA, June 20–24, 1998, pp. 267–276.

Shipman III F. M., Marshall C. C., Furuta R., Brenner D. A., Hsieh H.-W. & Kumar V. (1997) Using networked information to create educational guided paths. *International Journal of Educational Communications* **3**(4): 383–400.

Siviter P. (1997) Authoring tools for courseware on WWW: the W3Lessonware project. *WebNet'97, World Conference of the WWW, Internet and Intranet*, Toronto, Canada, November 1–5, 1997, pp. 513–516.

Smeaton A. F. & Crimmons F. (1997) Virtual lectures for undergraduate teaching: Delivery using real audio and the WWW. *ED-MEDIA/ED-TELECOM'97 – World Conference on Educational Multimedia/Hypermedia and World Conference on Educational Telecommunications*, Calgary, Canada, June 14–19, 1997, pp. 990–995.

Specht M. & Oppermann R. (1999) ACE – Adaptive courseware environment. *The New Review of Hypermedia and Multimedia* **4**: 141–161.

Specht M., Weber G., Heitmeyer S. & Schöch V. (1997) AST: Adaptive WWW-courseware for statistics. *Workshop "Adaptive Systems and User Modeling on the World Wide Web" at 6th International Conference on User Modeling, UM97*, Chia Laguna, Sardinia, Italy, June 2, 1997, pp. 91–95, http://www.contrib.andrew.cmu.edu/~plb/UM97_workshop/Specht.html.

Stanford Online Team (1998) Stanford Online, Stanford University, http://stanford-online.stanford.edu/.

Stern M., Steinberg J., Lee H. I., Padhye J. & Kurose J. (1997a) MANIC: Multimedia asynchronous networked individualized courseware. *ED-MEDIA/ED-TELECOM'97 – World Conference on Educational Multimedia/Hypermedia and World Conference on Educational Telecommunications*, Calgary, Canada, June 14–19, 1997, pp. 1002–1007.

Stern M., Woolf B. P. & Kuroso J. (1997b) Intelligence on the Web? In: Boulay B d & Mizoguchi R (eds.) *Artificial Intelligence in Education: Knowledge and Media in Learning Systems*. IOS, Amsterdam, pp. 490–497.

Stern M. K. (1997) The difficulties in Web-based tutoring, and some possible solutions. *Workshop "Intelligent Educational Systems on the World Wide Web" at AI-ED'97, 8th World Conference on Artificial Intelligence in Education*, Kobe, Japan, 18 August 1997, pp. 1–10, http://www.contrib.andrew.cmu.edu/~plb/AIED97_workshop/Stern.html.

Suthers D. & Jones D. (1997) An architecture for intelligent collaborative educational systems. In: Boulay B. d & Mizoguchi R. (eds.) *Artificial Intelligence in Education: Knowledge and Media in Learning Systems*. IOS, Amsterdam, pp. 55–62.

Synnes K., Lachapelle S., Parnes P. & Schefström D. (1998) Distributed education using the mStar environment. *Journal of Universal Computer Science* **4**(10): 803–823.

Tardif P.-M. & Zaccarin A. (1997) An interactive Java applet for teaching discrete-time system theory. *FIE'97, Frontiers in Education Conference*, Pittsburgh, PA, November 5–8, 1997, pp. 1517–1520.

Thimbleby H. (1997) Gentler: a tool for systematic web authoring. *International Journal on Human-Computer Studies* **47**(1).

Tilbury D. & Messner W. (1997) Development and integration of Web-based software tutorials for an undergraduate curriculum: Control tutorials for Matlab. *FIE'97, Frontiers in Education Conference*, Pittsburgh, PA, November 5–8, 1997, pp. 1070–1075.

Titus A. P., Martin L. W. & Beichner R. J. (1998) Web-based testing in physics education: Methods and opportunities. *Computers in Physics* **12** (Mar/Apr): 117–123.

Trigg R. H. (1988) Guided tours and tabletops: Tools for communicating in a hypertext environment. *ACM Transactions on Office Information Systems* **6**(4): 398–414.

Vassileva J. (1997) Dynamic course generation on the WWW. In: Boulay B. d & Mizoguchi R. (eds.) *Artificial Intelligence in Education: Knowledge and Media in Learning Systems*. IOS, Amsterdam, pp. 498–505.

Virtual-U Research Team (1997) Virtual-U, Simon Fraser University, Vancouver, http://virtual-u.cs.sfu.ca/vuweb/.

Wagner R. (1998) HyperStudio, Roger WagnerPublishing, Inc., http://www.hyperstudio.com/.

Warendorf K. & Tan C. (1997) ADIS – An animated data structure intelligent tutoring system or Putting an interactive tutor on the WWW. *Workshop "Intelligent Educational Systems on the World Wide Web" at AI-ED'97, 8th World Conference on Artificial Intelligence in Education*, Kobe, Japan, 18 August 1997, pp. 54–60, http://www.contrib.andrew.cmu.edu/~plb/AIED97_workshop/Warendorf/Warendorf.html.

WBT Systems (1999) TopClass, WBT Systems, Dublin, Ireland, http://www.wbtsystems.com/.

WebCT (1999) World Wide Web Course Tools, WebCT Educational Technologies, Vancouver, Canada, http://www.webct.com.

Weber G., Brusilovsky P. & Specht M. *Submitted* ELM-ART: An adaptive educational system on the WWW. *User Modeling and User-Adapted Interaction.*

Weber G. & Specht M. (1997) User modeling and adaptive navigation support in WWW-based tutoring systems. In: Jameson A, Paris C & Tasso C (eds.) *User Modeling*. Springer-Verlag, Wien, pp. 289–300.

Westhead M. D. (1996) EPIC: Building a structured learning environment. *WebNet'96, World Conference of the Web Society*, San Francisco, CA, October 15–19, 1996, pp. 480–485.

Whitehurst G. J. (1997) CourseNet, Worth Publishers, New York, http://129.49.71.236/coursenet.htm, http://www.whitehurst.sbs.sunysb.edu/Coursenet/CourseNet/default.asp.

Wright C. & Jones R. (1997) A Novel Architecture for WWW-based learning: the implementation of guided tours on the WWW. *5th Annual Conference on the Teaching of Computing*, Dublin, Ireland, August 1997, pp. 263—265.

Xiao G. (1999) WIMS, Université de Nice, Sophia Antipolis, http://wims.unice.fr/~wims/wims.cgi?lang=en.

Yang J. C. & Akahori K. (1997) Development of computer assisted language learning system for Japanese writing using natural language processing techniques: A study on passive voice. *Workshop*

"Intelligent Educational Systems on the World Wide Web" at AI-ED'97, 8th World Conference on Artificial Intelligence in Education, Kobe, Japan, 18 August 1997, pp. 99–103, http://www.contrib.andrew.cmu.edu/~plb/AIED97_workshop/Yang/Yang.html.
Yoon I., Mah P., Cho C. & Shin G. (1997) An adaptive visual programming tool for Web application. *WebNe '97, World Conference of the WWW, Internet and Intranet*, Toronto, Canada, November 1–5, 1997, pp. 576–581.

7

Digital Libraries and Virtual Universities

CHRISTINE L. BORGMAN

Introduction

Universities are places of learning and libraries are places to provide access to information in support of learning. How do these functions change when universities become virtual and libraries become digital? The tension between places and spaces is profoundly evident in university libraries. These are institutions charged with preserving the past, serving the needs of the present, and inventing new information service frameworks for the future. They are expected to be many more things to many more people, and often with few, if any, additional resources. Librarians find themselves living in "interesting times," confronting an array of new technical, economic, managerial, cultural, and political challenges.

As Tschang notes elsewhere in this volume, virtual universities have the potential for student-centered learning, access to almost unlimited information resources, and interaction with globally-distributed communities. Digital libraries are a mechanism for providing access to distributed information resources in support of these and other models for learning, instruction, and independent information-seeking. Digital libraries also have the potential to support the cycle of information creation, use, and seeking. In some senses, university libraries are becoming digital libraries because they are the agencies responsible for providing access to information resources in digital form.

Most of the literature specific to digital libraries is found in journals (e.g. special issues of the *Journal of the American Society for Information Science, Communications of the ACM, Information Processing & Management*) and conference proceedings (e.g. *ACM Digital Libraries, Advances in Digital Libraries*). The first volumes of a new book series on digital libraries and electronic publishing are appearing (Arms, 1999; Borgman, 2000a; Bishop, Buttenfield, & Van House, 2000), and many new books are being published on related topics such as the transformation of university libraries, electronic publishing, and scholarly communication (Criddle, Dempsey, & Heseltine, 1999; Dowler, 1997; Hawkins & Battin, 1998; Graubard & LeClerc, 1998; Meadows, 1998; Newby & Peek, 1996). These are areas of growing interest, and as is typical of new topics, definitions tend to be fuzzy and boundaries tend to be porous. The first section of this chapter explores the emergence of digital libraries as an area of research and practice, including a subsection on definitions of the term "digital libraries." Subsequent sections explore the relationship between digital libraries and virtual universities, the evolution of university libraries, and trends that will influence the usability of information technologies in educational settings.

Evolution of Digital Libraries

Scholarly and professional interest in digital libraries grew rapidly throughout the 1990s. In the United States, digital libraries (DL) were designated a "national challenge application area" under the High Performance Computing and Communications Initiative (HPCC), and a key component of the National Information Infrastructure (Office of Science and Technology Policy, 1994). The Digital Library Initiative (1994–98) involved three U.S. federal agencies: the National Science Foundation, Computer and Information Science and Engineering Directorate; the Advanced Research Projects Agency Computing Systems Technology Office and the Software and Intelligent Systems Technology Office; and the National Aeronautics and Space Administration. The Digital Libraries Initiative—Phase II (1998–2003) involves nine agencies (the original three agencies plus the National Library of Medicine, the Library of Congress, and the National Endowment for the Humanities, in partnership with the National Archives and Records Administration and the Smithsonian Institution. The Federal Bureau of Investigation joined the DLI partnership to provide additional funding after the initial call for proposals was published), indicating the expansion of interest and scope over this short period of time. The first Digital Libraries—Phase II awards were made in 1999, including some five-year grants that will continue until 2004. In addition, an International Digital Libraries Program managed by the (U.S.) National Science Foundation supports joint work between U.S. researchers and researchers in partner countries; the first awards were made in 1999. A directory of projects funded under DLI-1, DLI-2, and the International Digital Libraries Initiative can be found at (http://www.dli2.nsf.gov).

In the United Kingdom, the Electronic Libraries Programme (eLib) (http://ukoln.bath.ac.uk/elib/) is funded by the Joint Information Systems Committee. Their programmatic focus is on research and development of new information services. In the European Union, research on digital libraries falls under a variety of programs and directorates; one portion of the Second EU-US Conference New Vistas in Transatlantic Scientific and Technical Cooperation (Stuttgart, 1999) was devoted to telelearning and digital libraries, for example. A wide array of digital library research projects are under way in Europe, Asia, and elsewhere, whether under DL-specific funding initiatives or funding from other areas. Libraries are undertaking research and development projects in areas such as digital imaging, digital preservation, electronic publishing, and digital course reserves.

A substantial number of conferences on digital libraries were established during the 1990s. Two series of digital libraries conferences take place annually in the United States, as well as one in Europe, one in Japan, and yet another in southeast Asia. All of these meetings draw international audiences. Papers relevant to digital libraries are presented at meetings in fields as diverse as computer science, information science, librarianship, education, museum studies, archives, communication, sociology, engineering, medicine, geo-sciences, space sciences, and in the arts and humanities. Several new print and online journals on DLs have appeared (e.g. *D-Lib Magazine*, *Journal of Digital Information*). Online distribution lists with news of DL projects proliferate.

Why all of this interest and activity? Did an urgent research and development problem lead to large amounts of grant funding? Did the availability of grant funding create opportunities for a new research area? Did successful research lead to practical developments? Did practical concerns lead to research on solutions? Is digital library research and practice a definable area of interest, or has "digital library" merely become an umbrella term for a wide array of information and technology projects? And specific to the concerns of this book, how are digital libraries related to higher education and virtual universities?

What are digital libraries?

Despite its popularity, "digital library" remains an ambiguous term. Clifford Lynch (1993a) was prescient in noting that the term obscures the complex relationship between electronic information collections and libraries as institutions. Greenberg (1998, p. 106) comments that "the term 'digital library' may even be an oxymoron: that is, if a library is a library, it is not digital; if a library is digital, it is not a library." Battin (1998, p. 276–277) rejects the use of the term "digital library" on the grounds that it is "dangerously misleading." Indeed, a review of definitions reveals that "digital library" describes a variety of entities and concepts (Bishop & Star, 1996; Lesk, 1997; Levy & Marshall, 1995; Lyman, 1996; Lynch & Garcia-Molina, 1995; Waters, 1998; Zhao & Ramsden, 1995).

Of these many definitions, the most succinct one from the computer and information science community originated in a research workshop on scaling and interoperability of digital libraries (Lynch & Garcia-Molina, 1995):

A digital library is a system that provides "a community of users with coherent access to a large, organized repository of information and knowledge."

In contrast, the most succinct definition arising from the community of library practice is that set forth by the Digital Library Federation (Waters, 1998):

> Digital Libraries are organizations that provide the resources, including the specialized staff, to select, structure, offer intellectual access to, interpret, distribute, preserve the integrity of, and ensure the persistence over time of collections of digital works so that they are readily and economically available for use by a defined community or set of communities.

As discussed in more depth elsewhere (Borgman, 1999, 2000a), computer and information science researchers are focusing on digital libraries as networked information systems and as content collected on behalf of user communities, while librarians are focusing more on digital libraries as institutions or services. The interests and concerns of both communities are reflected in a broader, two-part definition that arose from the Social Aspects of Digital Libraries research workshop (Borgman, et al., 1996):

1. Digital libraries are a set of electronic resources and associated technical capabilities for creating, searching, and using information. In this sense they are an extension and enhancement of information storage and retrieval systems that manipulate digital data in any medium (text, images, sounds; static or dynamic images) and exist in distributed networks. The content of digital libraries includes data, metadata that describe various aspects of the data (e.g. representation, creator, owner, reproduction rights), and metadata that consist of links or relationships to other data or metadata, whether internal or external to the digital library.
2. Digital libraries are constructed—collected and organized—by [and for] a community of users, and their functional capabilities support the information needs and uses of that community. They are a component of communities in which individuals and groups interact with each other, using data, information, and knowledge resources and systems. In this sense they are an extension, enhancement, and integration of a variety of information institutions as physical places where resources are selected, collected, organized, preserved, and accessed in support of a user community. These information institutions include, among others, libraries, museums, archives, and schools, but digital libraries also extend and serve other community settings, including classrooms, offices, laboratories, homes, and public spaces.

The above definition moves beyond information retrieval to include the cycle of creating, searching, and using information. Rather than simply collecting content on behalf of user communities, it embeds digital libraries in the activities of those communities, and it encompasses information-related activities of multiple information institutions. This broad definition of the term "digital libraries" is assumed in this chapter and is appropriate to educational applications.

Research and practice

The availability of grant funding for research on digital libraries has attracted scholars and practitioners from a variety of backgrounds, some of whom have minimal prior knowledge of related areas such as information retrieval, computer networks, cataloging and classification, library automation, education, museums, archives, or publishing. Sometimes other research topics were simply relabeled "digital libraries," adding to the confusion. The rapid growth in computer networks, databases, and public awareness has contributed to a bandwagon effect in hot topics such as digital libraries, digital archives, and electronic publishing.

One reason for the confusion of terminology is that research and practice in digital libraries are being conducted concurrently at each stage of the continuum from basic research to implementation. Some people are working on fundamental enabling technologies and theoretical issues, others are working on applications, others are studying social aspects of digital libraries in experimental and field contexts, and yet others are deploying the results of earlier research. Their concerns and foci are understandably different.

The variety of concerns within the digital libraries research community reflects the interdisciplinary nature of the topic. Scholars based in computer science are largely concerned with enabling technologies and networks. Scholars based in library and information science are largely concerned with content, organization, user behavior, and publishing. Those based in education, sociology, or economics are more likely to concern themselves with learning, social context, and economic models, respectively. Topics such as human–computer interaction, interface design, and service delivery cross all of these disciplines and more.

Research libraries are concerned with the evolution of libraries as institutions, the changing nature of the university, the role that libraries play in serving the university community, and how that role is changing with the advent of digital collections and services. Librarians are faced with formulating visions for the future of their institutions and services while managing daily operations that may serve tens of thousands of users.

Research and practice have a symbiotic relationship. Interesting research problems often arise from practice. Scholars attempt to isolate problems for research purposes and then provide solutions to practitioners for implementation. Partnerships between researchers and practitioners are essential to the design of digital library applications for higher education.

Digital Libraries, Higher Education, and Virtual Universities

"Virtual Universities" is another new term that encompasses many meanings. Earlier in this volume, Tschang distinguishes between campus-based classroom environments, open learning environments that serve off-campus or part-time students, and campus-less universities that use Internet technology for their main delivery mode. Only the latter form is considered to be a "virtual university."

However, as he acknowledges, it is technology-based developments in the first two situations that is leading to the emergence of virtual universities. On-campus classrooms increasingly are enhanced with information technologies such as Internet connections, large-screen projection systems, and assorted devices such as video tape and disk players. Computer laboratories and classrooms merge when classrooms have workstations at every student desk. Some classrooms have Internet connections at every desk, enabling students to plug in their own machines to participate in the class session. Such classrooms serve "same time/ same place" courses, and can be adapted to open learning environments for "same time/different place," "different time/same place," or "different time/different place" courses (Besser, 1996; Johansen, 1988). Audio and video connections serving remote classrooms may supplement the technologies required for campus-based instruction. With the advent of Internet-based audio and video communication, individual homes and offices can be connected as well.

Tschang includes "telelearning" and "distance-independent learning" in the category of open learning environments that serve off-campus or part-time students. The term "telelearning" tends to be used in Europe, while "distance-independent learning" is more commonly used in the U.S. For the purposes of this chapter, the term telelearning is preferred both because it is better known internationally and because it is broader, encompassing the use of information technologies on campus as well as over wide-area distributed networks. Digital libraries, when viewed as distributed information systems, are a form of telelearning technology. When viewed in the larger sense of a new organizational form of library, as defined above, digital libraries are an organizational component of a virtual university.

Research on telelearning, virtual universities, and digital libraries has only recently begun to overlap. Much of the research on digital libraries, especially in the United States, has focused on technical aspects such as information retrieval algorithms, organization of collections, and user interfaces. Research and development conducted in university libraries has focused largely on information services aspects of digital libraries, such as electronic course reserves and networked access to databases. Some recent examples of research and development at the intersection of digital libraries and higher education include a project based at the University of California-Berkeley to teach the Catalan language at several campuses (Faulhaber, 1996) and the Alexandria Digital Library at the University of California-Santa Barbara to develop a digital library of geo-referenced information resources for use by students, faculty, and others (Buttenfield & McLafferty, 1996; Hill et al., in press; Smith, T. R. et al.). The latter project is being extended to construct the Alexandria Digital Earth Prototype (ADEPT; http://www.alexandria.sdc.ucsb.edu/adept/), whose use will be studied in undergraduate classrooms in several disciplines. Educational evaluation will be conducted at UCSB and at the University of California, Los Angeles (http://is.gseis.ucla.edu/adept/index.html). A list of other digital library research projects focusing on undergraduate education can be found at (http://www.dli2.nsf.gov).

As other chapters in this volume note, virtual universities are part of a general trend toward greater deployment of information technology in education at all levels. Technology is seen as an answer to several pressing issues. The costs of scholarly publications are spiralling, leading to the current "crisis in scholarly communication" (Hawkins & Battin, 1998). Electronic publishing, online pre-print services, and consortium contracts for access to digital collections are among the many technology-based responses to maintaining and improving library services. Universities also are concerned about population projections indicating that the number of students will soon exceed the capacity of existing institutions of higher education. This trend is variously referred to as "Tidal Wave II," the "Demographic Time Bomb," or the "Baby Boom Echo." Many university administrators are hoping that telelearning technologies can be employed to relieve some of the pressure on classroom space and teaching resources. The trend toward more off-campus learning is not universally welcomed by university faculty (academic staff, in British parlance) or students, as other chapters in this volume also note. Those skeptical of these developments often see telelearning technologies as means to increase teaching loads and devalue the university experience (Winner, 1998).

Telelearning technologies, including digital libraries, serve forms of instruction other than matriculated degree programs, however. The market for higher education is becoming more diverse and complex, as we experience Toffler's (1981) "third wave," in which people will have several careers over a lifetime. Continuing education is essential for staying current with one's field, for career advancement, and for changing careers. Online instruction may be especially well-suited for short courses on specific topics, in addition to its uses in degree programs. Universities increasingly are competing with private educational ventures that offer degrees and certification with extensive use of telelearning technologies.

Another reason for the increased interest in digital libraries and higher education is that the idea has achieved political popularity. A project led by the (U.S.) National Research Council proposed a "digital national library for undergraduate science, mathematics, engineering, and technology education" (Center for Science, Mathematics, and Engineering Education, 1998). President Clinton (January 29, 1999) also proposed funding for "A Digital Library for Education," saying "It is a time to build, to build the America within reach… An America where every child can stretch a hand across a keyboard and reach every book ever written, every painting ever painted, every symphony ever composed."

The political visibility of information technology in education, as well as the widespread deployment of computers and network access in workplaces and homes, has raised concerns about the usability of information technologies. Particularly notable are two recent (U.S.) National Academy of Science reports, *More than Screen Deep: Toward Every Citizen Interfaces to the Nation's Information Infrastructure* (Computer Science and Telecommunications Board, 1997), and

Being Fluent with Information Technology (Commission on Physical Sciences, Mathematics, and Applications, 1999). The first report outlines the difficulties of using today's technologies and lays out a research agenda addressing issues to be resolved for widespread deployment of computers and networks. The second report identifies the needs for information literacy and information technology skills at the undergraduate level, with the goal of a 21st century workforce that is "fluent with information technology."

Technology alone is rarely a solution to political, social, or institutional problems. Researchers, educators, administrators, system developers, and librarians are at the early stages of distinguishing means from ends and challenges from opportunities in the deployment of information technology in higher education. One goal of this book is to move these discussions further along. Research on digital libraries and higher education is only beginning to shift from technological concerns to broader pedagogical and institutional issues. For these reasons, the focus of this chapter is more on problem definitions and research questions than on solutions.

Digital libraries as enabling technologies

The continued expansion of information infrastructure and the penetration of information technology into more aspects of daily activities will require basic and applied research in many disciplines. Just as the frontiers of computing have moved from desktop to mobile computing to embedded systems, digital libraries are themselves becoming "enabling technologies" for many other applications. Contributing materials to DLs is a form of electronic publishing. Telelearning requires content associated with instruction, and DLs are a means to provide distributed access to that content.

Electronic publishing

Electronic publishing is essential to the growth of telelearning and virtual universities, since content must be collected, organized, preserved, and disseminated electronically. Electronic publishing is not universally popular, however. At one extreme, the strongest proponents of electronic publishing claim that it is less expensive than print, offers faster and wider distribution, and will continue to become less expensive and more accessible as technology gets cheaper and better and as computer network penetration grows. If such predictions are correct, electronic publishing could replace print publishing for all but niche materials. A corollary effect would be that libraries and publishers, as we know them, will no longer be needed, because people will acquire their own content directly from computer networks. In contrast, the strongest proponents of print publishing claim that electronic materials are unpleasant and uncomfortable to read on screen, and are at best a means to select content to be printed on paper for useful reading. Print is a proven technology and its portability, combined with the ability to flip through pages, offers advantages as yet unmatched by any electronic form. While even strong proponents of print publishing acknowledge that electronic forms will have an important niche, they note that print publishing continues unabated and that bookstores are proliferating, online and offline. In a hybrid world of print and electronic resources, libraries will continue to select, collect, organize, conserve, preserve, and provide access to content in many media. Similarly, publishers will be needed to provide editorial, marketing, distribution, and other services.

Both positions overstate their cases in defense of their preferred forms of publication. Most such scenarios are ahistorical, not recognizing the continuity of debate over forms of media. Gutenberg's improvements in movable type were decried by those who feared the loss of handcrafted manuscripts (Nunberg, 1996). As Odlyzko (1997) comments, even Plato decried the invention of writing, fearing the loss of oral tradition and of reliance upon human memory. Yet we still have oral traditions, handwriting remains a basic skill of literacy, and hand-crafted books are valued more than ever. Television did not replace radio or newspapers. Rather, these media co-exist with many other mass media, including cable and satellite television and online news and entertainment services, each of which has evolved into its own niche.

Underlying these debates are multiple meanings of the term "electronic publishing." Simple characterizations tend to focus on electronic distribution of content, such as electronic journals, monographs, newsletters, or documents in digital libraries. However, electronic technologies pervade all aspects of the publishing process. Authors write their texts on word processors and send them to their publishers online or on disk. Images, tables, and graphics are likely to be created on computers as well. Even if the authors do not create content initially in electronic forms, most publishers key, scan, or otherwise digitize content for production. Copy-editing, page layout, and other production tasks take place online, regardless of whether the final product appears in print or electronic form. The days of hand-setting type are largely gone, except for specialized presses in the book arts and publishing in countries that lack adequate computing and technical infrastructure.

In sum, most aspects of modern publishing are electronic. Manual production no longer is cost-effective for most mainstream trade and scholarly publishing. Given that so much of the publication process is electronic, why are so many publications, including this one, transferred to paper form for distribution? If the issues surrounding form of publication were merely technological, most content would be distributed and read in electronic form by now. Technology is only one determinant of the most appropriate form to produce and distribute content, however, and multiple forms of print and electronic publication will exist indefinitely.

Telelearning and digital libraries

Digital libraries are an enabling technology for telelearning, as they are a means to provide access to information resources associated with instruction (Faulhaber, 1996). They can support independent, guided, or collaborative learning (Criddle, Dempsey, & Heseltine, 1999; Twidale & Nichols, 1998a, b). Regardless of the mode of instruction, students can use digital collections provided by their university libraries. They can hold any type of content in digital form, whether "born digital" or digitized from another medium such as print, audio, or visual recordings. Their subject matter can be as diverse as any other form of library. Individual digital libraries also can be created in support of specific courses. These may contain materials to be read or used by students, software tools for course projects, other forms of assignments and examinations, and pointers to other resources online and offline.

Digital libraries are well-suited to learning that occurs in other than "same time/ same place" instruction because they support asynchronous interaction (available at anytime) over distributed networks (accessible from any place with a network connection). Rather than students having to visit a library or laboratory in person to use instructional materials, often competing for one or a few copies (or a limited number of laboratory work stations), one digital document can be accessible to multiple students at multiple places at all times. Documents that are independent in physical form can be interdependent in digital form, hyperlinked together. Students and instructors can follow paths between documents linked by citations, common terms, formats, or other relationships. Some of these links are machine generated, while others are created by instructors and students. The scope of linking is not limited to materials gathered for one course. Links can be followed from one digital library to another, following paths to materials in many countries, cultures, and languages. The possibilities for guided learning and self-directed learning are limitless.

Digital libraries can be accessible from the same computers on which students write their papers and compose their assignments, leading to a seamless transition between tasks. When working with paper documents, a student often will photocopy materials, annotate them, take notes on separate paper or note cards, and transcribe those notes before or during the process of writing a paper or assignment. When working with electronic documents, sections of interest can be clipped into a notes file, annotated, and edited into the final document. The completed document is not limited to text on paper. Students can include still and moving images, sounds, animated models, and other digital resources into their work. Far richer and more complex products can be produced, and they can be tailored to the subject matter of the product. Chemistry projects can include animated models of chemical bonding and dance projects can include films of dancers and animated choreography, for example. Students' creativity in using these new tools is evident in the National Digital Library of Theses and Dissertations (http://www.ndltd.org/), which includes masters' theses and doctoral dissertations containing various multimedia content, in addition to the usual textual documentation.

Along with these benefits of digital libraries for telelearning come disadvantages—or at least challenges that have yet to be overcome. The most obvious disadvantage is the expense of creating digital resources for instruction. Materials already in digital form often can be acquired for a fee that includes certain intellectual property rights. Usually such materials are acquired under contracts for specific uses. Rarely would a university obtain digital materials with unlimited rights to re-use the content in other products, for example. If materials of interest are not in digital form, the university or instructor must obtain permission to digitize them. Owning a physical copy of a book, audio recording, videotape, or photograph usually does not include the rights to digitize the content. Over and above the cost of acquiring intellectual property rights, the process of digitizing is expensive and time consuming. Every page to be scanned must be handled, colors must be calibrated, and text should be proofread (especially if text is converted to machine-readable form through optical character recognition). Similarly, audio and video must be played and calibrated. Content must be described, indexed, and organized so that it can be accessed. Parts of these processes can be automated, but considerable manual labor and supervision are required. Unless the digitizing process is contracted out, equipment must be purchased or leased, which also entails costs for space and maintenance.

The use of digital libraries makes instruction both more and less flexible. It is more flexible because of the variety of content that can be offered to students and the variety of ways in which they can use it. It is less flexible because of the overhead of creating digital resources and the need to stabilize and test them before using. Instructors who are accustomed to producing their instructional materials one class session at a time may find that telelearning technologies such as digital libraries constrain their ability to adapt quickly to changing needs. The effort and expense of creating a digital library in support of a course or curriculum is more easily justified when the materials can be reused multiple times. Ironically, digital libraries may be better suited, at least in economic terms, for courses with relatively stable content than for courses with content that evolves rapidly. However, in fast-moving areas of science, technology, and engineering, materials that are "born digital" are readily available. An intermediate approach is to support courses with digital libraries that consist largely of pointers to electronic resources. A list of pointers is essentially a bibliography, but the entries can be active hyperlinks to the full resources.

A growing problem for digital libraries is digital preservation. Given the rate of advances in information technology, maintaining content in a continuously viable form is a major challenge. Most paper documents can be set on a shelf and remain readable for centuries, under proper storage conditions. Magnetic media (computer disks; audio, video, and data tapes; etc.) must be copied every few years to maintain the readability of content, and must be stored properly to ensure long-term readability (Hedstrom, 1998; Van Bogart, 1995). Even if the medium remains viable, finding devices to read older formats can be problematic. Already it is difficult to locate operational devices to read media that were widely distributed only a few years ago, such as 5.25" floppy disks or 33-1/3 rpm phonograph records. Devices to read 8" floppy disks, 78 rpm records, Betamax videotapes, and reel-to-reel film are harder to find. Drives for 3.5" disks already have ceased to be a standard feature of new computers, thus reading these disks will soon be difficult.

Even if the media are readable, finding hardware with the necessary operating systems and application software to read older files can be impractical. Unless files are transferred to the subsequent generation of hardware and software quickly, it is unlikely they ever will be read again. All of the proposed data preservation strategies require active efforts to maintain the data in a readable form, rather than the passive strategies of putting a book on a shelf or microfilm in a storage vault. Thus when universities create digital libraries, they commit themselves to recurring expenses of maintaining electronic content. Digital preservation is an active research area and one of great import for the use of digital libraries in higher education (Foster and Kesselman 1999; Hedstrom, 1991, 1998; Moore, et al. 1999; Rajasekar, et al. 1999; Rothenberg, 1995).

Digital libraries as institutional framework

Digital libraries are an essential enabling technology for electronic publishing and telelearning. They are also an essential part of the institutional framework for universities, virtual and otherwise. University libraries are faced with being three institutions concurrently (Frye, 1997, p. 12–13): "the library of the past, with all of its traditional expectations about building comprehensive collections and providing direct access to printed materials; the library of the present, with the extraordinary added costs of inflation, automation, and preservation of decaying print; and the library of the future, with all the attendant costs of developing and implementing new concepts, prototypes, and technologies for publishing, acquiring, storing, and providing access to information through digital technology."

All three forms of libraries are needed concurrently; none of these roles can be abandoned. The extant physical collections must be maintained for the use of present and future scholars. Access to physical collections must be substantially improved through the use of information technologies. Online catalogs and indexes provide access superior to that of card catalogs and printed indexes, because they offer more sophisticated searching capabilities, are available via distributed networks, and can combine multiple functions. For example, online catalogs are combined with acquisition and circulation systems to identify what materials are owned, on-order, and available (whether "on shelf" or "on loan" to another user). Similarly, indexing and abstracting databases can be linked with full content databases, enabling users to search for materials of interest and display the full content for those materials that are available in other digital libraries to which the library subscribes. University libraries also must provide access to collections that exist in digital form, regardless of the location or ownership of those collections. Many are responsible for publishing electronic resources in addition to contracting for digital materials provided by commercial services.

The concept of a digital library as an organization (Waters, 1998) is converging with the concept of a "gateway library" (Dowler, 1997). A gateway library is variously defined as an information system that "provides a single, convenient, uncomplicated entry point to a carefully selected library of bibliographic, full-text, and numeric information" (Olsen, 1997, p. 124–125), "a constellation of services, the organization required for providing those services, and the spaces dedicated to student learning" (Dowler, 1997, p. 98), and as "a process rather than a place" (Rockwell, 1997, p. 112). The gateway concept is not new, however. One of the most essential (and perhaps least recognized) roles of university libraries is to select materials from the vast universe of published and ephemeral resources. Once selected, librarians are responsible for collecting and organizing these materials in ways most usable and accessible by the university community. What is new is that the library, as gateway, is no longer confined to a physical place.

Concerns for the role of libraries as learning centers and the tension between space and place are evident in the titles of recent books about the future of libraries in a digital age: *Books, Bricks, & Bytes* (1996), *Gateways to Knowledge: The Role of Academic Libraries in Teaching, Learning, and Research* (Dowler, 1997), *The Mirage of Continuity: Reconfiguring Academic Information Resources for the 21st Century* (Hawkins & Battin, 1998), and *Information Landscapes for a Learning Society: Networking and the Future of Libraries* (Criddle, Dempsey, & Heseltine, 1999).

Form and function, place and space

The simple definition of a library—an agency that selects, collects, organizes, preserves, conserves, and provides access to information on behalf of a user community—says little about how these activities are performed, the relative emphasis on each, or the relationship between them. Indeed, libraries come in so many types and sizes, encompass such a wide variety of activities, and vary so much by social context, that no single agreed-upon definition of "library" appears to exist. Libraries differ along such lines as type of institution (national, academic, school, public, and special are the usual categories), and by politics, such as the contrast between public library services in democratic and totalitarian systems. *Why* they differ is little studied, however (Buckland, 1988).

Libraries are social institutions that have evolved over a period of many centuries. They serve the information needs of their user communities, adapting collections and services as those needs change. Libraries tend not to be autonomous institutions. Rather, most are funded by universities or schools to serve students, teachers, and staff; by governments to serve a local, regional, or national constituency; or by businesses, hospitals, museums, or other organizations to serve their employees, clients, patients, or patrons. Most libraries have a mission statement and collection development plan that identifies who they serve and what they collect, thereby drawing boundaries around their responsibilities. Libraries thus are both institutions and functions. Much of what distinguishes them from other information organizations such as museums, archives, and commercial enterprises is their professional principles and practices.

Library buildings serve as gathering points for communities, bringing together people, information resources in physical forms, access to information in electronic forms, and professionals to assist people in their information-related activities (Agre, 2000; Kent, 1996; Lyman, 1996; Mason, 1996). In the last decade, grand new buildings or major additions have been constructed for public libraries in Los Angeles, San Francisco, Cleveland, and other major American cities. Britain, France, Germany, and Croatia recently have opened new showcase buildings for their national libraries, and the Library of Congress has reopened its historic Thomas Jefferson Building after a major renovation. These large investments of public monies signify a resurgence of interest in libraries as physical places (Dowlin & Shapiro, 1996; Lehmann, 1996). These new buildings are designed to support the latest technologies, while preserving existing materials in multiple formats. The new model for library services is neither a print nor a digital library, but a "hybrid library" with complementary print and digital collections. The presence of complementary resources and services is having a crossover effect. People visit a library building to borrow a book, and learn to use the Internet while they are on site. Conversely, they come to the building to use computer networks, then browse printed materials and leave with a book (Mason, 1996).

Many of the functions that libraries serve can be supported online, in a virtual space. People can search catalogs and databases from home or office, can consult with librarians or other advisors by email, and can convene meetings online. Libraries provide services to discover, locate, and obtain materials online, thus many are using libraries without even visiting the physical locations.

Digital libraries are bringing issues of space and place into conflict in some anticipated ways. One of the promises of a global information infrastructure is for individuals to have direct access to information resources located anywhere on the network, so that they can seek and use information on their own, and can create new resources for others to use. In many respects, however, individuals are becoming more dependent on institutions for information access, rather than less dependent as predicted.

In the world of print, individuals can gain access to many libraries, public and private, at home and abroad. Depending on local practices, prospective users may or may not need an affiliation with the institution to gain access, may or may not be required to identify themselves to gain entry, and may or may not be allowed to browse shelves or borrow materials. Normally users can browse catalogs and indexes, once admitted to the library. If they are not allowed to browse shelves, they will be able to request specific documents to be retrieved on their behalf.

In an online world, more credentials are being checked and fewer resources are available without authorization. Intellectual property rights and contractual arrangements result in libraries and universities making fine distinctions on access policies that are unnecessary or impractical with print materials. In a print world, access is largely controlled through entry to the building. In the online world, access rights often are resource-specific. In the hybrid libraries of today, the result is an odd mix of privileges that vary between on-site and off-site usage.

American public libraries and libraries in public universities, for example, typically allow anyone to enter their buildings. Once inside, anyone usually can use materials on open shelves or use computer terminals. Borrowing privileges, special collections, interlibrary loan, and other services may be restricted to members of the local community. Different access restrictions apply when entering the same library institution through an online connection. Many of the digital resources freely available in the building may be available online only to authorized members of the community. Contracts with vendors of digital libraries typically define the user community as institutional affiliates (e.g. employees; university students, faculty, and staff; residents of the jurisdiction of a public library) plus those physically in the building, due to the impracticality of controlling physical access on a resource-by-resource basis, each governed by a different contract. Defining institutional affiliation for national libraries is especially problematic, as national libraries typically are open to all residents of the country and to foreign visitors. With regard to digital materials, most national libraries are defining the user community as those people physically present in the building. Electronic documents acquired by the Library of Congress and the British Library as materials deposited under copyright law, for example, are available only within their respective buildings. Thus, in an ironic twist, individuals may now have less access to information resources via a global digital library than they do within a physical library.

Hybrid libraries

People need information resources that exist in a wide variety of formats. Some wish their libraries would collect a greater proportion of print materials than they do, while others would prefer a greater proportion of digital materials. These debates often degenerate into discussions of the appropriate portion of library collections that should be in electronic form—20%, 50%, 75%? Approaching collection development in terms of format begs the questions of content and community, however. Traditional approaches to collection development begin with questions of 'who are we collecting for?' (community) and 'what do they need?' (content). The choice of format (electronic, print, or other forms) usually follows from answers to the questions about community and content. However, these factors interact in complex ways. The materials selected must be in formats that are usable (readable, playable, etc.) on equipment available in the library or in the community, for example.

Even if the same content exists in multiple formats, consideration must be given to the ways in which print and electronic materials are used. Buckland (1992) draws a contrast between the uses of paper documents and electronic documents. Paper is a solo technology, typically best used by one person at a time, and is a localized medium, requiring that the document and the reader must be at the same place at the same time. Electronic documents can be used by many people at the same time, and can be located at a different place from the reader, making them suitable for telelearning environments.

Other characteristics of form are relevant in collection choices. For example, paper documents often are more effective than electronic forms for discussions by a small number of people. Paper is well-suited for purposes such as explaining diagrams, where people need to point and make hand motions over a document, and generally is more suitable for making annotations and for browsing by flipping through pages than are most forms of hypertext (Dillon & Gabbard, 1998). Technologies such as "electronic white boards" and "electronic books" try to replicate these features of paper (Schilit et. al, 1999; Silberman,1998; http://www.itl.nist.gov/div895/ebook99/,http://www.ebooknet.com). Paper is a very flexible medium, with many features that are still difficult to support in digital environments.

The economic tradeoffs between print and electronic materials are complex and evolving. One factor is the difference in business models for print and electronic publication. Print materials normally are sold as individual copies or annual subscriptions, while electronic materials are subject to a variety of elaborate pricing schemes that may include annual fees, fees per use, fees based on size of user population, combination fees based on acquiring sets of print and electronic resources from one publisher, etc. Other economic considerations for libraries include the continuing costs of maintaining the materials, whether stored on a bookshelf, mounted on local computers, or access provided to remote computers, and the conservation and preservation of the materials as technology evolves.

Implications for the Virtual University

In designing digital libraries for tomorrow's virtual universities, we can build upon what is known about electronic publishing, telelearning, libraries, and information-related behavior, but which of the many research findings should be applied? What findings are relevant to digital libraries that are even larger in size, collect an expanding array of digital content, employ increasingly sophisticated technical capabilities, and serve an increasingly diverse international population?

Rather than begin design afresh with each new generation of technology, we can build upon experience and research from prior technologies.

We discuss four trends in the design of digital libraries for virtual universities. In doing so, we make several assumptions about the future. One is that the number and variety of digital libraries will continue to grow. Another is that the number of people with access to computer networks will continue to rise at rapid rates (although access will continue to be unequal, especially between more and less developed countries). Similarly, network capacity will increase quickly, as will the price–performance ratio of computing technology. However, the actual trajectory of these developments is subject to many factors that cannot be predicted. The price of computer workstations is decreasing, but hardware and software represent only 4% to 30% of the total cost of ownership, depending upon how these figures are computed (Barnes, 1997; Bits, 1998; Strassmann, 1997), thus the true costs of owning, operating, and maintaining information technology are not necessarily dropping. We also assume that patterns of access will change as various new "information appliances," many of which will be mobile or otherwise "untethered" devices, supplement or supplant the use of desktop workstations. Similarly, economics will affect usage patterns in unpredictable ways. Network congestion is leading to differential pricing models for access, which will influence who uses what network resources, at what cost, and at what time (Lynch, 1998).

From metadata to data

Digital libraries are useful only to the extent that their existence can be determined and that the content within them can be identified. A pile of books shorn of covers, title pages, and other identifying data is of little value, as is a magnetic disk containing files that lack names, types, dates, or hierarchical relationships. If the books are shredded or the files merged, the problem is only worse. Digital libraries and their contents need to be described and organized in some way to facilitate access. The description and organization of resources is an essential part of network architecture (Arms, 1995; Dempsey & Heery, 1998; Weibel, 1995).

Metadata, or "data about data," is the basic mechanism for describing and organizing information resources. Metadata provide information that is necessary for understanding the content of resources (such as who created them and when), and for making use of them (such as the format in which they are stored, and the ownership of the intellectual property). Metadata can serve as the "Rosetta Stone" to decode information objects and to transform them into knowledge (Gilliland-Swetland, 1998). Metadata also are needed to describe entire digital libraries, as a means to identify and locate collections of resources on a topic, or relationships between resources.

Earlier generations of digital libraries consisted largely of textual metadata describing offline, printed materials such as journal articles, books, court cases, legislation, memoranda, and personal papers. The best known of these are online catalogs and finding aids for archival materials. Examples include the Melvyl catalog which incorporates holdings of the nine campuses of the University of California, and the Online Archive of California, which contains finding aids for a variety of public and private archives. Both are available to the public via the California Digital Library (http://www.cdlib.org). The primary thrust of current digital library design is to maintain full content online. New forms of organization, user interfaces, and functional capabilities are being developed to address the changing environment for information systems.

Digital libraries containing only metadata continue to be valuable for identifying and locating offline sources. Many resources will continue to exist only offline, such as the holdings of libraries and archives that may have been collected over a period of centuries. Even when digital representations of documents are available online, the metadata are essential for locating the originals. Most library collections are subject to the "80/20 rule": 20% of the materials satisfy 80% of the needs. The difficulty is predicting which 20% are most likely to be used, and how usage patterns will change over time. Popular materials fall into disuse, and obscure items later become essential resources. The only practical and economically feasible solution to providing comprehensive access is to maintain online digital libraries of metadata that describe offline sources.

The relationship between data and metadata is changing with the advent of full content digital libraries, the extensive penetration of computing networks, and the increasing variety of players creating digital libraries. Librarians will continue to create and manage digital libraries. They also will assist university faculty, staff, and students in creating their own digital libraries. At the same time, many other individuals and organizations are producing digital libraries. Some are following established principles for metadata creation and maintenance, while others are inventing their own techniques.

Information retrieval mechanisms

Until recently, information retrieval consisted almost entirely of textual searching on metadata. Even when digital libraries contain other formats such as images and numeric data, the primary access may be via textual metadata that describe and represent those objects. When searching full-text content, rather than metadata, the characteristics of the data and the search process change.

In principle, metadata consist of carefully chosen terms to describe and represent some document or object. Metadata on published materials, for example, usually consist of bibliographic descriptions of documents (e.g. author, title, date, place of publication, physical characteristics) plus subject or index terms, and may include an abstract summarizing the content. As a result, each term (with the exception of "stop words" such as articles and conjunctions) in the metadata record carries significant meaning about the document or object. When the full content is searched, this assumption becomes problematic because information retrieval is very sensitive to scale. One term among five in a title or among 200 in an abstract is more likely to be relevant than when the same term is found among 10,000 words in a journal article, much less among 100,000 words in a book. For the same reasons, Boolean operators are less useful when searching full content. If two or more terms appear in the same metadata record, the record is more likely to be relevant than if the same two or more terms appear somewhere in an article, book, or report.

Information retrieval also is subject to the principle of "garbage in, garbage out." The effectiveness of retrieval depends heavily on the quality and consistency of the metadata in the digital libraries being searched. Some metadata can be generated automatically, such as extracting terms from documents. Other metadata must be supplied manually, such as terms that describe the meaning of the content, or for information external to the document such as intellectual property rights or hardware and software requirements. When metadata are created by library catalogers, archivists, museum registrars, records managers, knowledge engineers, and other information professionals, established principles and practices are followed to assure quality and consistency. These principles and practices are learned through course work, apprenticeship, and experience. When metadata are produced by instructors, students, and other non-professionals, the resulting metadata will be less consistent and more idiosyncratic (Bowker & Star, 1999). Digital libraries for virtual universities will need robust searching mechanisms that can produce reasonable results from messy contents.

A growing amount of content in all media is being "born digital," and content that originated in other forms is being digitized. However, the technology for storing and displaying multimedia content is improving at a faster rate than is the ability to search for that content. While text is at least partially self-describing, sounds and images are not. Two general approaches are possible. One is to describe sounds and images with words, which then are searchable. This approach reduces a multimedia digital library to a textual database, failing to take full advantage of the richness of the content. The other approach is to apply media-specific searching mechanisms that can identify sounds (e.g. by speaking or playing them) or images (e.g. by matching similar objects or recognizing shapes and colors). Each approach has strengths and weaknesses. Regardless of the application, some amount of textual description usually is necessary for basic metadata such as titles, authors/creators, and ownership. Manual description may be essential for interpretation. Photos, for example, are sometimes of interest based on who is represented in the picture and sometimes for who took the picture, neither of which can be generated automatically in most cases.

Digital libraries that can match on sounds and images are at their early stages, but are promising (Bainbridge, Nevill-Manning, Witten, Smith, & McNab, 1999; Wactlar, Christel, Gong, & Hauptmann, 1999). However, retrieval mechanisms tend to be media- and application-specific. Methods that work for sound retrieval in one application are difficult to transfer to other applications, much less to video or graphics, for example. Few true multi-media searching capabilities exist as yet (Croft, 1995).

Dynamic documents

Most efforts in designing and maintaining digital libraries have addressed static documents (e.g. published materials; completed work products, whether text, audio, video, or other media). Digital libraries for virtual universities will include a wide range of dynamic documents, i.e. documents that are "mobile, malleable, and mutable" (Bishop & Star, 1996). Instructors will modify documents during and between offerings of a course. Students will modify their own documents continuously throughout a course, with links to other documents that also are changing continuously. Thus a substantial research and practical challenge is to describe, represent, and organize dynamic documents.

Metadata that describe dynamic documents somehow need to distinguish between versions, while identifying and maintaining relationships among versions of materials that are substantially the same. For example, when locating a document it is helpful to know that prior and subsequent versions may exist. It is also helpful to know if the content exists in substantially similar forms that may be equivalent for a given purpose (e.g. materials for related courses; other student papers on the same topic; conference paper and sequential versions of journal article; transcript and audio version of a speech; theater, broadcast, and airplane versions of a film). Determining whether versions of a document are substantially the same is often a matter of interpretation and may require information that is external to the documents themselves, all of which complicates the creation of standards and practices for document representation (Lynch, 1993a, b).

From independent to linked systems

If a digital library is to be designed for a specific course, to be taught once, by one instructor, and available only to a limited number of students, the system can be tailored explicitly to local needs and circumstances. However, this situation will be rare in virtual universities. More often, digital libraries will contain materials intended for a large audience (e.g. databases of discipline-specific papers and articles) or will be designed for courses that will be taught multiple times by one or more instructors. They will be accessible over computer networks, and will be used at home, office, library, dormitory, hotel, and other sites. Under these circumstances, digital libraries for virtual universities must be interoperable with a variety of hardware, software, and computer networks.

Determining the appropriate balance between tailoring systems to local circumstances and maintaining interoperability among systems is a fundamental challenge in designing digital libraries for virtual universities. On the one hand, the description, representation, and organization of digital libraries should be tailored to the students, course material, and technology environment. On the other hand, digital libraries should be interoperable over a global information infrastructure, so that what exists on a topic can be identified regardless of location, format, language, or other characteristics. At the extreme of tailoring, the result would be idiosyncratic systems of little value outside their local context. At the extreme of interoperability, the result would be highly-standardized systems that might serve no application well.

The set of digital libraries on distributed networks can be viewed collectively as a global digital library. In turn, a global digital library needs organizing mechanisms and searching capabilities that enable people anywhere to search this vast information space. Capabilities are needed to search multiple databases concurrently, to discover and locate individual digital libraries worthy of deeper exploration, to follow links between systems, and to trace paths between documents within individual systems. Materials also need to be transferrable between digital libraries and other applications for document creation and use. Despite the progress in standards for networked data exchange, however, full interoperability remains a distant goal (Lynch & Garcia-Molina, 1995; Paepcke et. al., 1998). Some of the problems are due to the existence of a wide variety of hardware platforms and software applications that span several generations of technology, some are due to the lack of appropriate software to translate between formats, some are due to inadequate user interfaces, and others are due to lack of adequate skills on the part of users. Yet others are due to social, cultural, and policy differences that are well beyond the scope of technical solutions (Agre, 2000).

Organizing within and between digital libraries

The goal of most organizational schemes is to impose internal consistency upon individual digital libraries. Concepts are mapped to a common term or category ("collocated"); broader, narrower, and related term relationships may be identified; and facets or other relationships may be specified as well. Most organizational schemes were designed to link content relationships between documents, or between the metadata that describe documents, within a digital library. These schemes are being extended and others created to represent relationships among digital items such as whole/part, same origin of content in different medium (e.g. book, script, film, play), multiple instances of an artifact, original and translation, similar content in multiple formats (text, numeric, images, audio, video), and to support hybrid digital libraries that combine metadata and data (Borgman et al., 1996; Smiraglia & Leazer, 1999; Tillett, 1991, 1992).

Organizing information between digital libraries takes place at two levels: Mechanisms that enable multiple digital libraries to be searched at once, bridging the organizing structures of each; and mechanisms that enable links between documents to cross system boundaries. Many models exist for organizing materials in a single collection, but no similar model exists for organizing resources across multiple collections (Lynch, 1993a). New methods are needed to bridge the organizational schemes that provide consistency within digital libraries but that may produce inconsistencies between them. Many factors introduce inconsistencies, one of which is perspectives inherent in schemes applied by different information professions such as librarians, archivists, and curators. Digital libraries organized from each of these perspectives may contain related information, yet the information is structured and organized so differently that it can be hard to identify related documents. When materials are indexed by students and other non-information professionals, even more inconsistencies will be introduced.

The same content may be pertinent to multiple audiences, each of whom uses different terminology, has different levels of knowledge about the content, and has different requirements for organizing and navigating that content. For example, scientific data collected by and for scientists may be of considerable value for instruction at the high school, undergraduate, and graduate levels. Much scientific data, especially in the geo-sciences, also is valuable for business and government applications. Scientists require a finely-detailed organizational structure for these data and employ highly-specialized terminology in describing content. For the same digital libraries to be useful to students, teachers, business people, and government officials, simpler analytical structures, more common vocabulary, and user interfaces that require minimal domain knowledge are needed. Similarly, legal information systems are predicated on technical knowledge of the law, yet non-lawyers need legal materials too. The same is true of most digital libraries containing medical and health information.

Digital libraries need to span community boundaries both for reasons of resource allocation and for resource discovery. Digital libraries are expensive to construct and maintain, thus it may be more economical to make one system available to multiple user communities than to create multiple systems. Conversely, when people are seeking information on a topic, they may need to discover information resources located in digital libraries designed for a different user community, and which may be organized and represented in unfamiliar ways.

Legacy data and migration

The contents of digital libraries for virtual universities will be drawn from a variety of sources that will exist in a variety of physical and electronic formats. Constructing a digital library from multiple sources usually requires content to be restructured, reorganized, or otherwise converted to a common form. The means depends upon the originating and target formats. Clean text can be scanned and then converted by optical character recognition; degraded or handwritten text may need to be rekeyed. Sounds recorded on vinyl can be played and re-recorded digitally, etc.

Any data not in a desirable form for a current application is considered "legacy data." Paper records are legacy data when computer-readable records are required for a digital library. Automated records are legacy data when they are in a structure or format that is incompatible with the requirements of a new system on which they must operate. Converting and restructuring data can be expensive, depending upon the amount of manual labor required. Once in digital form, the data may be migrated to other hardware and software multiple times, because data often outlive the technology by many years.

The economics of digital library applications for virtual universities are complex. University libraries usually make decisions about investments in legacy data, such as metadata, content representation, intellectual access, and organization, with a long view in mind. In the case of general applications, such as digital libraries of published discipline-specific materials (e.g. science, social science, arts, humanities databases) the usual considerations of conserving and preserving content are likely to apply. In the case of course-specific digital libraries, the considerations are somewhat different. The content of these digital libraries may have a shorter life span, so less investment in the quality of data and metadata might be justified. Yet unless these digital libraries meet international standards for metadata and interoperability they will be of little value in distributed applications.

From searching to navigation

At present, digital library technology is more advanced for creating electronic resources than for searching them. Most digital libraries are still difficult to use (Borgman, 1986, 1996, 2000a, b) due to weaknesses in user interfaces, to the lack of adaptation to user tasks, and to users' knowledge and skills with information technology (Commission on Physical Sciences, Mathematics, and Applications, 1999). Furthermore, searching mechanisms designed for independent digital libraries that contain only metadata (e.g. local online catalogs), have limited effectiveness when applied to networked digital libraries of full content. Navigation and browsing are more flexible and adaptable means of searching than are submitting queries, and are better suited to distributed environments. The term "browsing" has many definitions (Marchionini, 1995); dimensions include scanning, intention, goals, and knowledge (Chang & Rice, 1993). "Navigation" is used here as a collective term for browsing, for other forms of scanning through information, and for following paths within and between systems.

Information-related problems frequently are ill-defined, especially in educational settings. Part of the learning process is understanding concepts and issues well enough to formulate questions about them. For these reasons, people typically approach digital libraries with "anomalous states of knowledge" (Belkin et al. 1982a, b). The knowledge they bring to a system often is better applied as a starting point than as a statement of their problem. Navigation features can enable users to follow paths from one related resource to another, whether from metadata to data, from one version of a document to another, or from one document to another related by subject, creator, owner, price, rights availability, features, or any other attribute that can be described and represented.

Despite the advantages of navigation, the paradox of information retrieval remains: the need to describe the information that one does not have. The searcher still must establish a starting point for the navigation process, which requires selecting terms and features that describe the information resources sought. The searcher also must specify a starting point in one or more digital libraries or the characteristics of target collections.

Navigation behavior

Navigation appears to be a natural form of information-seeking behavior, based on principles of cognition that people learn by making associations with prior knowledge and that people find it easier to recognize information presented to them than to recall it from memory (Anderson, 1990). Systems that require the user to enter keywords rely on recall knowledge, since the user must generate the terms. Systems that present an organized framework of terms from which to select rely on recognition knowledge. Recognition-based systems are particularly effective in subject areas that are not part of everyday knowledge, as we found with a digital library of science materials for children (Borgman et al., 1995).

Navigation also is effective because it makes system features more visible. Rather than examining documentation to identify search capabilities, users can select from an array of choices, whether functions, terms, related documents, or related digital libraries. However, navigation approaches are not a direct substitute for query-based searching. These approaches appear to be more effective in small, focused collections than in large and diverse systems in which the choices may be overwhelming. Navigation also is better suited for ill-defined problems where exploration is helpful than for well-defined problems in which a query can be more efficient.

Among the greatest difficulties in searching for information is reframing the search when too many, too few, or irrelevant results are retrieved. Navigation provides some assistance by enabling users to follow paths rather than submitting queries that return results sets. Even so, searchers still may need to browse through large sets of documents or long lists of options they have found by following links. Hypertext is a mechanism that enables people to follow non-linear paths, such as the links embedded in the World-Wide Web. Early research on hypertext revealed that people can become lost in the system (Conklin, 1987; McKnight et al., 1991). Various approaches to improving navigation in hypertext have been tried, including geo-spatial metaphors such as maps and signposts (McKnight et al., 1991). More recent research suggests that the effectiveness of hypertext for learning and comprehension is limited, and its value depends on individual abilities and learning styles (Dillon & Gabbard, 1998).

User interface design

Query-based interfaces often are simple and terse, while navigation-oriented interfaces are more richly structured, typically with multiple windows, menus, and other options. Color and screen placement can be used to distinguish among features. The manner in which information is presented or delivered will influence the way it is received and interpreted. Navigation interfaces work best when they are tailored to the content of the digital library and to the purposes of those who search it. In principle, such interfaces should be effective for telelearning applications, especially for guided and exploratory learning modes.

User interface design for navigation-oriented systems must balance the tailoring vs. interoperability tradeoffs noted earlier. Tools for navigating through text should take advantage of textual characteristics, while tools for navigating through music or videos should take advantage of their format and content characteristics. However, the more that interfaces are tailored to content, the more likely that searchers will encounter a new, different, and often foreign interface for each digital library. Similarly, people vary widely in their information-seeking habits, preferences, skills, and abilities. For example, those with strong spatial skills are likely to be adept at navigating spatially organized interfaces, while others with weak spatial skills may be confused by them (Egan, 1988). A single, simple user-interface that will support effective navigation through text, images, audio, video, and numeric data is not yet feasible. A larger challenge is to design user-interfaces for diverse communities with a wide array of interests, purposes, and skills to navigate through diverse content.

From individual to group processes

Most of what is known about information-related behavior, in both in print and electronic environments, concerns the activities of individuals in creating, using, and seeking information. Until recently, most research in library settings was confined by entry and exit from a physical library or logon and logoff from a digital library. Despite the historical focus on individuals, information-related activities typically involve group processes. People learn in classrooms; consult with teachers, librarians, fellow students, experts, and colleagues; work in teams; and correspond with friends, families, colleagues, and strangers. Questions posed to digital libraries often arise from interaction within a group.

Incorporating group processes into the design of digital libraries is a paradigm shift that is just now under way (Twidale & Nichols, 1998a, b). Support for collaborative work is one of the greatest benefits of telelearning technologies, as Gloria Mark discusses elsewhere in this volume. She emphasizes the importance of establishing social context, or a sense of place, in collaborative learning environments. In virtual universities, the metaphor for a gathering place may be a virtual classroom, with its associated instructor, fellow students, and mechanisms to share artifacts and knowledge. In digital libraries, the metaphor for a gathering place perhaps should be a virtual library, with its associated librarians, collections, organizational mechanisms, devices to read, search, or otherwise use resources, and even some equivalent for sitting in "comfy chairs" to pore over documents.

Bishop and Star (1996, pp. 369–372) propose a number of research questions related to issues of the social context of digital libraries, such as, How do creators, librarians, and users collaborate in creating, finding, and using digital library documents? How does work change with the introduction of digital library access in the work place and the home as well as in the traditional library? How do digital documents structure social interactions (e.g. make, maintain and differentiate social groups)? and How do organizations make digital libraries usable or unusable for their members? These are but a few of the issues to be addressed in making digital libraries more suitable to group work.

Sharing documents

Users of digital libraries will share documents in many ways, for many purposes, especially in virtual university applications. Some sharing is direct and intentional, as when multiple people author a document. Collaborative writing tools can assist in this process. However, the task of writing over distance with two or more authors making changes to a document is complex and idiosyncratic; current tools are of limited value (Bruce & Sharples, 1995; Twidale & Nichols, 1998a, b). Users also share documents without being aware of each other, as when multiple people read the same document. In paper environments, individuals annotate their own copies of documents. In electronic environments, it is possible for users to annotate documents in a shared space, or to annotate them in personal spaces but make their annotations available to others.

Annotations and other records of how and when documents are used can be indicators of value. These indicators can be utilized in searching, for example to request documents that are rated highly by known experts or to request documents whose frequency of use is above a given threshold (Twidale & Nichols, 1998a, b). Another technique that relies upon the community basis of documents is "collaborative filtering" (*Filtering and Collaborative Filtering,* 1997; Green, Cunningham, & Somers, 1998). The underlying principles are most easily explained by example. If two people like the same set of documents, such as the same three novels or movies, then each of them probably will like other novels or movies that the other party likes. To conduct collaborative filtering, the group of participants would be offered a list of novels or movies and asked to select their favorites. When all have made their selections, people with similar tastes are clustered. Each participant is then given a list of novels or movies he or she has not yet read or seen, but which are favorites of people with similar tastes to theirs. Online bookstores are using collaborative filtering to recommend books, music, and movies to customers based on commonalities of taste. Collaborative filtering also is being used with automated intelligent agents to improve information retrieval.

Modern collaborative filtering techniques have much in common with citation analysis, a long established method of clustering documents and following paths between related documents (Borgman, 1990). Authors cite other documents for a variety of reasons, and if the sample is large enough, the influence relationships are reasonably reliable. Citation data can be utilized alone for retrieval and analysis or in combination with other measures to assess trends and relationships (Lievrouw et al., 1987).

The role of intermediaries

In physical libraries, professional librarians play an important mediating role between users and collections. Users pose vague questions typical of the early stages of problem formulation. Librarians interview users to clarify and expand upon questions, using this information to rephrase or restructure user queries in ways more readily searchable. In doing so, librarians draw upon expertise in information-seeking behavior and in interpersonal communication, as well as expertise in knowledge organization and in the subject disciplines of the collection. They also draw upon expertise of other librarians. Librarians share their experiences about how they solved difficult problems. They also share tips on searching specific information systems. Some of this knowledge is gathered into "ready reference" files of frequently asked questions and tips about solving certain types of queries. In these and other ways, librarians add value to library collections and services.

As Agre (2000) comments, designers of networked information technologies tend to discount the importance of human intermediaries. They argue that once networks "connect" buyers and sellers, or searchers and digital libraries, then intermediaries are no longer needed. Agre counters this argument by noting that intermediary services are among the most successful new businesses on the Internet (Agre, 2000; Sarkar et al., 1995). Intermediaries serve many purposes in addition to simply "connecting" people. Few would argue that virtual universities could operate without instructors, for example. Rather, instructors are taking on expanding roles such as designing online courses, developing electronic course materials, supervising "chat rooms," and facilitating online interaction between students. Similarly, librarians' roles are expanding to include "networked learner support" (Fowell & Levy, 1995; Twidale & Nichols, 1998a, b), part of which is digital library development and management. University instructors and librarians will be working together yet more closely in the future if virtual universities are to provide resource-rich instruction.

Summary and Conclusions

Digital libraries are essential to the development and deployment of virtual universities. "Digital libraries" has many meanings, two of which were highlighted here. One is an advanced form of information retrieval system, usually containing the full content of electronic documents, often in multiple media, and available via distributed computer networks. The other is an institution or organization that provides information resources or services in digital form, has specialized staff, and is responsible for selecting, collecting, organizing, conserving, preserving, and providing access to information resources. Both meanings are relevant to higher education and virtual universities. In the first sense, digital libraries are a technical means to provide access to information resources in support of instruction. In the second sense, they are an institutional framework within which to provide information services.

As information systems, digital libraries are an enabling technology for electronic publishing and for telelearning. Materials that are "born digital" or are digitized are collected and organized in digital libraries. Students and faculty alike can publish materials in digital formats. Digital libraries can support many modes of telelearning, whether local or remote, and whether lectures, guided learning, or self-directed learning. University libraries provide generalized digital libraries to be used for all forms of instruction and research. Instructors, alone or in partnership with librarians, can produce course-specific digital libraries. Digital libraries offer many benefits for instruction, such as facilitating the cycle of information seeking, creation, and use; the ability to follow links between related materials; and the ability to support collaborative work by students. They also have several disadvantages, such as the cost of creating and maintaining them, the continuing costs of digital preservation, and difficulties in searching and use.

As an institutional framework, digital libraries serve a variety of functions and operate in virtual spaces. University libraries are at least three institutions at once. They are charged with maintaining their physical collections, with improving access to print and digital resources through automation, and with developing new information services for the future. The present and future model for university libraries is that of the "hybrid library," which provides a growing array of resources and services. Older materials and formats continue to be useful. New information technologies supplement, rather than supplant, traditional libraries.

This chapter identified four trends that will influence the future of digital libraries and virtual universities. The first, from metadata to data, addresses the transition from digital libraries that consist primarily of metadata (typically pointers to offline materials), to digital libraries of full content. New forms of retrieval algorithms will be needed, especially as students, instructors, and other non-information professionals construct their own digital libraries. Digital libraries also will require new mechanisms for managing documents that are dynamic, such as versions of student projects, rather than static, such as published articles.

The second trend is the transition from independent to linked systems, in which users can follow links from one document to another, within or between digital libraries. The challenge here is to tailor digital libraries to individual user communities while making them interoperable with libraries constructed for other classes, universities, disciplines, countries, cultures, and languages, scattered across a global information infrastructure. A related challenge is to transfer content from one format to another as needed for a specific system or generation of technology, which involves continuous efforts in data migration. The third trend is the transition from query searching to navigating through digital libraries. While navigation appears to be a natural form of information-seeking behavior, new forms of user interfaces to assist in navigating, filtering, and displaying information are needed.

The fourth trend identified is the shift from focusing on the individual user of digital libraries to a concern for the group processes involved in information-related behavior. Research on collaboration in digital libraries is in its early stages. We identified sharing documents and the role of intermediaries as concerns. Users share documents directly, such as when co-authoring a book or an article, and indirectly, such as when reading and using the same documents. Some technological assistance is becoming available for these tasks. Intermediaries, especially instructors and librarians, will continue to be essential to the growth of digital libraries and virtual universities.

While this chapter focuses largely on identifying problems in the design, development, and deployment of digital libraries for virtual universities, we conclude by emphasizing that the most significant challenges are related to social issues and policy, rather than to technology. Digital libraries will be used and will be useful only to the extent that they fit into the social context of instruction, research, and practice. To achieve wide acceptance, they must become easier to create, to maintain, and to search than is presently the case. Revolutionary new designs rarely work in operational settings, because people need opportunities to shape the technology to their own needs and practices. "Radical incrementalism" is a more reliable design method, because adaptation requires "extensive fine tuning in the real world," as Brown (1996, p. 30) puts it. Designing digital libraries for virtual universities must build upon what is currently known about learning, instruction, and information-related behavior. To identify the best practices for the design of digital libraries for higher education, it is essential that a diverse array of field studies of digital library use be conducted. Librarians, educators, and scholars of many disciplines will need to work together if digital libraries are to achieve their promise for virtual universities.

References

Agre, P. A. (2000). Information and institutional change: The case of digital libraries. In Bishop, A. P., Buttenfield, B. P., & Van House, N. (Eds.). (2000). *Digital Library Use: Social Practice in Design and Evaluation*. Cambridge, MA: MIT Press.

Anderson, J. R. (1990). *Cognitive psychology and its implications*. San Francisco: W.H. Freeman.

Arms, W. Y. (1995, July). Key Concepts in the Architecture of the Digital Library. *D-Lib Magazine*. http://www.dlib.org/dlib/July95/07arms.html.

Arms, W. Y. (1999). *Digital libraries*. Cambridge, MA: MIT Press.

Bainbridge, D.; Nevill-Manning, C. G.; Witten, I. H.; Smith, L. A.; & McNab, R. J. (1999). Towards a digital library of popular music. In E.A. Fox & N. Rowe (eds). *Digital Libraries 99, the fourth ACM conference on digital libraries.* August 11–14, 1999, Berkeley, CA. New York: ACM. Pp. 161–169.

Barnes, M. (1997). The costs of systems: What we're spending. *Bobbin*, **38**(11), 25–26.

Battin, P. (1998). Leadership in a transformational age. In B.L. Hawkins & P. Battin (eds.), *The Mirage of Continuity: Reconfiguring Academic Information Resources for the 21st Century.* Washington, DC: Council on Library and Information Resources and the Association of American Universities. Pp. 260–270.

Belkin, N. J., Oddy, R. N., & Brooks, H. M. (1982a). ASK for information retrieval: Part I. Background and theory. *Journal of Documentation,* **38**(2), 61–71.

Belkin, N. J., Oddy, R. N., & Brooks, H. M. (1982b). ASK for information retrieval: Part II. Results of a design study. *Journal of Documentation,* **38**(3), 145–164.

Besser, H. (1996). Issues and challenges for the distance-independent environment. *Journal of the American Society for Information Science*, **47**(11), 817–820.

Bishop, A. P., Buttenfield, B. P., & Van House, N. (Eds.). (2000). *Digital Library Use: Social Practice in Design and Evaluation*. Cambridge, MA: MIT Press.

Bishop, A. P.; & Star, S. L. (1996). Social informatics for digital library use and infrastructure. In M. E. Williams (ed.), *Annual Review of Information Science and Technology,* **31** Medford, NJ; Information Today, pp. 301–401.

Bits. (1998). *Eye-opening Reports About Desktop Computing Costs.* http://www.lanl.gov/Internatl/divisions/cic/bits/archive/98june/Marcia_REDI.html.

Books, Bricks, and Bytes. (1996). *Daedalus, Journal of the American Academy of Arts and Sciences; Proceedings of the American Academy of Arts and Sciences,* **125**(4).

Borgman, C. L. (1986). Why are online catalogs hard to use? Lessons learned from information retrieval studies. *Journal of the American Society for Information Science,* **37**(6), 387– 400.

Borgman, C. L. (Ed.) (1990). *Scholarly Communication and Bibliometrics.* Newbury Park, CA: Sage Publications.

Borgman, C. L. (1996). Why are online catalogs *still* hard to use? *Journal of the American Society for Information Science*, **47**(7), 493–503.

Borgman, C. L. (1999). What are digital libraries? Competing visions. *Information Processing & Management,* **38**(3), 227–243. In G. Marchionini & E. Fox (eds.), Special Issue: Progress Toward Digital Libraries.

Borgman, C. L. (2000a). *From Gutenberg to the Global Information Infrastructure: Access to Information in the Networked World.* Cambridge, MA: MIT Press.

Borgman, C. L. (2000b). Uses, Users, and Usability of Digital Libraries. In Bishop, A. P., Buttenfield, B. P., & Van House, N. (Eds.). (2000). *Digital Library Use: Social Practice in Design and Evaluation*. Cambridge, MA: MIT Press.

Borgman, C. L.; Bates, M. J.; Cloonan, M. V.; Efthimiadis, E. N.; Gilliland-Swetland, A.; Kafai, Y.; Leazer, G. L.; Maddox, A. (1996). *Social Aspects Of Digital Libraries*. Final Report to the National Science Foundation; Computer, Information Science, and Engineering Directorate; Division of Information, Robotics, and Intelligent Systems; Information Technology and Organizations Program. Award number 95–28808. http://dlis.gseis.ucla.edu/dl/.

Borgman, C. L., Gallagher, A. L., Hirsh, S. G., & Walter, V. A. (1995). Children's Searching Behavior On Browsing And Keyword Online Catalogs: The Science Library Catalog Project. *Journal of the American Society for Information Science,* **46**(9), 663–684.

Bowker, G. C., and Star, S. L. (1999). *Sorting things out: Classification and its consequences.* Cambridge, MA: MIT Press.

Brown, J. S. (1996). To dream the invisible dream. *Communications of the ACM*, **39**(8), 30.

Bruce, B. C., & Sharples, M. (eds.). (1995). Special Issue on Computer-supported collaborative writing. *Computer Supported Cooperative Work (CSCW).* **3**(3–4), 225–404.

Buckland, M. K. (1988). *Library services in theory and context, 2nd ed.* Oxford: Pergamon.

Buckland, M. K. (1992). *Redesigning Library Services: A Manifesto.* Chicago: American Library Association. Available at http://sunsite.berkeley.edu/Literature/Library/Redesigning/html.html.

Buttenfield, B. P., & McLafferty, S. (1996). Spatial data in the classroom: A vision for education. *Proceedings of NAS Mapping Committee and FGDC Conference on The Future of Spatial Data and Society*, Washington, DC, April 24–25, 1996. Washington, DC: National Academy of Sciences.

Center for Science, Mathematics, and Engineering Education, National Research Council. (1998). *Developing a Digital National Library for Undergraduate Science, Mathematics, Engineering, and Technology Education: Report of a Workshop.* Washington, DC.: National Academy Press.

Chang, S-J. & Rice, R. E. (1993). Browsing: A multidimensional framework. In M. E. Williams (ed.), *Annual Review of Information Science and Technology,* **28**. Medford, NJ: Learned Information, pp. 231–277.

Commission on Physical Sciences, Mathematics, and Applications; Computer Science and Telecommunications Board; Committee on Information Technology Literacy, National Research Council. (1999). *Being Fluent with Information Technology.* Washington, DC: National Academy Press.

Computer Science and Telecommunications Board; Commission on Physical Sciences, Mathematics, and Applications; National Research Council. (1997). *More than Screen Deep: Toward Every-Citizen Interfaces to the Nation's Information Infrastructure.* Washington, DC: National Academy Press.

Conklin, J. (1987). Hypertext: An introduction and survey. *IEEE Computer,* **29**(9), 17–41.

Criddle, S. Dempsey, L. & Heseltine, R. (1999). *Information Landscapes for a Learning Society. Networking and the Future of Libraries,* **3**. Bath, UK: UKOLN, the UK Office for Library and Information Networking and London: Library Association.

Croft, W.B. (1995). What do people want from information retrieval? (The Top 10 Research Issues for Companies that Use and Sell IR Systems). *D-Lib Magazine,* November. Http://www.dlib.org/November98/11croft.htm.

Dempsey, L. & Heery, R. (1998). Metadata: a current review of practice and issues. *Journal of Documentation*, **54**(2), 145–172.

Dillon, A. & Gabbard, R. (1998). Hypermedia as an educational technology: A review of the quantitative research literature on learner comprehension, control, and style. *Review of Educational Research.* **68**(3), 322–349.

Dowler, L. (ed.). (1997). *Gateways to Knowledge: The Role of Academic Libraries in Teaching, Learning, and Research.* Cambridge, MA: MIT Press.

Dowlin, K.E. & Shapiro, E. (1996). The centrality of communities to the future of major public libraries. In Books, Bricks, and Bytes, *Daedalus, Journal of the American Academy of Arts and Sciences; Proceedings of the American Academy of Arts and Sciences,* **125**(4), 173–190. Republished in S. R. Graubard & P. LeClerc (eds.). (1998). *Books, Bricks, and Bytes: Libraries in the Twenty-First Century.* New Brunswick, NJ: Transaction Publishers.

Egan, D. E. (1988). Individual differences in human-computer interaction. In M. Helander (ed.), *Handbook of Human–Computer Interaction.* Amsterdam: Elsevier, pp. 543–568.

Faulhaber, C.B. (1996). Distance learning and digital libraries. *Journal of the American Society for Information Science*, **47**(11), 854–856.

Filtering and Collaborative Filtering. (1997). *Fifth DELOS Workshop.* Budapest, 10–12 November 1997. European Consortium for Informatics and Mathematics. Le Chesnay, France: ERCIM. Http://www-ercim.inria.fr.

Foster, I. & Kesselman, C. 1999. *The Grid: Blueprint for a New Computing Infrastructure.* Morgan Kaufmann.

Fowell, S. & Levy, P. (1995). Developing a new professional practice: A model for networked learner support in higher education. *Journal of Documentation*, **51**(3), 271–280.

Frye, B. E. (1997). Universities in transition: Implications for libraries. In L. Dowler, (ed.). *Gateways to Knowledge: The Role of Academic Libraries in Teaching, Learning, and Research.* Cambridge, MA: MIT Press. Pp. 3–16.

Gilliland-Swetland, A. (1998). Defining metadata. In M. Baca (ed.), *Introduction to Metadata: Pathways to Digital Information.* Los Angeles: Getty Information Institute.

Green, S.; Cunningham, P. and Somers, F. (1998). Agent mediated collaborative Web page filtering. In Klusch, M.; Weiss, G. (eds.), *Cooperative Information Agents II. Learning, Mobility and Electronic Commerce for Information Discovery on the Internet. Second International Workshop, CIA'98.* Proceedings, Paris, France, 4–7 July 1998. Berlin, Germany: Springer-Verlag, 1998. Pp. 195–205.

Greenberg, D. (1998) Camel drivers and gatecrashers: Quality control in the digital research library. In B.L. Hawkins & P. Battin (eds.), *The Mirage of Continuity: Reconfiguring Academic Information Resources for the 21st Century*. Washington, DC: Council on Library and Information Resources and the Association of American Universities. Pp. 105–116.

Graubard, S. R. & LeClerc, P. (eds.). (1998). *Books, Bricks, and Bytes: Libraries in the Twenty-First Century*. New Brunswick, NJ: Transaction Publishers.

Hawkins, B. L. & Battin, P. (eds.) (1998). *The Mirage of Continuity: Reconfiguring Academic Information Resources for the 21st Century*. Washington, DC: Council on Library and Information Resources and the Association of American Universities.

Hedstrom, M. (1991). Understanding electronic incunabula: A framework for research on electronic records. *The American Archivist,* **54**(3), 334–354.

Hedstrom, M. (1998). Digital preservation: A time bomb for digital libraries. *Computers and the Humanities,* **31**, 189–202.

Hill, L. L.; Carver, L.; Larsgaard, M.; Dolin, R.; Smith, T. R.; Frew, J.; & Rae, M. A. (2000). Alexandria Digital Library: User evaluation studies and system design. *Journal of the American Society for Information Science*, **51**(3), 246–259.

Hill, L. L., Dolin, R., Frew, J., Kemp, R. B., Larsgaard, M., Montello, D. R., Rae, M.A., & Simpson, J. (1997). User evaluation: Summary of the methodologies and results for the Alexandria Digital Library, University of California at Santa Barbara. *Proceedings of the American Society for Information Science Annual Meeting*, **34**, 225–243. http://www.asis.org/annual-97/alexia.htm.

Johansen, R. (1988). *Groupware: Computer Support for Business Teams.* New York: Free Press.

Kent, S. G. (1996). American public libraries: A long transformative moment. In Books, Bricks, and Bytes, *Daedalus, Journal of the American Academy of Arts and Sciences; Proceedings of the American Academy of Arts and Sciences,* **125**(4), 207–220. Republished in S. R. Graubard & P. LeClerc (eds.). (1998). *Books, Bricks, and Bytes: Libraries in the Twenty-First Century*. New Brunswick, NJ: Transaction Publishers.

Lehmann, K-D. (1996). Making the transitory permanent: The intellectual heritage in a digitized world of knowledge. In Books, Bricks, and Bytes, *Daedalus, Journal of the American Academy of Arts and Sciences; Proceedings of the American Academy of Arts and Sciences,* **125**(4), 307–330. Republished in S. R. Graubard & P. LeClerc (eds.). (1998). *Books, Bricks, and Bytes: Libraries in the Twenty-First Century*. New Brunswick, NJ: Transaction Publishers.

Lesk, M. E. (1997). *Practical Digital Libraries: Books, Bytes, and Bucks.* San Francisco: Morgan Kaufman.

Levy, D. M.; & Marshall, C. C. (1995). Going digital: A look at the assumptions underlying digital libraries. *Communications of the ACM,* **38**(4), 77–84.

Lievrouw, L. A., Rogers, E. M., Lowe, C. U., & Nadel, E. (1987). Triangulation as a research strategy for identifying invisible colleges among biomedical sciences. *Social Networks,* **9**, 217–238.

Lyman, P. (1996). What is a digital library? Technology, intellectual property, and the public interest. In Books, Bricks, and Bytes, *Daedalus, Journal of the American Academy of Arts and Sciences; Proceedings of the American Academy of Arts and Sciences,* **125**(4), 1–33. Republished in S. R. Graubard & P. LeClerc (eds.). (1998). *Books, Bricks, and Bytes: Libraries in the Twenty-First Century*. New Brunswick, NJ: Transaction Publishers.

Lynch, C. A. (1993a). *Accessibility and Integrity of Networked Information Collections* (Background Paper No. BP-TCT-109). Washington: Office of Technology Assessment.

Lynch, C. (1993b). Interoperability: The standards challenge for the 90s. *Wilson Library Bulletin,* March, 38–42.

Lynch, C. (1998). The evolving Internet: Applications and network service infrastructure. *Journal of the American Society for Information Science,* **49**(11), 961–972.

Lynch, C. & Garcia-Molina, H. (1995). *Interoperability, Scaling, and the Digital Libraries Research Agenda.* http://www.hpcc.gov/reports/reports-nco/iita-dlw/main.html.

Marchionini, G. (1995). *Information Seeking in Electronic Environments.* NY: Cambridge University Press.

Mason, M. G. (1996). The yin and yang of knowing. In Books, Bricks, and Bytes, *Daedalus, Journal of the American Academy of Arts and Sciences; Proceedings of the American Academy of Arts and Sciences,* **125**(4), 161–171. Republished in S. R. Graubard & P. LeClerc (eds.). (1998). *Books, Bricks, and Bytes: Libraries in the Twenty-First Century*. New Brunswick, NJ: Transaction Publishers.

McKnight, C.; Dillon, A. & Richardson, J. (1991). *Hypertext in Context.* Cambridge: Cambridge University Press.

Meadows, A. J. (1998). *Communicating Research.* San Diego: Academic Press.

Moore, R., Baru, C., Gupta, A., Ludaescher, B., Marciano, R. & Rajasekar, A. 1999. *Collection-Based Long-Term Preservation,* GA Report GA-A23183 submitted to National Archives and Records Administration, June 1999.

Newby, G. B. & Peek, R. M. (eds.). (1996). *Scholarly Publishing: The Electronic Frontier.* Cambridge, MA: MIT Press.

Nunberg, G. (ed.). (1996). *The Future of the Book.* Berkeley, CA: University of California Press.

Odlyzko, A. M. (1997). Silicon dreams and silicon bricks: the continuing evolution of libraries. *Library Trends,* **46**(1), 152–167.

Olsen, J. (1997). The gateway library: Point of entry to the electronic library. In L. Dowler, (ed.). *Gateways to Knowledge: The Role of Academic Libraries in Teaching, Learning, and Research.* Cambridge, MA: MIT Press. Pp. 123–134.

Paepcke, A. Chang, C-C. K. Garcia-Molina, H. & Winograd, T. (1998). Interoperability for digital libraries worldwide. *Communications of the ACM,* **41**(4), 33–43.

Rajasekar, A. Marciano, R. & Moore, R. (1999). Collection Based Persistent Archives, *Proceedings of the 16th IEEE Symposium on Mass Storage Systems,* March 1999.

Rockwell, R. C. (1997). The concept of the gateway library: a view from the periphery. In L. Dowler, (ed.). *Gateways to Knowledge: The Role of Academic Libraries in Teaching, Learning, and Research.* Cambridge, MA: MIT Press. Pp. 109–122.

Rothenberg, J. (1995). Ensuring the longevity of digital documents. *Scientific American,* **272**(1), 24–29.

Sarkar, M. B., Butler, B., & Steinfield, C. (1995). Intermediaries and cybermediaries: A continuing role for mediation players in the electronic marketplace. *Journal of Computer-Mediated Communication,* **1**(3).

Schilit, B. N. Price, M. N. Golovchinsky, G. Tanaka, K. & Marshall, C. C. (1999). As we may read: The reading appliance revolution. *IEEE Computer,* **32**(1), 65–73.

Silberman, S. (1998). Ex Libris: The joys of curling up with a good digital reading device. *WIRED,* **6**(07), 98–104.

Smiraglia, R. P. & Leazer, G. H. (1999). Derivative bibliographic relationships: The work relationship in a global bibliographic database. *Journal of the American Society for Information Science,* **50**(6), 493–504.

Smith, T. R., et al. (1996). A digital library for geographically referenced materials. *Computer*, **29**(7), 54–60.

Strassmann, P. A. (1997). Will big spending on computers guarantee profitability? *Datamation,* **43**(2), 75–85.

Tillett, B. B. (1991). A taxonomy of bibliographic relationships. *Library Resources & Technical Services,* **35**, 150–159.

Tillett, B. B. (1992). Bibliographic relationships: An empirical study of the LC machine-readable records. *Library Resources & Technical Services,* **36**, 162–188.

Toffler, A. (1981). *The Third Wave.* London: Pan.

Twidale, M. B. & Nichols, D. M. (1998a). *A Survey of Applications of CSCW for Digital Libraries.* Technical report CSEG/4/98, Computing Department, Lancaster University, U.K. Http://www.comp.lancs.ac.uk/computing/research/cseg/projects/ariadne/docs/.

Twidale, M. B.; & Nichols, D. M. (1998b). Computer supported cooperative work in information search and retrieval. In Williams, M. E. (ed.), *Annual Review of Information Science and Technology*, **33**, pp. 259–319.

Van Bogart, J. W. C. (1995). *Magnetic Tape Storage and Handling: A Guide for Libraries and Archives.* Washington, DC.: Commission on Preservation and Access, and St. Paul, MN: National Media Laboratory.

Wactlar, H. D. Christel, M. G. Yihong Gong & Hauptmann, A. G. (1999, February). Lessons learned from building a terabyte digital video library. *IEEE Computer,* **32** (2), 66–73.

Waters, D. J. (1998). What are digital libraries? *CLIR (Council on Library and Information Resources) Issues,* No. 4. http://www.clir.org/pubs/issues/issues04.html.

Weibel, S. (1995, July). Metadata: The foundations of resource description. *D-Lib Magazine*, http://www.dlib.org/dlib/July95/07weibel.html.

Winner, L. (1998). Report from the Digital Diploma Mills Conference. *Science as Culture,* **7**(3), 369–377.
Zhao, D. G.; & Ramsden, A. (1995). Report on the ELINOR electronic library pilot. *Information Services & Use,* **15**, 199–212.

8

Social Foundations for Collaboration In Virtual Environments

GLORIA MARK

Introduction

The astounding growth of Web technologies is profoundly impacting collaborative work and learning. Whereas less than five years ago few people outside of the field of computer science had ever heard of a World-Wide Web, the Web is now influencing which information people will encounter, the form that it is presented in, who one will collaborate with, and how the results will be worked on and shared.

The development of new forms of Web-based multimedia and collaborative software environments has extended the possibilities for interacting with others and viewing information. These technologies are enabling information to be transmitted fast and world-wide so that at any time and place, people who can access an Internet connection can find out about past, current, and future events anywhere in the world, can meet others, and can respond from their computer with a wide selection of media choices. Such technologies have tremendous potential for enhancing education.

It is perhaps not surprising to see that the use of global networks for education are influencing the development of new models for teaching and learning. New models of active individual computer exploration are replacing old models of classroom lectures, access to experts on-line is replacing the single teacher, learning from up-to-date on-line information is replacing outdated textbook information, and discussion about course work becomes continuous, as opposed to being limited to face-to-face meetings (Hiltz and Turoff, 1993). New paradigms in distance learning are employing synchronous and asynchronous participatory models, whereas formerly they were based on non-interactive live video transmissions. Although many of these models are still based on individual learning, cooperative learning models using Web technology are beginning to emerge. Experiments with virtual classrooms and universities using collaborative software and multimedia have for sometime already proved to be successful for motivating students (e.g. Hiltz, 1993; Harasim et al., 1995).

Global education can be distinguished from other cooperative work activities, since it involves aspects such as curriculum planning, framing the role of teacher, assembling course materials, and determining performance measures. On a finer-grained level, activities in a virtual education environment can be structured for questions and responses, reading documents, selecting topics, viewing grades, and taking exams. Yet another component in virtual education is collaboration, which spans a wide range of types of interaction, from a project group working together, to collaborative writing in a class, to communicating in seminars, or even discussing course work outside of the virtual classroom. The potential to discuss coursework and work collaboratively in groups is considered an essential aspect of the learning process in virtual educational environments (Hiltz and Turoff, 1993).

In this chapter, I focus on that aspect of virtual education which concerns collaboration. To support collaborative processes involved in teaching and learning, it is necessary to reconceptualize the direction of the design of Web technologies from that of supporting single users searching for and contributing information in isolation, to the collective and social processing of information. To be truly effective, collaborative learning and work must enable rich social interaction, both inside and outside of the virtual classroom. This means not only providing a means for students and teachers to express themselves within a classroom environment, but also to find others and form relationships outside of the classroom setting to enrich the learning experience. To achieve this, we need to move beyond a task-oriented approach toward technology design, and take account of the social aspects involved in collaboration. We need to support processes involved in the formation and maintenance of groups: finding others, establishing channels for formal and informal communication, and developing trust and group identity. It is the premise of this chapter that the current technology developments make networked computers a promising candidate and baseline for the development of a new on-line social infrastructure for people.

In the next section I begin by presenting an overview of technologies already available to support Internet-based collaboration, and discuss their strengths and shortcomings in enabling interaction. In Section 3, I will discuss how social information can help people not only to collaborate, but to understand their work context. Communication on-line occurs through mediums ranging from asynchronous e-mail to virtual environments, and unless one is a member of the "in-group" of expert users of such systems (Donath, 1998), users of Internet technologies will need to develop new models of communication and interaction, involving new forms of representations, awareness information of others' activities, and conventions. How we can reconceptualize the design of Web technology to support the development of such models will be discussed in the last section.

Current Means for Internet-based Collaboration

The number of on-line systems available to support collaboration and communication across distance has grown rapidly in the last few years. In general, collaborative systems differ in the way that interaction and communication with them is structured. Unlike face-to-face interaction, where social information about cooperating partners is rich, and which we are experts at encoding, the communication channels in on-line collaborative systems are in comparison quite limited. In this section I present a brief survey of systems available currently for collaboration through the Internet. The different systems support different aspects of collaboration, and correspondingly, different types of social information are transmitted.

TABLE 8.1

Synchronous and asynchronous Internet and Intranet systems

Primarily Synchronous Collaboration	Telephony Chat systems Video conferencing Video snapshots Shared virtual environments Data sharing
Primarily Asynchronous Collaboration	Electronic mail Electronic bulletin boards Asynchronous conferencing List servers Shared workspaces
Combination Environments	Collaboratories Team rooms Virtual classrooms

Table 8.1 shows different classes of Internet and Intranet-based systems sorted according to a typology of those supporting same time (synchronous) and different time (asynchronous) collaboration. Other typologies have been used to classify CSCW systems using same and different place categories as well (see Baecker et al., 1995). I have confined the discussion here to systems used across distance since the lack of social information is more critical when people are not meeting face to face, and do not share the same environmental context when they collaborate.

Synchronous collaborative systems

Synchronous systems enable people to collaborate at the same time, regardless of where they are located geographically, using audio, video, text mediums, and combinations thereof. An advantage of synchronous communication in terms of social information is the immediate feedback that partners can receive, as well as the ability to continually confirm implicit assumptions in conversation (Clark, 1996). Data sharing, available in conjunction with audio, text-based, and video communication systems, enables people on-line to collaboratively view, write, and edit documents.

Due to increased bandwidth possibilities and data compression techniques, video systems can provide digital visual information on conversation partners, either via continuous film or up-dated snapshots. The introduction of video on the Internet has been hailed by many for its potential in providing social information, yet various studies have failed to find more than marginal effectiveness, especially when audio is available (Edigo, 1988; Gale, 1991). The disadvantage of video snapshots, while saving on bandwidth requirements, is that the images often do not coincide with the audio data; this limits their aid in interpreting fine nuances in conversation, by making it difficult to associate facial expression with speech.

A more informal means of communication on the Internet is through chat rooms, which appeared on the early bulletin boards and which became very popular for multi-person conversations. Chat rooms are essentially 'living' transcripts of events: who is present, who enters, who leaves, who has said what to whom. One types into a scrolling text window next to one's on-line alias. People cruise chat rooms looking for the right atmosphere. Private rooms are available when the conversation tone becomes more intimate. In some sense, chat rooms simulate the idea of chance meetings as in a cocktail party; the windows are rooms filled with simultaneous conversations, and the private rooms are corners that people can retreat to.

Discussion forums have expanded the original notion of Usenet newsgroups to offer threaded discussion around a number of themes. Usenet began in 1979 and has since grown to over 20,000 different newsgroups at the time of this writing. Users can filter contributions of other users, and can receive up-dates on what is new since their last visit. The use of appropriate metaphors, such as meeting rooms, helps users distinguish different discussion topics. One method used in some discussion forums of enhancing the social experience is through facilitators, which have been proven effective in coordinating face-to-face meetings. Their use in discussion forums is to integrate new members and guide the discussion. However, despite such aids, the problem still remains that on-line conferencing lacks many cues that verbal and visual communication offer. This lack of cues has been shown to affect group processes such as decision-making, effect of status, and participation (Sproull and Kiesler, 1991; Hiltz and Turoff, 1993).

Chat room functionality is also offered in MUDs, (multi-user dungeons) which provide spatial navigation and text-based interaction with other users and the environment (e.g. operations with the objects to be found in the diverse rooms of the dungeon). While chat rooms, MUDs, and MOOs (MUD Object-Oriented) offer a certain amount of awareness information not normally present on a Web page, they are restricted to text-based communication. People establish MUD identities, and the extent to which people identify with their characters is exemplified in reported cases of genuine grieving when their character dies (Turkle, 1997). A spin-off of these are MUSEs (Multi-User Simulation Environment), such as that used at MIT for interactive experiments in a virtual museum setting. There are well over 200 MUSEs in existence for educational purposes, enabling children and adults to create, and share results with others (Hunter, 1995).

The multi-user interactive MUDs have evolved into 3-D graphical environments, which offer communication using audio and/or text. These virtual environments are often structured in terms of a geography of places; one can navigate or teleport between different rooms or spaces. In the case of text-based MOOs, the user must retain a mental model of the geography; with 3-D graphical environments, one can readily see the spatial relations of objects and people in the environment. People use certain social conventions in these environments that involve spatial relations, such as facing one another when they speak and maintaining a social distance when they converse (Mark and Becker, 1998).

Asynchronous systems

Whereas synchronous environments make it easier to construct context for interaction, with asynchronous systems, the context in which the social information is presented can be minimized or lost due to the time delay. Compared with asynchronous interaction in the physical environment, where place and artifacts provide cues for a context, interaction in an asynchronous electronic medium (and often synchronous, too, for that matter), is not unlike working in a void, where language and results are produced in at best, a vague context. Context must be reproducible for conversation and documents to take on a fuller and clearer meaning, such as in providing design rationale.

Standards for the coding of media information and the availability of multimedia computers helped e-mail to emerge as the primary asynchronous transport media for the exchange of multi-media information. Although e-mail is considered an asynchronous medium, its technical basis enables transmission to be so fast that if two people exchange messages rapidly, it can be perceived as a synchronous medium. The simplicity of e-mail is a great advantage, yet for many users it is difficult to filter information, and structure conversations, problems which were addressed by systems such as Information Lens (Malone et al., 1987) and The Coordinator (Winograd, 1987/88).

Distribution lists are a means for people to broadcast their business, or opinions, as may be the case. William Mitchell (1996) in *City of Bits* describes lists as electronic Speakers Corners, as in Hyde Park: anyone can speak to the gathered crowd. Being on the right list is like being in the right place; it means access to the most current up-to-date information from the people most informed.

The emergence of newsgroups on the Internet began to unite users into virtual communities. Recently the number of newsgroups that provide a forum for leisure, social or political topics has increased. However, as a social forum, newsgroups are also deficient. It is difficult to know exactly who and how many participants are involved in the forum. Further, issues such as identity deception and trust in the integrity of the information are concerns for members who must devise collective means to solve this problem (Donath, 1998).

Shared workspaces are repositories of information where people can collectively organize and store documents. Documents are stored in public shared folders, and then retrieved and worked on, either in the public workspace, or on one's private desktop. The first shared workspace based purely on Internet and World Wide Web technology was the BSCW system (Bentley et al., 1997). An important consideration in cooperation with shared workspaces is the tradeoff between personal and group work styles. This is reflected in individual differences in data organization, naming conventions, and work protocols (Mark et al., 1997). This suggests the need for social information about others' work habits that can facilitate users in a process of developing congruent work styles. Group members who work virtually have even more difficulty than face-to-face groups in achieving such congruency. The hard task is that they must gradually accommodate their work styles to that of a common group procedure while receiving limited information about others' activities.

Some Web-oriented applications and toolkits are emerging to provide users with the means to see and meet each other, such as surfing together through the Web using a text-base chat mechanism (Ubique, 1995), and viewing common Web pages with application sharing and telepointers, writing in comments, and collaboratively navigating (Greenberg and Roseman, 1996). However, these systems are limited by restricting awareness information to the current Web page. A notable exception is MetaWeb, which provides a basic infrastructure for deploying social information about Web users by presenting mutual awareness about other coworkers' current activities and current locations (Trevor et al., 1997).

Combination environments

Although I distinguish between synchronous and asynchronous collaboration support, the traditional concepts of time and place are changing as people become networked to each other. The borders between distinct times and different places are becoming blurred as communication takes place almost independently of the locations and time in which partners' messages have been sent. For this reason, some environments have emerged using metaphors that provide people with a sense that they are collocated in the same media space. Collaboratories, or global laboratories, are such environments, and they enable scientists to extract samples and view data from sites thousands of miles away (Johnston and Agarwal, 1998).

Team rooms are another type of such combination environment which often rely on a room metaphor to provide workplaces for both group and individuals, using both asynchronous and synchronous means (e.g. Greenberg and Roseman, 1998). The persistence of objects and awareness mechanisms that inform who is working in which room contribute to the sense of shrinking the distance between collaborating partners in time and space. DIVA was a prototype virtual office environment which employed metaphors of rooms and desks (Sohlenkamp and Chwelos, 1994). Another prototype, the GroupDesk system, based on a shared desk metaphor, was designed to integrate the users with the objects of work (Fuchs et al., 1995).

Virtual classrooms convey the sense of a virtual place for learning without walls. They employ a number of different software, combining both asynchronous and synchronous conferencing systems and shared workplaces. In addition, there is program support for keeping records of completed activities and structuring questions and responses among students. Students from any place at anytime can select and view lecture materials, and at scheduled times, can see presentations and participate in seminar-type discussions.

As a consequence of such environments, physical place becomes less important, and the notion of temporal order and time-dependent speaking turns is vanishing. As new concepts of time and place evolve, so must corresponding new behaviors develop in order to enable cooperation and coordination, such as new social conventions and etiquettes. And designers must also encourage the development of these behaviors by introducing mechanisms that can convey and facilitate the transmission of social information.

A Social Web

The evolutionary direction of these Internet technologies is heading closer toward supporting more comprehensive collaborative behaviors. Yet despite these advances, current technologies still do not adequately account for two fundamental social aspects of work and learning: its situated and complex nature.

According to the notion of situated behavior, work and learning are embedded in social practice. Reconceptualizing work and learning from individual cognitive activities to behavior within a social context is a paradigm shift (Suchman, 1987; Lave and Wenger, 1991). According to this view, activities are not performed in isolation; they are performed within an intricate system of social relations. The process of learning is described by Lave and Wenger (1991) as being intertwined with gaining a sense of identity within a particular community of practice. A newcomer in a learning field strives to be integrated into this community, developing identity as one moves closer into the field. This idea presupposes that the community of practice is clear and identifiable for the newcomer.

Such a theory of situated learning relies heavily on making the community of practice transparent and accessible. This poses a particular challenge for global networks of people who form communities of practice independent of the geographical location of the members. This calls for the requirement that the social context and relations of the members of a work and learning community be salient, which extends beyond making the results of cooperation accessible. In other words, in on-line collaboration, people must be represented as social actors.

A shortcoming also exists with current technologies in representing another social aspect of collaboration: namely, the fluidity and lack of precise boundaries for working spheres. What characterizes most work is that it is continuous and complex. On any given workday, people continually move from one task to the next. Additionally, people move in and out of different working spheres involving distinct individuals. In a large study of work activity, Reder and Schwab (1990) found that at a Fortune 500 company, sales managers engaged in approximately 15 different tasks per day (defined as discrete work objectives), involving about 15 different persons in interactions. Senior managers participated in even more work spheres, performing roughly 30 different tasks a day, with about 24 non-redundant interactions. Not all tasks directly involved cooperative behavior, but it is difficult to distinguish among those tasks in which solitary activity was directed toward some form of collaboration. People might move from a staff meeting to a telephone conference, to calling home, to working on a document in a shared workspace, to meeting a colleague in the coffee room, to searching the Web for information.

The empirical results of Reder and Schwab confirm that people are members of different working spheres; moreover, people are also involved in spheres that extend beyond their daily work: members of professional specialties, invisible colleges and social spheres. The same is true whether we consider work or university life; in the course of their daily routines, people experience a continual transition from one sphere to the next, engaging in a mixture of informal and formal interaction: attending classes and seminars, working on joint projects, seeking information from libraries, collaboratively writing, and participating in social functions. Social information is an important component that helps define and establish the spheres and helps people move seamlessly in and out of them.

Virtual team members are not part of the same working spheres that we share with our colleagues who are in physical proximity. Supporting the means to communicate, and to transfer and receive data across distance, is not enough to support the complexity of work. Additionally, there is a requirement for technological support to ease the transition between work in virtual and physical spheres. The same types of requirements that exist when we interact with face-to-face teammates, become even more crucial in a digital environment: the means to communicate precisely, establish trust, and authenticate identity. In face-to-face environments, through direct and peripheral awareness, we are aware of the activities of others. Through this we can discern implicit conventions about coordination. To add to the complexity, virtual collaboration occurs not only when one is on-line, but coordination begins before system use, during, and after. A social infrastructure is necessary to aid people in finding cooperating partners, planning all phases of cooperation, and sustaining the relationships once the formal system use has ended.

The technologies described in the last section support components of work. That is, they are designed to support a wide range of tasks such as collaborative writing, synchronous and asynchronous communication, and even informal meetings. Although they provide the functionality for combining experiences and producing results, they only weakly support the social nature of work. If we consider on-line environments for collaborative work or study, we must provide an infrastructure to enable people to maintain a coherent and continual sense of work life. The technology must enable people to find others to engage in rich and rewarding interaction, and must promote seamless transitions in and out of different work and social spheres.

We need to view collaboration not just in its immediate context, but in a broader and more permanent sense. The notion of a *Social Web* refers to an infrastructure that people can use to build an integrative on-line experience through the provision and use of social information, even when they are not formally collaborating (The Social Web, 1998). Through the integration of a number of mechanisms, some which already exist, some which need to be developed, the social cohesion for collaboration and developing communities of practice can be provided. In the next section design considerations that could transform the Web into a Social Web will be presented.

Some Issues and Challenges in Designing a Social Web

Developing a Web environment that supports the inherent social nature of collaboration involves many research challenges. To begin with, a breakdown of the social processes involved in collaboration can clarify requirements for an on-line social infrastructure. It is intended that such an infrastructure should serve as a complement for other collaborative systems, i.e. it should be application independent. This infrastructure should support the social aspects of collaboration through the following goals: establishing presence of the cooperating partners, building a sense of community even when members are not formally collaborating, managing collective information of the community, and creating a cohesive environment with appropriate cues that can facilitate social interaction. Table 8.2 presents a summary of these.

TABLE 8.2
Classification of Social Information to beIncluded in a Social Web Infrastructure

Social Component	Means
Establishing presence	Sensory information
Building sense of community	Determining group parameters
Self-organization	Norms and conventions
	Identity persistence
Social processing of information	Social construction of knowledge
	Agents as mediators
Creating a cohesive environment	Appropriate metaphors
	Interface as environment
	Awareness
	Design of virtual space

Establishing presence

The study of presence in media environments has received a lot of attention recently; undoubtedly the development of new Internet communication technologies has played a significant role. Although different conceptions of presence have been proposed, the commonality underlying them is suggested by Lombard and Ditton (1997), as the perception of non-mediation in a communication environment.

Face-to-face communication is rich with social information, and we are experts at encoding and interpreting verbal inflections and non-verbal behavior during interaction. But communicating through media leads to a reduction of these social cues. Media differ in their capability of conveying social cues and fine nuances in communication (Rice, 1992). For example, media research suggests that visual media facilitate more presence than audio media, which in turn would facilitate more presence than text media (Short et al., 1976).

A high sense of presence has also been defined as actors feeling that they are together and sharing the same space. This has been used to describe the immersive feeling that people can experience when they are in a virtual environment (Slater and Usoh, 1993). With shared virtual environments, presence in this sense can mean that one feels "transported" to a common location with others (Benford et al., 1996). A shared media space then refers to that location that all members agree upon that is common among them, and separate from their physical environments. A high degree of presence should lead to a higher degree of involvement in the environment, and with the other actors who share the experience, even though they do not share the same physical surroundings and corresponding traits, such as common culture, and language. It should be the goal of a Social Web to include media to convey sensory information that enhances a non-mediated communication experience.

The role of sensory information

Much information that we gain through our senses in the real world is lost when entering an electronic environment. We lose the sense of touch, of background noise, the scent of odors, kinesthesis (the sense of bodily movement), the sensation of taste, and visual cues which give us a peripheral awareness of the presence of others. With the loss of such sensory impressions, the design of an electronic environment takes on an added importance. In the same way that a blind person compensates for the loss of one sense by acutely developing other senses, we must also consider how an electronic environment can provide new kinds of sensory information to make up for the loss of what we know as physical sensory information.

The lack of sensory information affects how we perceive ourselves in a virtual world, and how we perceive others. In real life, there is a primary contact between ourselves and others through our bodies; in virtual communication, the physical body is either portrayed in words in a text-based environment, or in simplistic avatars in a graphical environment. Both representations imply a lack of non-verbal communication strategies and social cues: gesture, facial expression, body language, age, gender, etc. Sensory information should provide social cues so that members in a digital community can portray and express their affiliations, beliefs, and interests and comprehend a wide range of nuances in communication. Although there are current bandwidth limitations on how much video and audio information can be portrayed on-line, we can envision alternative modes of information transmission, such as sensors, which can capture information about people's physical actions and show them in an on-line environment (Benford et al., 1997). Symbolic acting is a method developed to portray people's on-line activities through an electronic visual representation (McGrath, 1998). Social information can also be shown through directional gaze during an on-line conversation (Donath, 1995).

Establishing a sense of community

In his article *Design Principles for Online Communities*, Peter Kollock (1998) discusses attributes that contribute toward the success of an on-line community: the existence of identity, rituals, an internal economy, ownership of property, a recorded history, coherent sense of space, and the opportunity for casual interaction. In addition, in real communities the existence of social conventions are fundamental to the stability of a social community (Giddens, 1990). When we consider community in a real campus or work life, many of these attributes apply as well. In university life, people belong to a number of communities; in each they have a sense of identity. Each university has its own set of conventions unique to its history and environment. There is intellectual ownership of ideas. Moreover, the opportunity for casual interaction is one of the richest offerings of university life; one knows gathering points, and has expectations of who one might meet depending on whether one goes to the library, a lecture, a cafe, or sprawls on the university lawn. In such a learning environment, the sense of community plays an integral role to education. Who we meet and the relationships formed between the university community members essentially determines a large part of our education.

Many different communities already exist on the Web: in Usenet groups, Internet Relay Chat, on-line MUDs and virtual environments. This points to an evolution in Web usage not only for information seeking, but also for developing relationships (Wellman and Gulia, 1998). While some critics argue that so-called on-line communities lack a sense of cohesion (Poster, 1995), other empirical research suggests that among users in on-line environments there is indeed a sense of cohesion and belonging (Roberts, 1998).

Determining group parameters

A vital aspect of a real-life community, whether it be a city, an enterprise, or a university, is the users' abilities to dynamically form and reform relationships, groups, and networks of people, according to their interests and goals. A challenge for a digital community to sustain itself is for its members to effectively organize and govern themselves, without formal "steering mechanisms", devising rules well-matched to the group (Ostram, 1990). Optimal group size is essential to reduce coordination costs and the risk of free-riding (Kollock and Smith, 1998). This means that groups need the means to restructure and find new relationships as the group's needs change.

A self-organizing principle also means that a community is aware of what its resources are and can effectively utilize and manage them. Collective resources of an on-line community can include technical knowledge, programming experience, and emotional support. To avoid falling into social dilemmas, members must be motivated to contribute their share of resources (e.g. answering queries, contributing information) without excessive use of mechanisms that draw upon limited bandwidth (Kollock and Smith, 1998). On the other hand, because of the potential to be overrun with information resources of the group's members, the need for technical support for group and individual information filtering becomes necessary.

Norms and conventions

Successful collaboration and interaction depends on people developing the appropriate norms and conventions. This is not only true in real working and social life, but is especially true in an electronic environment such as the Web, where the lack of social information makes it difficult for people to establish appropriate norms of behavior and conventions for carrying out work and interactions. We have learned from experiences with groupware that conventions for using a collaborative system cannot be ignored. Groupware conventions are vast, unpredictable, dynamic, unique to the group, and are often violated (Mark et al., 1997).

Social norms and conventions evolve as a consequence of interacting with others. Conventions in an on-line environment are not easy to establish. First, feedback and communication may be restricted compared to real life. In a face-to-face group, communication, feedback such as back-channel responses, and visual information about body language and facial expressions, are rich, which facilitates the negotiation and agreement of appropriate behavior. In an on-line environment, restricted communication makes it more difficult for people to see others' actions and to understand them. Second, a digital environment lacks appropriate cues for behavior. One may walk into a conference room or party in real-life, and know at once how to behave. Cues such as type of dress, posture, or tone of voice, immediately inform one as to whether the group is formal or informal, serious or funny. Such cues in an on-line environment are either lacking or not obvious. Third, representations, artifacts, and movement may be totally new experiences for many on-line users, making it difficult to apply analogy of conventions from real-life. Technical support, in the way of awareness mechanisms, sensory information, and agents is needed to help groups form and maintain conventions.

Social norms will also differ according to the group. In an electronic environment, new forms of norms and conventions must be developed to ensure that the community thrives: these may include new language, new styles of communication, a new basis for trust, ways for preserving privacy of information, and even greeting rituals, among many other unforeseen conventions.

Identity persistence and representation

Identity persistence and authentication of identity is a requirement in an on-line environment to establish trust and security of information exchange (Donath, 1998). Trust in virtual groups is especially difficult to achieve since it is generally believed to be formed through personal relationships and common membership in social networks (Powell, 1990). Yet it is an important component of development in a group's life-cycle. Jarvenpaa and Leidner (1998) found that trust in virtual teams may form fast, but may not be substantial. This points to the need for identity persistence to substitute for knowledge of identity gained through personal contact.

Currently in on-line environments people can be represented on-line by text, audio, video images, or avaters. Each of the forms supply incomplete visual information about a person. However, the potential with electronic information exists, not yet fully taken advantage of, to supply a representation with unique attributes about a person, such as one's activities, interests, moods, and experiences. Such a comprehensive representation that is persistent could be expected to facilitate the development of trust among virtual group members.

Social processing of information

At the heart of a Social Web is the shared knowledge among the members. The Social Web would require that we change our conception from individual electronic stores of knowledge to a common, community constructed knowledge base. Community knowledge consists not only of the information itself, but also of the characteristics of the knowledge: its centrality to the group, its rating, and relationships (Nakata et al., 1998). Building a knowledge base is intricately tied to the development of a community in an on-line environment, described above: the norms and conventions, authentication of contributors, and social willingness of the members to contribute. Thus, once knowledge is reconceptualized from individually owned to community-based, the social properties of the group must be considered as influences upon that body of knowledge. Agents are a means to decrease the complexity of managing this information.

Social construction of knowledge

Reading information alone, searching for information, collecting it, and editing, are not social activities. Currently individual collections of information are owned by Web users: e.g. data files, bookmarks, and pointers to relevant databases. Instead of a vast collection of separate pieces of information of Web users, the notion of a shared information and knowledge base belonging to a group or community can be valuable (Nakata et al., 1998). Depending on the sphere of people, information should have a meaningful structure, which can be achieved by collecting and structuring the information socially. Examples of how shared knowledge bases can be collectively constructed is by contributing personal information and knowledge, links, commentaries, cross-references, and keywords to facilitate querying and filtering for others. An example of a system designed to gather and facilitate the information flow between people is the Knowledge Pump (Glance et al., 1998). What is interesting about this system is that knowledge flow and use is supported within a community of people based on social properties of the group, namely their interactions and recommendations.

Once gathered, social knowledge must be kept alive. An important property of collective knowledge is for the community to be aware of what knowledge is being socially used and what is abandoned (Maeda et al., 1997). To insure that knowledge is kept robust in the community, members must establish appropriate conventions, and must guarantee the reliability and authenticity of the knowledge.

Agents as mediators

Inhabiting a social web involves cognitive complexity. In addition to managing information and navigation, we need to process social information and interactions as well: recognizing appropriate on-line behaviors, finding attractive people to interact with whose interests match ours, constructing common knowledge bases, managing communication, and representing ourselves to others.

Agents can provide people with a data base of knowledge, experience, and events. Agents designed to reduce information overload has already been shown to be a promising research direction (Mitchell et al., 1994; Maes, 1994). Personal agents can act as representatives of individuals, and specify their owner's interests to other agents, they can help their owner find information and expertise, they can perform enquiries, they can relay information, and can make contacts with other agents on behalf of their owner. Communal agents can act as mediators on a higher level, identifying commonalities among clusters of people and knowledge, and connecting groups in different disciplines who have overlapping interests.

Creating a cohesive environment

While links are technical connections between Web sites at present, we need to recognize that places are meaningfully connected and grouped into distinct clusters, and reflect spheres of different activities: e.g. classrooms, campuses, neighborhoods, and larger cities. We can learn from architects and urban designers those principles of design that are used to communicate behavioral expectations and facilitate social constellations of people. Yet we still need to apply these principles to virtual space. We need to reconceptualize the notion of interface as inter-*active*-face to convey the sense that people are truly in an interactive shared environment.

Just as environmental factors in the physical world shape culture such as, e.g. climate, terrain and natural resources available, we should also expect the technological environment to shape the culture of its inhabitants as well. For example, in newsgroups where text is the sole communication channel, many linguistic conventions have developed, ranging from the use of capital letters which indicate shouting, to determining ways for authenticating user identity and information through writing conventions (Donath, 1998). Thus, we should expect the environment created by the technology to shape culture and influence behavior among virtual actors.

Interface as environment

The content and services of a social digital environment will influence its users; but additionally, the design of the interface will also influence the inhabitants to feel comfortable and motivated to participate. The interface design must integrate the different Social Web functions so that they are cognitively connected for the user. We must consider which metaphors will make the presentation of a Social Web environment, as well as its interface, intuitively easy for its users: orientation and navigation must be simple. For example, teamrooms and many MUDs use a metaphor of interconnected rooms that are containers of information and people. A broader metaphor is needed in a learning environment to convey the wide range of social spheres of interactions possible, as well as to make it cognitively easy for the members to differentiate their smaller social spheres. The metaphors must present the dynamic nature of a Social Web environment: its changes, its events, and its processes. The types of artifacts in a digital environment influence how people will use the environment. Objects which are persistent, and which can be shared among users, will play an integral role.

Awareness

A comprehensive environment must also present awareness information about collaborating partners: not only to heighten the social information about people during interaction, but also to indicate which activities other colleagues are engaging in, and where they are at, when they are not present in common working spheres. Awareness, as defined by Dourish and Bellotti (1992), enables one to construct a context for one's own work, by providing an understanding of the activities of the other participants. In the same way that one is aware of peripheral activities of others in the physical environment, reconstructing this information on-line enables one to broaden the work context so that others' become a part of the virtual setting. In order to create a perception of an integrative on-line social environment, the goal of an awareness mechanism should be to provide activity information independent of the application one is working in (Prinz 1999). In other words, cooperation is not bounded by immediate working spheres; awareness needs to cut across the current working spheres to promote a sense of seamlessness in cooperation.

Design of virtual space

When people move and interact in virtual spaces, many similarities exist to their social behaviors in physical spaces: in the social distances kept between individuals, in positioning within a group, in behaviors used to signal privacy, and even in the way that avatars turn to face one another when they interact (Jeffrey and Mark, 1999). Such behaviors suggest that people construct virtual space as a social space. Capitalizing on this idea, designers need to consider the design of virtual space to support and promote social interactions. Spatial distance can be used to convey group membership, to represent similarities among people; and to show people's relationships and interests. Architectural concepts can be applied to the design of virtual space in the same way that sociopedal environments promote the formation of constellations of people, and in how landscape design can facilitate traffic flow. The design of the environment needs to support easy transitions between spaces, conveying the notion upon entering of what type of place it is, and the behaviors and information expected there.

Types of Settings for a Social Web

Thus far I have described social aspects that must be considered when designing a technological infrastructure to support collaboration. In this section I describe potential settings that could benefit from such a Social Web infrastructure. This describes not only an enhancement of technological applications that already exist, but also some visions of what could be.

There are three objectives associated with a Social Web:

- Enrichment of social life in a community: a Social Web can offer the opportunity for people to co-enjoy new forms of culture, entertainment, and leisure.
- Creating and sharing knowledge in a community: a Social Web can help to organize the exchange of the wealth of knowledge and experience in local and world-wide communities outside of market mechanisms.
- Reducing social isolation in a community: a Social Web can support people in finding others with similar interests, needs, and goals, thereby expanding a person's social radius independent of geographical bounds.

A Social Web could be expected to provide benefits for a broad range of people including children and senior citizens, as well as for heterogeneous groups in schools, universities, cities, industries, hospitals, government, and professional and political organizations. The following are some proposed collaborative application areas where such a social infrastructure could prove beneficial:

- Global Schools and Virtual Universities: that support group learning over wide-area distances, including on-line learning and cross-cultural exchange.
- Scientific Communities: that help professionals make contact, keep up with the development in the field, and prepare or hold on-line conferences.
- Learning Factories, that allow the building of social networks within enterprises to utilize the organizational knowledge of a company.
- Virtually Collocated Teams, that enable people within geographically disperse organizations to be virtually collocated to work in project teams and task forces.

Global schools and virtual universities

The new advances in computer-supported learning provide students with the opportunity to utilize and build upon other students' results. Students can gather information globally and document their projects and results in a reproducible and attractive form. Schools distributed geographically could cooperate in discussing and adding their experiences.

Additionally, global networks offer students in remote areas the opportunity to overcome intellectual isolation. Students can contact others with similar goals who may be located geographically far away, and they can collaborate together on projects. They can seek teachers who have expertise to match their interests, even though they may be long-distance, such as the experience of gifted students in the far outreaches of Montana who received instruction from MIT professors, in the Big Sky Project (Uncapher, 1998). The opportunities arising for teachers to extend globally and find highly motivated students can be equally rewarding for them as well.

Students can also participate in interactive group experiences worldwide. Students in North America studying German can have dialogues with students in Germany studying English. Studying Geography can be made more alive for students; they can ask each other directly about their country, culture, politics, and religion, seeing video images and hearing different languages. Engaging in common projects with students around the world can be a highly motivating and exciting experience for children and young adults.

What distinguishes networks for higher education from elementary and secondary school networks is that with the former, a significant, if not total amount, of the education can take place on-line (Harasim et al., 1995). Harasim et al. reports that already on-line peer reviews, collective ratings on assigned readings, seminars, and discussion groups have shown positive results in learning. In a comparative study of face-to-face versus on-line course participation, students reported higher interest, involvement, motivation, and time spent in on-line coursework.

With a Social Web infrastructure, the virtual university experience can be enhanced by providing not only the components for collaborative learning but also a means to form and maintain different communities in the universities relating to interests, or even fields of specialty. Additionally, students can become members of social communities, as one might find on a physical campus. This could include forming clubs and extracurricular activities, competitions, support groups—activities that occur outside of the bounds of the virtual classroom. In this way, an on-line education could be enriched by becoming more of a social experience.

Scientific communities

Members within a scientific discipline have specialized requirements for obtaining up-to-date knowledge which might be widely distributed. While collaboratories are being developed, currently most scientists use e-mail, distribution lists, and newsgroups for receiving information, and coordinating and collaborating in their work. For specific themes, directories, bookmark collections, and home pages are sources that collect distributed knowledge.

Conferences are going online: papers and reviews are exchanged, focus sessions and workshops are prepared and results are made available via the Internet. Real-world conferences can also occur simultaneously on-line in a Social Web. Video cameras will enable people to see who is present, and are even now beginning to enable remote participants to contact real-life conference participants for discussions. Conference participants can ask questions, and even give lectures when they are unable to travel.

The Internet accelerates research. Paper publications are often superseded by on-line publications. Even on-line journals, due to time involved with reviewing processes, cannot always keep up with the presentation of information as authors become their own private publishers and promoters presenting papers, project descriptions, and proposals on the Web. On the other hand, researchers need to find information from others. New research themes may lead to the creation of new groups, whose members soon develop their own terminology. This makes it more difficult to become aware of parallel work in other research communities, and to initiate an exchange of ideas and results. In a Social Web, word-of-mouth processes could help to bridge such gaps. Researchers need the means to detect what their colleagues consider relevant. Social rating systems could also help in selecting material; researchers could rate, annotate, and comment on papers taken from the Web. This volunteer process could play a similar function as organized reviewing procedures for traditional hardcopy media.

To serve the needs of professionals, a Social Web can offer the means to construct collective knowledge bases, to socially process them, to find others with similar backgrounds and visions, and form collaborations. Invisible colleges can become more "visible". People can be informed both on- and off-line of important events filtered to their needs: upcoming conferences, journal publications, and job openings.

Learning factories

Formerly, many organizations lacked documentation of rules, traditions, and procedures. Finding the relevant information was especially difficult for new members entering the organization, or even for those who might be long-time members. Organizational memory has been developed as a means to bring past experience and knowledge to share among the employees in a number of contexts: small business team support across time and project, product design, large projects with distributed participants, continuity in series of meetings, answers on a range of topics, access to documents, and assistance in interacting with the organizational environment.

It is envisaged that this knowledge will lead to a higher organizational efficiency since decision processes become transparent and reconstructable, redundant work is avoided, and new decisions and developments benefit from previous experiences. Although these attempts appear reasonable, experiences show that they often fail due to the difficulties in collecting and organizing the appropriate information from their members. We also see that the process of creating the organizational knowledge cannot be regarded as a formal data collection, but as a social collaborative process.

A Learning Factory can aim at the development of organizational knowledge, expertise, and best practices within an organization in a Social Web. It can do this through the provision of that knowledge to the organization members, and the contribution to that knowledge base by members as well. By the provision of services to find, meet, interact and cooperate with people who have similar expertise, problems, or who can provide help, a Social Web facilitates the development of relationships and groups within an organization independent of predefined organizational hierarchies or process structures. A best-practice workshop could offer the exchange of experience with existing processes, and a place for the development of proposals for new procedures, products, and services. An interactive help desk can offer aid for organizational procedures and technical problems. As a way to preserve tribal knowledge in an organization, users can also have access to the valuable knowledge and experience of retired employees and they can even connect with them remotely.

Virtually collocated teams

As organizations become more geographically distributed, the challenges of supporting global virtual teams are being increasingly recognized. Reorganization can now mean forming teams independent of their physical location, and dynamically redefining them as the company's needs change.

A virtual team, like any physically-collocated team, starts out without its own culture. It needs to define goals, roles, assignment of tasks, criteria for success, and what Schein (1990) describes as a common communication system. Forming a culture occurs as a gradual process; the group members need to accumulate information about each other. Trust and security in information are developed in this process and must ideally be established in the early stages of the group life-cycle.

A social infrastructure can help groups in these early stages of development by facilitating informal communication and awareness. An environment which contains "public" places such as lobbies, mail rooms, and coffee rooms can facilitate chance encounters. Private areas such as offices would provide the chance for more personal conversations with colleagues. A team visibility room would be a place for groups to store their information and establish a sense of identity by posting their accomplishments: similar to a Web page but instead is an interactive environment. Team-building can be promoted through events that are outside of the formal group meetings: brown-bag discussions, seminars, or learning about the different members' home turfs, e.g. through virtual tours of a work place.

Developing New Models of On-line Interaction

Trust provides the basic seeds for cooperation. In order to build solid and fulfilling on-line collaborative relationships, people need to achieve the level of trust where information about people, communication, and information is of the highest integrity. Social information about members is needed to transform a digital environment into an environment where one can receive abundant awareness information, where communication channels are rich with sensory information, where agents can sift through the mire and find others with similar interests and help people construct collective knowledge bases. Moreover, appropriate norms and conventions must be developed, especially to encourage reciprocity of pro-social behavior (Putnam, 1995). These mechanisms need to exist independent of, and complementary to, collaborative applications.

Transforming the Web into a Social Web involves the *integration* of these different mechanisms. New concepts for the design of places and events should lead to new *cultural models* of on-line environments. Web sites will become more like places. A "place" has behavioral expectations and cultural meanings, similar to a home, library, railroad station, museum, or pub. We see signs of such types of places in collaboratories, virtual classooms, and team rooms that are emerging. On-line places should be unique, but also recognizable as certain types, being invested with appropriate cues for behaviors and cultural understandings: roles, purpose, and function. They should have an atmosphere that can convey these aspects to the visitor upon entering and while staying in a place, whether the place is a virtual classroom, a study lounge, or a student center. Although presently on the Web there exist "places" such as shopping malls and libraries, we believe that while functional, they only weakly (if at all) convey social aspects. Others, while alluding to a metaphor, fail to provide functionality, as well as the means for social interaction. New types of places will come into existence on the Web that may have no analogy to real-world sites, in the same way that FAQs have arisen as a new form of literature on the Internet.

The pages in a Social Web should be considered as places for interaction: a foundation for supporting social processes. While events on the Web are already appearing, like art exhibits, concerts, and political forums, we have still to develop schemas associated with new types of on-line events that will emerge, such as: expert interactive advice sessions, guided tours, Chataquaa festivals, new forms of games, such as on-line Olympics, unique forms of expression for political opinion, among many others. New forms of on-line traditions are also emerging, such as language and rituals; collective ratings and seals of approval; and collective publishing of memoirs, photo albums, commentary books, and hypertext novels.

The new models of on-line places, events, and social processes will have even more of a profound impact on virtual education than what we are seeing today. Not only is the global village expanding, but it will become more personal and intimate as well. I will soon share my early morning coffee break with a fellow researcher in Sydney, who is having her after-dinner espresso. And I may enjoy the smell of her coffee as well.

References

Baecker R., Grudin J., Buxton W., & Greenberg S. (1995) *Readings in Human computer Interaction: Toward the Year 2000.* Morgan-Kaufmann, San Mateo, CA.

Benford S., Greenhalgh C., Snowden D., & Bullock A. (1997) Staging a public poetry performance in a collaborative virtual environment. *Proc. of the 5th European Conference on CSCW*, Kluwer Academic Publishers, Dordrecht, pp. 125–140.

Benford S., Brown C., Reynard G., & Greenhalgh C. (1996) Shared spaces: Transportation, artificiality, and spatiality. In: Ackerman M (ed.) *Proceedings of CSCW 96,* November 16–20, Boston, MA. New York: ACM Press, pp. 77–86.

Bentley R., Appelt W., Busbach U., Hinrichs E., Kerr D., Sikkel K., Trevor J. & Woetzel G. (1997) Basic Support for Cooperative Work on the World Wide Web. *International Journal of Human-Computer Studies* Special issue on Innovative Applications of the World Wide Web **46(**6): 827–846.

Clark H. H. (1996) *Using Language*. Cambridge University Press, Cambridge, U.K.

Donath J. (1998) Identity and Deception in the Virtual Community, In: Kollock P. & Smith M. (eds). *Communities in Cyberspace*, University of California Press, Berkeley.

Donath J. (1995) The Illustrated Conversation. *Multimedia Tools and Applications,* Vol. **1** March 1995.

Dourish P. & Bellotti V. (1992) Awareness and coordination in shared workspaces. *Proceedings of ACM 1992 Conference on Computer-Supported Cooperative Work*. Oct. 31–Nov. 4, Toronto, pp.107–114.

Edigo C. (1988) Videoconferencing as a technology to support group work: A review of its failure. *Proceedings of CHI'88*, ACM Press, New York, pp. 13–24.

Fuchs L., Pankoke-Babatz U. & Prinz W. (1995) Supporting cooperative awareness with local event mechanisms: The GroupDesk system. *Proceedings of the Fourth European Conference on Computer-Supported Cooperative Work*, Sept. 10–14, Stockholm, pp. 247–262.

Gale S. (1991) Adding audio and video to an office environment. In: Bowers J. M. & Benford S. D. (eds.). *Studies in Computer Supported Cooperative Work,* Elsevier, Holland, pp. 49–62.

Giddens A. (1990) *The Consequences of Modernity*. MIT Press, Cambridge, MA.

Glance N., Arregui D., & Dardenne M. (1998) Knowledge Pump: Supporting the Flow and Use of Knowledge. In: Borghoff U. & Pareschi R. (eds), *Information Technology for Knowledge Management*, Springer Verlag, Berlin.

Greenberg S. & Roseman M. (1996) GroupWeb: A WebBrowser as Real-Time Groupware. *Proceedings of CHI'96*, April 13–18, Vancouver, Canada, ACM Press, pp. 271–272.

Greenberg S. & Roseman M. (1998) Using a room metaphor to ease transitions in groupware. Research report 98/611/02, Dept. of Computer Science, University of Calgary, Calgary, Canada.

Harasim L., Hiltz S. R., Teles L. & Turoff M. (1995) *Learning Networks: A Field Guide to Teaching and Learning Online*. MIT Press, Cambridge.

Hiltz S. R. (1993) *The Virtual Classroom: A New Option for Learning*. Ablex Press, Norwood, NJ.

Hiltz S. R. & Turoff M. (1993) *The Network Nation: Human Communication via Computer*. The MIT Press, Cambridge, MA.

Hunter B. (1995) Learning and teaching on the Internet: Contributing to educational reform. In: Kahin B. & Keller J. (eds.) *Public Access to the Internet.* The MIT Press, Cambridge, MA.

Jarvenpaa S. L. & Leidner D. E. (1998) Communication and trust in global virtual teams. *Journal of Computer Mediated Communication* **3**(4).

Jeffrey P. & Mark G. (1999) Navigating the Virtual Landscape: Coordinating the Shared Use of Space. In: Munro A. Benyon D. & Hook K. (eds.) *Footprints in the Snow: Personal and Social Navigation of Information Space,* Springer Verlag, Berlin.

Johnston W. E. & Agarwal D. (1998) Issues in architectures and technologies for on-line, global-scale scientific environments. Presented at America in the Age of Information, Ernest Orlando Lawrence Berkeley National Laboratory, University of California, Berkeley, CA 94729 (http://www-itg.lbl.gov/~johnston/CIC.Paper.1.fm.html).

Kollock P. (1998) Design principles for on-line communities. *PC Update* **15**(5): 58–60, June, 1998.

Kollock P. & Smith M. (1998) Managing the virtual commons: cooperation and conflict in computer communities. In: Kollock P. & Smith M. (eds) *Communities in Cyberspace*, University of California Press, Berkeley.

Lave J. & Wenger E. (1991) *Situated Learning: Legitimate Peripheral Participation.* Cambridge University Press, Cambridge.

Lombard M. & Ditton T. (1997) At the heart of it all: The concept of presence. *Journal of Computer-Mediated Communication* **3**(2).

Maeda H., Kanjihara M., Adachi H., Sawada A., Takeda H., & Nishida T. (1997) Weak information structures for community information sharing. *International Journal of Knowledge-Based Intelligent Engineering Systems,* **1**(4): 225–234.

Maes P. (1994) Agents that reduce work and information overload. *Communications of the ACM,* **37**(7): 31–40.

Malone T. W., Grant K. R., Lai K. Y., Rao R. and Rosenblatt D. (1987). Semi-structured messages are surprisingly useful for computer-supported coordination. *ACM Transactions on Office Information Systems*. **5**, 115–131.

Mark G. & Becker B. (1998) Designing believable interaction by applying social conventions. In Dautenhahn K (ed.) *Applied Artificial Intelligence,* Special Issue on Socially Intelligent Agents **13**(3): 297–320.

Mark G., Fuchs L., & Sohlenkamp M. (1997) Supporting groupware conventions through contextual awareness. In Prinz W., Rodden T., Hughes J., & Schmidt K. (eds.) *Proceedings of ECSCW'97*, Sept. 7–11, Lancaster, Kluwer Academic Publishers, Dordrecht.

McGrath A. (1998) The Forum. *Siggroup Bulletin* **19**(3): 21–24.

Mitchell T., Caruana R., Freitag D., McDermott J., & Zabowski D. (1994) Experience with a learning personal assistant, *Communications of the ACM,* **37**(7): 81–91.

Mitchell W. J. (1996) *City of Bits: Space, Place, and the Infobahn.* The MIT Press, Cambridge, MA.

Nakata K., Voss A., Juhnke M. & Kreifelts T. (1998) Concept Index: Capturing emergent community knowledge from documents. The 7th Le Travail Humain Workshop, *Designing Collective Memories.*

Ostrom, E. (1990) *Governing the Commons: The Evolution of Institutions for Collective Action.* Cambridge University Press, Cambridge, U.K.

Poster M. (1995) *The Second Media Age*, Polity Press, Cambridge, U.K.

Powell W. W. (1990) Neither market nor hierarchy: Network forms of organization. *Research in Organizational Behavior,* **12**, 295–336.

Prinz W. (1999) NESSIE: An awareness environment for cooperative settings. *Proceedings ECSCW'99*, Copenhagen, 12–16 September.

Putnam R. D. (1995) Tuning in, tuning out: The strange disappearance of social capital in America. *Political Science and Politics,* **28**: 664–683.

Roberts T. L. (1998) Are newsgroups virtual communities? *Proceedings CHI'98,* April 18–23, Los Angeles, ACM Press, New York, 360–367.

Reder S. & Schwab R. G. (1990) The temporal structure of cooperative activity. *Proceedings of the Conference on Computer Supported Cooperative Work, CSCW'90*, Oct. 7–10, Los Angeles. ACM Press, New York.

Rice R. E. (1992) Task analyzability, use of new medium and effectiveness: A multi-site exploration of media richness. *Organization Science,* **3**(4): 475–500.

Schein E. (1990) *Organizational Culture and Leadership*. Jossey-Bass, San Francisco.

Short J., Williams E., & Christie B. (1976) *The Social Psychology of Telecommunications.* Wiley, London.

Slater M. & Usoh M. (1993) Representations systems, perceptual position, and presence in immersive virtual environments. *Presence*, **2**(3): 221–233.

The Social Web (1998) http://orgwis.gmd.de/projects/SocialWeb/.

Sohlenkamp M. & Chwelos G. (1994) Integrating Communication, Cooperation, and Awareness: The DIVA Virtual Office Environment. *Proceedings ACM 1994 Conference on Computer Supported Cooperative Work.* Oct. 22–26, Chapel Hill, NC, 331–343.

Sproull L. & Kiesler S. (1991) *Connections: New Ways of Working in the Networked Organization.* The MIT Press, Cambridge, MA.

Suchman L. (1987) *Plans and situated actions*. Cambridge University Press, Cambridge, U.K.

Trevor J., Koch T. & Woetzel G. (1997) MetaWeb: Bringing synchronous groupware to the World Wide Web. In Prinz W., Rodden T., Hughes J. & Schmidt K. (eds.), *Proceedings of ECSCW'97*, Sept. 7–11, Lancaster, Kluwer Academic Publishers, Dordrecht.

Turkle S. (1997) *Life on the Screen.* Touchstone, New York.

Ubique (1995) *Virtual Places.* http://www.vplaces.com/vpnet/index.html.

Uncapher W. (1998) Electronic Homesteading on the Rural Frontier: Big Sky Telegraph and its Community. In: Smith M. & Kollock P. (eds.) *Communities in Cyberspace,* Routledge, London.

Wellman B. & Gulia M. (1998) Net surfers don't ride alone: Virtual communities as communities. In: Kollock P. & Smith M. (eds.) *Communities in Cyberspace*, University of California Press, Berkeley.

Winograd T. A. (1987/88) A language/action perspective on the design of cooperative work. *Human-Computer Interaction* **3**(1): 3–30.

Part III

Moving Towards the Virtual University: Institutional Experiences from Different Regions

9

Institutional Models for Virtual Universities

ROBIN MASON

Background

A growing sense of crisis in the Higher Education sector over the last decade has given rise to a multiplicity of new organizational structures designed for meeting the new challenges of post-secondary education and training. Almost all of them exploit the developments in information and communication technology to design, manage and to deliver courses and training modules. The purpose of this chapter is to explore the range of these emerging structures, to consider the technologies they are using to meet the new challenges and to examine some of the market leaders in detail.

In order to understand why these new organizations are forming, it is important to consider what is driving the need to re-conceive the traditional university. The steady reduction in funding of public universities, as well as the falling population of traditional 18–20 year old students, at least in western countries, are the primary drivers. Related to these are the larger changes in the labor market as well as social and domestic changes which together have created the requirement for lifelong learning. The need for post-secondary education and training has never been greater, but the population requiring it has altered. Simply scaling up existing physical institutions would not meet this need even if resources were available to build new campuses, because physical attendance at specified times is ill suited to working adults.

Whether it is cause and effect, or mere chance, the fact is that as the need for open and flexible learning has exploded, so the technology to support new methods has been developing at an equally fast pace. The term "information and communication technology" (ICT) encompasses all the computer-based teaching systems, such as CD-ROM, as well as all the telecommunication systems, such as computer conferencing, the Web and videoconferencing. ICTs can support all aspects of teaching and learning from course development, presentation, delivery and support, to administration, registration, assignment handling and marking, even when the student body is widely dispersed and never meets face- to- face. Such has been the growth of ICT in higher education, that not a popular magazine nor an academic journal is free of references and articles describing and explaining ways in which new technologies are delivering new curricula to new kinds of students in new kinds of ways.

Arguments against virtual education

Although the enthusiasts for this technologization of education are more prominent in the media, they are probably equaled in number by the quieter voices whimpering for a halt to this headlong expansion of virtual education. The cognitive argument is based on the fact that the new delivery mechanisms for most virtual education are electronic and rely largely on the digitization and computerization of knowledge. Many people decry the cognitive effects of learning from screen-based information rather than from traditional text-based material, pointing to the breakdown of linear, narrative structures associated with the book, and the resulting fragmentation and superficiality induced by the hyperlinked structures of the Web and multimedia CD-ROMs. One of the more eloquent apologists for the culture of books is Sven Birkerts, who tots up the cognitive losses we are incurring with the rise of an electronic culture:

> In the loss column are (a) a fragmented sense of time and a loss of the so-called duration experience, that depth phenomenon we associate with reverie; (b) a reduced attention span and a general impatience with sustained inquiry; (c) a shattered faith in institutions and in the explanatory narratives that formerly gave shape to subjective experience; (d) a divorce from the past, from a vital sense of history as a cumulative or organic process; (e) an estrangement from geographic place and community; and (f) an absence of any strong vision of a personal or collective future. (Birkerts, 1994, p. 27)

Birkerts acknowledges that these are enormous generalizations, but he feels that they accurately reflect the comments of his students about their own experience.

The educational argument against technology-delivered education centers on the undesirable aspects of consumerism, wherein learning ceases to be about analysis, discussion and examination, and becomes a product to be bought and sold, to be packaged, advertised and marketed. While there are some who can argue the advantages of competition in course provision, there are those who view the growing competitive, consumer spirit amongst educators as detrimental to the learning outcomes (see, for example, Moore, 1996). This marketplace philosophy of higher education is particularly associated with distance education, and more so with technology-delivered, open and flexible education.

One consideration which is rarely discussed is whether students are prepared for the change to virtual, technology-delivered education. Those who definitely are, tend to be the technology-literate and the self-motivated who welcome the independence of student-directed approaches to teaching. However, there are very many students who find the technology off-putting, who do not have the learning skills for coping with large amounts of digital information, and who have not "learned how to learn".

Most students are not prepared for the need to take responsibility for directing their own learning, for pacing and motivating themselves, for managing and using the information available, and for constructing the knowledge, which previously the lecturer carried out for them. (Mason, 1998, p. 157)

Arguments in favor of ICT

So on the one hand, there are these voices against changing the status quo, and on the other, there are the financial pressures to compensate for falling government revenues. Are there educationally valid arguments in favor of the move to technology-based teaching and learning at post-secondary level? Many practitioners speaking from personal experience claim that ICT brings pedagogical benefits irrespective of cost savings and new markets. The most visionary proponents talk of a movement away from the bounded classroom, a retreat from the real world, self-contained and static, to a dynamic synergy of teachers, computer-mediated instructional devices, and students collaborating to create a window on the world. Interaction with learners across the country and even the world leads to an increased awareness of the extraordinary complexity of interrelations and a relativistic comprehension and tolerance of diverse approaches to understanding.

> The diversity of participants made for a far richer course than I could ever teach myself. Take, e.g., the time our correspondent in Istanbul reported on a lecture given there on medieval Christian philosophy by a Franciscan priest to the faculty of the (Islamic) University of the Bosporus. Well and good, the faculty opined when it was over, but it's too bad Christianity is not a truly rational religion, like Islam. Leaving aside the question of comparative rationality of religions, I think it is undoubtedly good for my students at Penn [University of Pennsylvania], taking a course on a very traditional, "western" figure, to be reminded that the whole picture looks quite differently if you happen to be in a different seat. (O'Donnell, 1996, p. 113)

The taste of online interaction given to this lecturer by his first venture onto the information superhighway, left him concluding:

> I cannot imagine ever passing a semester in the classroom again without the umbilical cord to the network to energize, diversify, and deepen what we do. (op cit., page 114)

O'Donnell's experience is echoed by many others who have discovered the educational benefits of a global student body, of interacting directly with students rather than lecturing to them, or of providing students with an opportunity to reflect on the course issues and engage in debate and collaborative activities with other students.

Technologies for Virtual Teaching

Virtual education is dependent on ICT for contact with students. There are three broad categories within which current technologies can be divided:

- text-based systems, including electronic mail, computer conferencing, real time chat systems, MUDS/MOOs, FAX and many uses of the World Wide Web
- audio-based systems such as audio conferencing and audiographics, and audio on the Web
- video-based systems such as videoconferencing, one-way and two-way, video on the Internet with products like CUSeeMe, Web-casting and other visual media such as video clips on the Web.

The implication of this list is that text, audio and video are discrete media. While this is partially true today, the evolution of all these systems is towards integration—of real time and asynchronous access, of resource material and communication, of text and video, in short, of writing, speaking and seeing. A good example of this integration is the CD-ROM, which combines elements of text, audio and video and consequently has tremendous potential as a stimulating learning resource. However, on its own, it lacks the significant component of person-to-person interaction. Furthermore, CD-ROMs are difficult and costly to update, and problematic as a global distribution medium. Recent developments overcome these shortcomings by integrating the CD-ROM with the interactive and updating capabilities of telecommunications. For this reason, I am including multimedia CD-ROM within the purview of my fourth the ultimate integrating medium and the rising star of global education delivery:

- the World Wide Web, which integrates text, audio and video, both as pre-prepared clips and as live interactive systems, both real time and stored to be accessed later and, furthermore, provides text-based interaction as well as access to educational resources of unprecedented magnitude.

Text-based systems

Without a doubt the most commonly used technology for communicating with students at a distance, is electronic mail and various forms of group communication. Text-based interaction, whether many-to-many in conferences, or one-to-one in electronic mail, is practiced at most institutions of higher education, whether students are geographically remote, or actually on campus. In fact, with the change towards the new majority (students who are older, or have some kind of part-time employment, or have family commitments or other barriers to attending campus full-time), many face-to-face teaching institutions use text-based interaction as a means of communicating with students, thereby reducing their dependence on physical attendance on campus at specific times.

Some institutions use standard electronic mail systems (which include the facility for sending messages to a group) to communicate with students at a distance. Those accessing from abroad usually use the Internet; those living locally may use a modem over telephone lines. The primary use is for students to ask questions of the tutor, but an additional use is the electronic submission of assignments, as an attachment to a mail message. This is the simplest and most accessible of all the telecommunications technologies, with the possible exception of FAX.

More commonly in distance education, a proprietary computer conferencing system is used. FirstClass is a very successful product amongst educators, and Lotus Notes is common particularly in Business Schools and in training organizations. Web bulletin boards are also becoming very popular and most of the proprietary systems have integrated with the Web, such that conferences can be accessed from a Web browser. Computer conferencing systems allow students on one course to share discussion areas, to have sub-conferences for small groups, and to have easy access to all the course messages throughout the length of the course. Computer conferencing systems are slightly more complex than email, and they may require the student to have client software. A faster modem may be necessary as well, or at least highly desirable. Generally, the more extensive the system and the more advanced the facilities it offers (e.g. frames on Web bulletin boards), the more powerful the clients' machine needs to be.

Most text-based communication systems are used primarily to support students (with the *contents* of the course delivered through some other medium); however, some educators run "online courses" in which the primary content of the course is the discussions and activities taking place amongst the students.

Audio systems

Straight audio conferencing using ordinary telephone lines is a "low-tech" solution to supporting students in the developed world, due to the near ubiquity of the telephone in these countries. Many print-based distance education programs use audio conferencing to help motivate students, and it has also been used for small group collaborative work at postgraduate level (Burge and Roberts, 1993). Nevertheless, there are few uses of this technology in group discussion mode (as opposed to simple student-to-tutor phone calls) in international programs of distance education. Audio conferences are difficult to manage with more than half a dozen sites, although it is possible to increase the number of participants by having groups of students at each site.

An extension of pure voice interaction is audiographics: voice plus a shared screen for drawing or sharing pre-prepared graphics. This technology has had more extensive use in distance education, and there are many applications of it being used between two sites in different countries. As with audio conferencing, audiographics use with more than two sites requires an audio bridge to connect all the lines together. There is no technical barrier to doing this internationally; cost is the primary deterrent. The term, "audiographics', will probably die out completely as shared screen and multiway audio are now possible on the Web.

Audio on the Web is a developing technology which many institutions are beginning to take seriously as an educational tool, especially when combined with various forms of real time, text-based interaction. RealPlayer, for example, is a product which allows real time lectures on a global scale (http://www.realplayer.com/). Many distance education systems have involved sending audiocassettes out to students through the post. With audio on the Internet products like RealPlayer, it is possible for large numbers of students around the world to access these "broadcasts" in both real and delayed time.

Video systems

Some educators feel that videoconferencing is not necessary in supporting students at a distance, and that audio, especially audiographics works better because it concentrates attention on content rather than distracting the learners' attention with the visual image of the speaker. For them, the significantly higher cost of providing video is not justified by the educational benefits. Others feel that we live in a visual age in which it makes no sense to restrict the learner to audio exchange. Video, when well used, contributes to the motivation of the student, makes the learning environment more social, and facilitates the delivery of exceptional learning materials in almost every area of the curriculum (Mason, 1994).

One-way video with two-way audio has widespread application in North American distance education and training in national and international companies. Many of these systems use satellite delivery to extend coverage. The educational paradigm of most programs is the lecture at a distance, with students watching either from smaller colleges, in the workplace, or (most commonly) later on recorded video at home. A very great deal of distance training is carried out by videoconferencing, some of it internationally, using ISDN.

Despite the popularity of such systems, there are many educational technologists who disparage their use. The following is a particularly incisive critique:

> The widely held view that face-to-face teaching is inherently superior to other forms of teaching has spawned a major industry worldwide. It is difficult to believe that videoconferencing would have become such a major influence, especially in North America, without the intellectual complacency associated with the tyranny of proximity. The investment in videoconferencing has been quite staggering despite the widely held view that the lecture is a process whereby the notes of the lecturer are transmitted to the notes of the student, without passing through the minds of either ...
>
> The apparently unwavering enthusiasm for the proliferation of videoconferencing systems for the purpose of enhancing teaching and learning represents "the tyranny

> of futility". If most lectures are relatively futile from a pedagogical perspective, why spend vast sums of money promoting expensive futile exercises? A reasonable explanation is related to the rate of change, or lack thereof, in the educational context—a phenomenon known as the "tyranny of eternity". Educational paradigms tend to change direction with much the same agility as an ocean-going oil tanker. (Taylor & Swannell, 1997)

Bates also concludes that instructional TV in which video is used to lecture, rather than to supplement, enhance and enliven course content delivered in another medium, is ultimately not cost-effective (Bates, 1995, 115).

Streaming video on the Internet, however, is developing technology, which may have more lasting potential, precisely because it is more closely tied to the model of personal computing, just-in-time learning and Web-based resources, than it is to the notion of teaching and learning through traditional lectures.

The Web

There is little doubt that the Web is the most phenomenally successful educational tool to have appeared in a long time. It combines all the media described above: text, text-based interaction, audio and video as clips, and, with somewhat less robustness, multi-way interactive audio and video. Its application in virtual education is unquestioned. Although access to the Internet is hardly universal, and large segments of the global population are more remote from access to it (whether through cost, or through unavailability at any cost) than they are to print- and post-based systems of distance education, nevertheless, vast numbers of people worldwide do have access, many from their home, and this access is growing exponentially.

Now that it is possible to download programs from the Web along with data, and to receive the appropriate software to handle the program automatically, the institution wanting to deliver course material can manage the maintenance and updating of any software required for the course. The student no longer needs to struggle with the installation of massive packages, and, furthermore, can use a relatively low cost machine—even a portable.

One of the more exciting developments which the Web supports is a truly interactive, multimedia environment consisting of tailor-made course materials (including graphics, simulations, video clips etc.) enhanced and extended by links to external Web sites containing articles, journals etc., as well as integrated conferencing media for both synchronous and asynchronous discussion, in text, audio and/or video. One such program is the UK Open University's Masters in Open and Distance Education (http:// iet.open.ac.uk). Collaborative activities involving Web searches, debates, and joint projects form the central content of several courses. The students are physically resident in many countries of the world. This is virtual education *par excellence.*

Synchronous versus Asynchronous Distance Education

As is apparent in this description of different technologies for delivering education at a distance, some of the systems rely on real time interaction, while others can be accessed asynchronously. This difference has major implications for the design and delivery of distance education, as well as for the study requirements of the learner. There are advantages to both forms and in the end, personal learning styles and the larger educational context determine what is most appropriate.

The following list details the major benefits of each mode in an educational context.

Asynchronous delivery

There are four crucial advantages to the asynchronous media, arranged in descending order of significance:

- flexibility—access to the teaching material (e.g. on the Web, or computer conference discussions) can take place at any time (24 hours of the day, seven days a week) and from many locations (e.g. oil rigs)
- time to reflect—rather than having to react 'on one's feet', asynchronous systems allow the learner time to mull over ideas, check references, refer back to previous messages and take any amount of time to prepare a comment
- situated learning—because the technology allows access from home and work, the learner can easily integrate the ideas being discussed on the course with the working environment, or access resources on the Internet as required on the job
- cost-effective technology—text-based asynchronous systems require little bandwidth and low end computers to operate, thus access, particularly global access is more equable.

Synchronous delivery

There are four equally compelling advantages to synchronous systems:

- motivation—synchronous systems focus the energy of the group, providing motivation to distance learners to keep up with their peers and continue with their studies
- telepresence—real time interaction with its opportunity to convey tone and nuance helps to develop group cohesion and the sense of being part of a learning community
- good feedback—synchronous systems provide quick feedback on ideas and support consensus and decision making in group activities, both of which enliven distance education
- pacing—synchronous events encourage students to keep up-to-date with the course and provide a discipline to learning which helps people to prioritize their studies.

Multi-synchronous course design

There are many technology-mediated virtual teaching programs which are entirely asynchronous (for example, those using print plus computer conferencing, or those using the Web for both course delivery and interaction), and others which are (almost) entirely synchronous (for example, those using videoconferencing for delivery and interaction). However, the trend is very much towards combining synchronous and asynchronous media in an attempt to capitalize on the evident benefits of both modes. The various permutations of media use and the amount of synchronous interaction included are almost as varied as the number of institutions providing virtual education.

New Models for New Needs

I have briefly reviewed some of the issues and the technologies which form the backdrop to the growth of new structures for delivering higher education and training. Let me turn now to a consideration of these structures and examine the way in which they have responded to the changes in the socio-economic context of post-secondary education and training.

In the typology of structures which follows, I have used evocative images to categorize existing and planned organizations. As with all such terms, the analogy is only appropriate up to a point. Furthermore, there is inevitably some overlap such that any organization belonging primarily in one category, may have some attributes of another. Bearing in mind these provisos, I think it is useful to describe some models in order to understand the issues, to consider alternatives, and to stimulate thinking about new solutions.

Brokerage

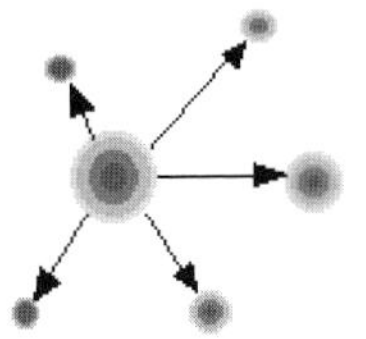

FIGURE 9.1
Brokerage

The brokerage model (Figure 9.1) consists of a new organization set up to provide courses for the life-long learning sector, but using the teaching and course resources of existing institutions. The new organization has a very small number of permanent staff, mostly administrators, but also possibly editors, educational technologists and curriculum designers. The broker puts students in touch with course providers: it may do this through managing study centers where students come for tutorials or access to the technologies which deliver the course; alternatively, it may be a largely invisible structure which draws together and markets a list of courses, registering students by phone or electronically. Frequently, the broker commissions courses which fit into a pre-determined curriculum.

A classic example of this model is the Open Learning Agency in Australia. Existing universities bid to provide courses requested by OLA, and all providers agree to give credit transfer to courses within the program.

The proposed University for Industry (Ufi) in the UK is also likely to work on this model, with the added element of catering for pre-university level courses as well. Furthermore, there is the notion of a pick and mix system with short modules which can be combined to fit with the existing credit points system. The aim is to meet employers' demands for just-in-time training of just-the-right-amount.

> The Ufi will connect those who want to learn with ways of doing so. It will act as the hub of a brand new learning network, using modern communication technologies to link businesses and individuals to cost-effective, accessible and flexible education and training.
>
> People and companies will be able to contact the University for Industry by telephone, letter, fax, email (through the Ufi's website) or by calling at a Ufi enquiry desk in, for example, a supermarket, high street shop, college, TEC or Business Link. The Ufi will tell you what learning is available and offer advice if you need it, and provide you with a course that meets your needs, whether full-time, part-time, or through study at home, at work or at a local learning center. For example, it could deliver a learning package on a CD-ROM to your home or send it by email, or contract with a college for an evening class, or broadcast an interactive TV program, or provide a course over the radio or on the Internet. Students will not need to be tied to one particular location. (The Learning Age, a renaissance for a new Britain, http://www.lifelonglearning.co.uk/greenpaper/index.htm)

Another example is the Jones Education Company (formerly Mind Extension University). JEC markets and registers students on courses designed and tutored by staff in associated universities. Most courses have a strong technology-delivered element—cable originally and increasingly the Web. Over the last few years JEC has steadily marched across the globe creating partnerships and agreements to market their educational TV enterprises in Asia, Europe, Latin America and soon, Africa, India and South America.

In theory there are many advantages to this model: it builds on existing resources in existing institutions; it helps to focus existing providers on the demands of the marketplace; it has low start-up costs. In practice, however, there are difficulties centered around ownership of courses, institutional loyalties, lack of prestige and in some cases, staying solvent!

Partnership

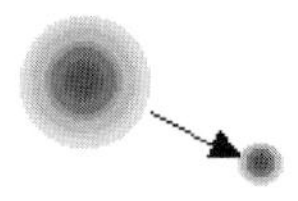

FIGURE 9.2
Partnership

The partnership model (Figure 9.2) does not involve the setting up of a new organization, but relies on agreements amongst existing universities, often in other countries. Typical partnerships consist of one large or prestigious university from a developed country working with smaller or newly forming universities in developing countries. Existing courses are offered to the partner with or without modifications, and may be tutored locally or by staff from the providing institution. Two-way partnerships (i.e. with each partner both providing and receiving courses) are rare, but some partnerships within national boundaries may be more reciprocal.

The UK Open University is a large and visible example of the partnership model in practice. Agreements exist with very many developing countries in Eastern Europe, Africa, Asia, and the Middle East. Increasingly partnerships are being formed with North American institutions, drawing on the OU's vast resource of high quality print and video-based distance education materials. Specific examples include:

- In Singapore about 6000 students are studying with the Open University, in conjunction with its partner, Singapore Institute of Management which carries out the marketing, local pricing and fee collection, registration, record keeping, employment of tutors, organization of local support services and tutorial meetings. The UK OU retains responsibility for the content of courses, for tutor training, development and monitoring, for assessment and examination standards generally and for determining awards and appeals.
- A variant on the Singapore model operates in the countries of Central and Eastern Europe, where similar contractual relationships are in place—albeit mainly with commercial partners—but where the courses are offered in the local language. Special arrangements are therefore in place for monitoring academic quality (i.e. translation of some scripts).
- In Hong Kong, the OU operates through the Open University of Hong Kong, which has about 20,000 students on courses for their own degrees and awards, the majority of them on courses which were acquired from the UK OU. In this model the local institution makes the adaptations to the assessment questions, tutor training material or course content to suit their own needs.

- A number of aid-funded projects are underway as part of the University's commitment to enabling access to higher education by those with very limited means. So, for example, the MBA program has been delivered to 115 heads of state in Ethiopia and to a similar number of ministers and senior government officers in Eritrea.
- The undergraduate program has been available to students across Continental Western Europe (CWE) and Ireland for some years. Over 6000 students are registered, studying primarily in the Arts, Business School, Social Science and Mathematics. The majority live in Ireland, Belgium, Germany, the Netherlands, Switzerland and France. All of the students in this program study the same course at the same time as their UK counterparts—where numbers allow, local tutors are hired and local tutorials are held. Otherwise students join UK-based tutor groups. Some of the courses have a compulsory online component in which CWE students interact with students in the UK.
- Joint course development features in the collaboration with the University of Waterloo in Canada, and in addition, the use of OU TV programs broadcast on nation-wide cable TV extends the OU's opportunities in Canada.

The positive aspects of this model derive from the way in which new institutions can "stand on the shoulders" of existing expertise to get started in the field, building up their own infrastructure gradually before becoming self-sufficient. There is also an efficiency gain in extending the use of existing teaching materials to a wider student population before they become out-of-date. The limitations of this model are fairly obvious: accusations of imperialism, imposing an already dominant cultural perspective, and lack of reciprocity.

Umbrella

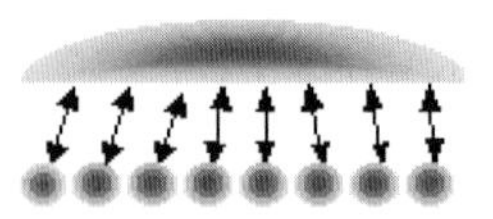

FIGURE 9.3
Umbrella

The notion of an umbrella model (Figure 9.3) is one where existing institutions pull together (instead of competing with each other) under the aegis of a slim superstructure, to provide new courses in new ways. This model has many similarities with the brokerage model, but the difference is in the vision and the intentions and outlook of the institutions which choose to come under the umbrella. The fundamental thinking is often one of "individually we are small and vulnerable; together we can be more than the sum of the parts".

A good example of this model is the newly emerging University of the Highlands and Islands in Scotland. The "superstructure" is located in Inverness, but the existing organizations—one university and many colleges—are spread throughout the area. With a newly installed telecommunications infrastructure, UHI plans to deliver new tailor-made courses as well as existing courses to new audiences. The vision is one of empowerment of small colleges, of preparing courses about local issues (Scottish history, archeology, language), and of using technology to bridge distance and disadvantage.

Another example of the umbrella model is the Western Governors University (WGU) which has formed to address the need for accessible, relevant lifelong learning. It acts both as broker and as course provider:

> Unlike other distance learning efforts, the WGU will bring together and act as a broker for both traditional and non-traditional educational providers, from universities, to corporations that train employees for specific skills. In addition to brokering courses, WGU itself will initially offer an Associate of Arts Degree and workplace certification.
>
> Competency Based Credentialing—This "anywhere, anytime" education system is also significant in that WGU certificates and degrees will be based on demonstrated competencies. Students will undergo a rigorous assessment to determine what they know and can apply. Where learning takes place will no longer be as important as what a student actually learns. Students will be able to use WGU credentials as proof of what they know, whether they are seeking jobs or further educational opportunities.
>
> *The Nerve Center.* The smart catalog/navigator has been called WGU's nerve center. While this Internet-based catalog will list various courses offered by traditional and non-traditional providers, it will be much more than a course list. Most important, it will map out the skills that need to be mastered in order to receive a WGU credential. Students will be able to use the catalog to assess their existing skills and knowledge to help determine what courses they need and are prepared to take. They will also use it to create a profile, including convenient times for taking courses and the types of technologies they prefer, e.g. the Internet, computer software, videotapes or satellite. The catalog will use the profile to offer students learning options, enabling them to seek a certificate of competency, a professional certification program, an academic degree or an individual course. The catalog will also include jobs and career information and assessment services.
>
> *Local/Regional Centers.* Each of WGU's participating states must establish at least one center that will provide one-stop shopping for WGU services, including access to the delivering technologies. Centers will be located at existing organizations such as public libraries, school extension sites or companies training their workforce through the WGU. The centers will conduct assessments of specific competencies; offer counseling; provide access to information technology, including computers, audio and video classrooms and Internet connections; and identify unmet education and training needs in the area. (http://www.wgu.edu)

The similarities with the UK University for Industry are obvious and underline a common need and a common solution for meeting the need in both the U.K. and U.S..

There is little to dislike about this model except the difficulties of carrying it off. Institutional politics, market protection, collaboration hostilities can only be overcome by near life-threatening crises, which command people to work together against a common threat.

Greenfield

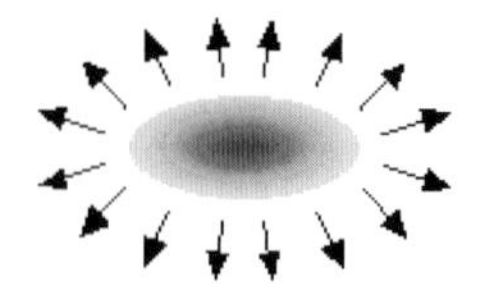

FIGURE 9.4
Greenfield

A greenfield site (Figure 9.4)—setting up an entirely new organization—has considerable appeal in the face of the difficulties of the other models. A new organization is not hampered by existing systems, outlooks or procedures. It can be designed to exploit new technologies for management, administration as well as course design and delivery. It can target the students and the areas of the curriculum most ready for technology delivery. It can focus on teaching and hire dedicated and outstanding teachers, rather than compromise teaching quality in favor of research credibility. Not surprisingly, there are many such organizations springing up around the world. Most of them have "virtual" in their title, their mission statement or their implementation.

One whole class of examples are the computer and telecommunications companies: Deutsche TeleKom's Global Learning network, Microsoft's Online Learning Institute, IBM's Global Campus. In most cases the name is grander than the actuality, but the impetus and the direction are clear. Whether they can seriously undercut existing universities is not the issue here; what is relevant is the model—technology-mediated, short modules, just-in-time, just-the-subject-you-want. There is usually no face-to-face component in the courses at all; tutoring is online, as are the course materials.

Another, smaller class of greenfield organizations are the new "open universities" e.g. in Greece, various Indian states, in the Arab states etc. These can build on the experience of existing open universities, but start with a technology base of the 21st Century. The funding for these as well as the mission is very different from that of the industry-sponsored, virtual organizations. The latter can "cherry pick" the most lucrative areas to exploit, whereas the former usually have to provide for broad undergraduate education across the curriculum and across the social spectrum.

A good example of one of the new, for-profit corporations is Magellan University, set up with the guiding principle, "Excellence in education, anywhere, anytime". The distinguishing feature of the Magellan environment is the division of course content (consisting of video recordings of outstanding teachers) and course support (consisting of classes of about 15 students managed by a tutor using Lotus/Domino software).

As a new institution, Magellan is free to find its faculty from around the world, selecting them on the basis of their teaching excellence. The tutors are paid on a per student basis, and are expected to interact asynchronously with students, as well as to mark their assignments. The Electric Library consists of over 400 online lessons developed initially by the Plato Laboratory at the University of Illinois. Each lesson takes between one and two hours, and includes examples and self-tests. They are aimed at life-long learners wanting to brush up their knowledge of particular subject areas, or get a taste of new subjects (Witherspoon, 1997, p. 161).

British Aerospace is in the process of setting up a corporate model of the virtual university, but though it is a "greenfield" in some ways, it is also looking to form partnerships with existing universities (e.g. the UKOU).

Network

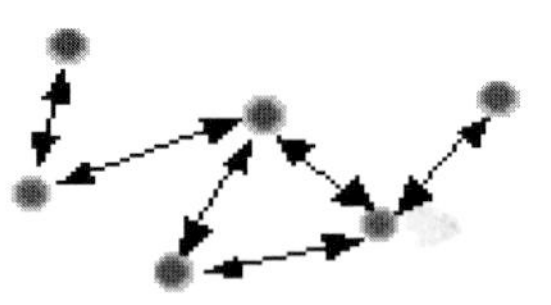

FIGURE 9.5
Network

A network model (Figure 9.5) is one in which existing universities and education providers collaborate in a variety of combinations to produce courses for the life-long learning market. This model is the most loosely organized, as there is no central driver of the program. However, it has the advantage of being founded on natural working relationships, rather than the forced arrangements which other external systems impose. In short, it is led from the grass-roots.

One minor example of this model is a European Commission funded project which calls itself the Virtual University for Europe:

EuroPACE 2000 is a trans-European network of universities and their partners in education and training, i.e. private enterprises, regional and professional organizations and public authorities. Approximately 60 member organizations (45 of them universities) participate in this network throughout Europe. (www.europace.be/).

The collaborating partners in the project agree to produce some joint courses which are delivered by ISDN videoconferencing. The model expands and contracts as the partners respond to their perceived need for courses, and new partnerships can be formed for each course.

This model is the least innovative and could be a development of existing arrangements and working relationships, with the injection of some extra resource for expansion: a "do it more" approach.

Dual Mode

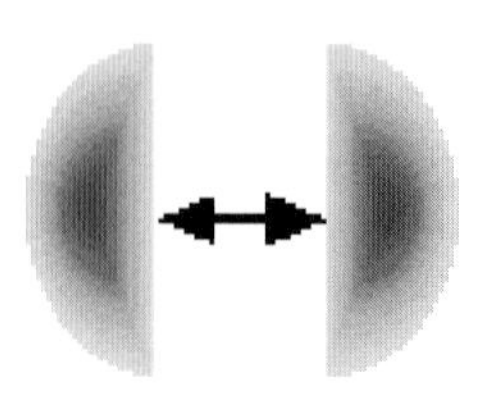

FIGURE 9.6
Dual Mode

The idea of teaching the same courses both face-to-face and off campus has a ring of efficiency about it which has led to the very widespread practice known as dual-mode (Figure 9.6). The Australians are most noted for this, and have found that there have been benefits for both categories of students, e.g. the printed and Web materials developed for the distance students are a help to the campus students and, in some cases, videos of the campus lectures are made available to the distance students.

In fact, there has been an explosion of dual-mode teaching in higher education over the last ten years. Figures from 1994 indicate that over half of UK universities, 42 out of 69 Canadian universities and "almost all" US universities teach at least one program of study by distance learning. Since then the proportion is closer to 80% teaching in both modes (Jenkins, 1995).

> In universities, a new generation of dual mode operations is growing up, after a period when open universities seemed to be the fashion. At first sight, the emphasis appears to be on growth—more institutions offering more distance learning to attract more students. But closer examination reveals a different picture. Many of the new distance learning programs are intended for small numbers of students. They are often tailored to highly specific needs—those of ethnic or linguistic minorities, of people in sparsely-populated remote areas, or specialist career requirements. (op cit., p. 427)

One of the more successful examples of this model is the University of Phoenix (www.uophx.edu). In addition to their extensive face-to-face program running from campuses and learning centers in ten states, University of Phoenix offers three forms of distance education:

- Group-Based Instruction through Online Campus, for small group interaction with the tutor and eight to thirteen other students
- One-on-One Instruction through the Center for Distance Education, for individual correspondence by email, phone or post with the tutor
- Self-Paced Professional Training through CPE Internet, for self-paced reading, study and testing via the Internet.

The guiding philosophy is characterized by choice and flexibility for the working adult, by short courses (five to eight weeks) which are taken sequentially, and by active and interactive study materials. Judging by the healthy enrollment, this is a very successful model!

There are many more venerable examples of dual mode teaching worldwide: the University of Southern Queensland, Penn State University and University of Wisconsin, to name only a few. All of these programs now use a variety of telecommunications technologies to delivery the distance education courses: computer conferencing, audio and audiographics, video and of course, the Web.

Some of the perceived advantages of dual-mode institutions are:

- they can provide a wide spectrum of courses—usually more than the dedicated distance teaching institutions, which tend to have large numbers on a few courses;
- they have a wider choice of teaching strategies than dedicated distance teaching institutions: they can mix print and video lectures, or campus-based evening classes with print, Web and computer conferencing, or resource-based learning using both campus facilities and network access;
- they can offer options to join the campus-based courses for certain periods and in many cases use the same staff to teach in both modes, thus providing parity of academic standards.

Two of the main disadvantages of dual mode teaching are:

- having to put in place systems to manage and administer both kinds of courses, students, and assignments, with the inevitable perception that one mode will be secondary, less desirable and less well resourced;
- maintaining cross fertilization between the two programs, so that each benefits from the other, at all levels of provision—the students, staff and administration.

Much distance teaching is added onto campus teaching usually by slow accretion. Division is inherent in the very notion of dual-mode. Yet with all the talk about convergence of the two modes as long as 15 years ago, it would be preferable to re-think, or perhaps merely re-badge course provision as multi-mode. The extracts from Ufi and Western Governors give an indication of this new thinking: client-centered, pick and mix, multi-accessible, lifelong learning approach.

I turn now to a discussion of the key elements which any educational structure or provider must consider in setting up a new organization or collaborating with an existing provider.

Key Elements of Course Provision in New Structures

In the following "checklist" of the key elements of higher education delivery, I consider a range of different ways of providing what traditional, campus-based institutions have been doing for centuries through lectures, seminars, written and oral examinations and paper based administration. Each of the new structures described above has had to re-think how they will manage these elements in order to meet the demands of the new market.

Curriculum definition

A very early decision to be made involves defining the areas of the lifelong learning curriculum to be addressed, e.g. pre-undergraduate, full undergraduate, postgraduate, professional updating, leisure and/or vocational markets. The next issue is how to generate courses: by commission, by tender, by collaborative arrangement, by re-badging from existing materials. Much of the nature and structure of the whole venture will be defined by these decisions.

There is considerable interest at the moment in adapting existing course material, e.g. to make short learning modules, to revise materials for a different culture, and to transform courses for Web delivery. The emphasis is on re-use and on targeting a new audience.

Course preparation

The most significant issue for any educational organization is course content, and many new ways of preparing it have been devised. All of them are technology-mediated in some way.

Videoconferencing is the least innovative, merely replicating the lecture at a distance. Despite the hype about two-way interaction, most videoconferencing systems are not used for small group seminars or other interactive dialogues, but for extending the reach of the lecturer. There have been a number of studies of costs related to videoconferencing (see for example Bates, 1995), and there is some evidence that it can save money depending on the context of use. However, the trend is definitely one of a movement away from lecturing as the means of delivering course content, and towards separating content from support. Content is delivered by print, CD-ROM, video, the Web, or combinations of these; support is offered through any of the interactive media: email, computer conferencing, realtime conferencing (audio or video), Web conferencing, face-to-face, or combinations of these.

It is generally agreed that technology plays only a small part in the ultimate success of a program. The corollary to this is that no one technology will be *the* solution. In fact, all the evidence points to increasing diversity in the nature and type of technologies used on any course or program. Combining synchronous and asynchronous media on the same program is very common (e.g. computer conferencing plus RealPlayer events; videoconferences plus CD-ROM). Another example involves the use of old technology and new (e.g. print plus computer conferencing; videos plus the Web). Finally, some courses are designed specifically to include both passive and interactive modes of learning (e.g. broadcast TV and realtime chat).

Technological diversity is expensive, but has many educational advantages: it caters to different learning styles; the variety maintains student interest, and each medium offers new opportunities for creating a rich learning environment.

Supporting students

Once the delivery of the course content is separated from the provision of support to the student, there arises the possibility of different staff carrying out these different tasks. Tutors can be hired to interact with students, mark their assignments and answer queries about the course. Just as it will probably involve a team of people to produce the course content (at least one academic, IT staff, an editor, perhaps a course manager), so a team of people are needed to interact with students (a group of tutors, perhaps one academic from the content team, IT staff to help students, counselors).

Depending on the technologies used, students may be accessing the course and/or the course support from home, using their own equipment; from work, using workplace equipment; or from telecenters, whether associated with the teaching organization, or a public or independent facility.

Each of these locations has implications for what can be delivered electronically (e.g. equipment, bandwidth) and what kind of support is necessary: e.g. a telephone help desk, an IT facilitator in the remote center etc.

Assessment and accreditation

Methods of assessing students' work are increasingly moving online: either through automated marking shells on the Web, collaborative work through computer conferencing, Web-based activities, or electronic submission of assignments. All of these free the student from physical attendance and speed up the turnaround time for feedback compared to postal arrangements.

Accreditation issues continue to dog many of the otherwise amicable partnerships, defining who is ultimately "in charge". Credit transfer amongst the various component parts is another hurdle to overcome. Arrangements for monitoring and evaluating the courses need to be in place, as well as proctoring for examinations. In theory, these issues relate to quality and quality assurance. In reality, they often reflect political and institutional defense mechanisms.

Ultimately the questions can be defined as simple ownership issues:

- Whose students are they?
  (the host organizations? the local colleges?, the one who accredits?, the one who tutors?)
- Whose courses are they?
  (who holds copyright? who updates the materials? whose logo is on the front? who will provide credit transfer?)

Organizational and administrative issues

Another aspect of the ownership debate relates to the marketing of courses and the billing and registering of students. The part of any system which handles these procedures and communicates directly with students commands loyalty from them. New systems for administering courses have been developed for the Web: Virtual-U and Web-CT are two Canadian products which have facilities for marking assignments and recording grades. Global Campus (http://www.ibm.com) and eCollege.com (http://www.ecollege.com) are products offering "cradle to grave" online campus solutions.

Conclusions

It is apparent from this survey of existing models and key elements of lifelong learning provision, that there are several elements common to all structures:

- flexibility—in meeting the needs of adults, fitting learning in with a range of other commitments
- interactivity—in supporting students in their learning processes
- accessibility—in making the courses available to students in the location, the amount and the subject they want.

There are several themes common to all approaches as well:

- re-conceptualizing the nature and delivery of learning materials
- re-using existing course materials and sharing resources
- re-organizing the structure of existing institutions to meet the requirements of the new learner.

References

Bates A. (1995) *Technology, Open Learning and Distance Education*. Routledge, London.
Birkerts S. (1994) *The Gutenberg Elegies. The Fate of Reading in an Electronic Age*. Fawcett Columbine, New York.
Burge E. & Roberts J. (1993) *Classrooms with a Difference. A Practical Guide to the Use of Conferencing Technologies*. Ontario Institute for Studies in Education, Toronto.
Jenkins J. (1995) Past Distance. In: Sewart D. (ed.) *One World Many Voices*. ICDE Conference, Birmingham, June 1995, The Open University, Milton Keynes, pp 427–430.
Mason R. (1994) *Using Communications Media in Open and Flexible Learning*. Kogan Page, London.
Mason R. (1998) *Globalising Education. Trends and Applications*. Routledge, London.
Moore M. (1996) Is There a Cultural Problem in International Distance Education? In: Thompson M (ed.) *Internationalism in Distance Education: A Vision for Higher Education*. ACSDE Research Monograph No. 10, Pennsylvania State University, PA, pp 187–194.
O'Donnell J. (1996) Teaching on the Infobahn. In: Corrigan D (ed.) The *Internet University. College Courses by Computer*. Cape Software Press, USA, pp 111–114.
Taylor J. & Swannell P. (1997) From Outback to Internet: Crackling Radio to Virtual Campus. In: *Proceedings of InterAct 97*. International Telecommunications Union, Geneva.
Witherspoon J. (1997) *Distance Education: A Planner's Casebook*. Western Interstate Commission for Higher Education, Boulder, Co.

10

From Traditional Distance Learning to Virtual Distance Learning in Higher Education in Africa: Trends and Challenges

MAGDALLEN N. JUMA

Introduction

This chapter examines the African Virtual University (AVU) within the historical context of African distance education (DE), and shows how the emergence of a new form of DE, the AVU, provides some of the initial building blocks for Africa to engage in the emerging global knowledge economy.

Distance education has had a long history, dating back to the mid-nineteenth century particularly in Britain and Germany, but until recently, it had been seen as a second best alternative to conventional education. In the Soviet Union, where there was a rapid expansion of part-time study opportunities in the 1950s, the use of correspondence methods, in conjunction with work experience, was initially perceived as a stop-gap measure (Kaye, 1988).

During the last several decades, a number of factors have contributed to a major change in the status of distance education (or distance learning) as to its being an appropriate and effective method for the education and training of adults. An important early indicator of this change was probably the publication by UNESCO of Open Learning: Systems and Problems in Post-Secondary Education, (MacKenzie et al, 1975), drew an analysis of many of the early experiences of the use of multi-media methods for distance education, and helped to redefine the field in broader terms than the simple provision of correspondence courses.

The importance of distance learning was emphasized by UNESCO in its recent commission for the vision of education in the twenty-first century (Delors et al, 1996). It has therefore become obvious that distance education is now seen as an effective, appropriate, and acceptable method of extending educational opportunities in many countries and contexts.

Taken together, these considerations suggest that distance learning is an appropriate means for developing countries to improve workforce skills, provide open and lifelong learning opportunities, thereby supporting the prospects for economic growth. The next section examines the history of African distance learning. The question is, how best can distance learning address the needs of knowledge-based economies? The effort to do this with the AVU is examined in the following sections.

Distance learning initiatives in African higher education

The 1960s introduced a number of basic changes in the direction of distance learning throughout the world, particularly pertaining to higher education. Distance learning has since become one of the fastest growing developments in higher education. Paralleling these worldwide developments, the landmark UNESCO Conference on the Development of Higher Education in Africa, held in Tananarive in September 1962, (UNESCO, 1963), laid down recommendations intended to guide the development of African higher education to the year 1980. The report was conceptually deficient in its ability to focus on and devise distance learning methods in higher education programs within the concerns of national development. Notwithstanding this, the Tananarive document significantly represents an achievement in support of it, as the document illustrates:

> The African universities must play a new role in order to prevent any widening of the already considerable gap between key university men and the masses and to ensure that higher education participates to the full in the economic and social development of the African countries... it is essential that higher education establishments draw up large scale programmes of extra-mural studies and university extension activities together with plans for peoples activities... including extensive participation in the education of the general population... enabling students who are not enroled in the university to obtain degrees... the popularization of science and considerable efforts in connection with applied science. The university has an important role to play in adult education outside its own walls, both in the cultural awakening of the people and in assisting them through extension classes and study groups... towards their own economic improvement... especially directed towards the vast majority who live and work on the land (UNESCO, 1963 pp. 39 and 50).

In the aftermath of independence, many African countries seemed to embrace the philosophy of universal equal opportunities in mass education via correspondence learning. Through its meetings in 1967 and 1968, the Dag Hammarskjold Foundation in Uppsala, Sweden, realized a growing African interest in adult and correspondence education, and thus sponsored seminars on the use of correspondence instruction in adult education: means, methods and possibilities. These seminars saw a majority of participation by individuals from the English speaking African countries.

The first seminar on correspondence education in Africa was finally held in Abidjan, Ivory Coast in April 1971, under the auspices of UN Economic Commission for Africa, in collaboration with the Ivory Coast National School of Administration. It had support and financial assistance from the Swedish International Development Authority (SIDA). The Abidjan conference culminated in the publication of the book *Correspondence Education in Africa*, which was a collection of the papers presented during the proceedings (Kabwasa and Kaunda 1973).

The Abidjan conference also marked an important step in the setting up of the Association of African Correspondence Schools. The conference, however, also addressed other issues: the promotion of correspondence education in Africa, and the evaluation of the current state of correspondence education in Africa and its future prospects. Other issues included the use of correspondence methods in specialized training in agriculture, teacher training and up-grading, public administration and others. The final aim of the conference was to consider the possibility of introducing new techniques to reinforce correspondence education in Africa (Kabwasa and Kaunda, 1973).

Since then, there have been a large number of conferences and organizations in Africa which have addressed issues in correspondence education, and sought to broaden the consideration of all aspects in distance education.

Several organizations have also given strong endorsement to the expansion of distance education in Africa as a way of promoting access, especially for secondary and post-secondary education. Among them has been the World Bank. In its famous study *Education in Sub-Saharan Africa* (1988), it observed that the new extramural study programs were perhaps the only viable way to address the massive problems of access to secondary and tertiary education for students and to continuing education for teachers. Accreditation examinations, it was noted, allow the certification as to a student's "equivalency" in having completed a conventional programme. Students can thus acquire diplomas and degrees by independent study, typically guided by correspondence materials supplemented by radio broadcasts. The replicability of high-quality teaching materials allows high performance standards to be set and maintained in the equivalency system. This is important since extramural programs conventionally utilize existing campus facilities during evenings or vacation months for tutorial sessions (World Bank, 1988).

The issue of lower costs of distance learning was also stressed in the World Bank Report: it noted that the unit cost of instruction in extramural programmes are typically 20–40% of the unit costs in conventional instruction; additionally, there are often substantial savings in student transportation and public budget savings in living expenses. It was further emphasized that during the present period of adjustment and austerity in Africa, a country's rationale for incurring the high construction and incremental costs of the "bricks and mortar" approach to educational expansion deserves to be scrutinized (World Bank, 1988:99).

Partly as a result of these policy trends, many African universities have established distance learning programs. Amongst Anglophone countries, the University of South Africa (UNISA) has had the longest running distance education program, tracing its roots to the founding of the University of Good Hope in 1873. Although it has always offered some print-based distance learning courses, it was not a uniform correspondence university until 1964 (Roberts and Associates, 1998).

In Sub-Saharan Africa, the Congo Brazzaville Service d'Enseignement par Correspondence is perhaps the oldest distance education program. The correspondence service of the Center d'Enseignement Superieur was set up in 1962, within the Fondation de I'Enseignement Superieur en Afrique Centrle (FESAC), which became part of the University of Brazzaville and modeled on the Centre National de T'ele Enseignement in Paris. It provided degree-level education by correspondence that was equivalent to conventional university praxis, offered to students in the Congo, Brazzaville, Gabon and the Central African Republic. Areas of study included law, economics, modern literature, modern foreign languages, history and geography. Students received general guidance and tutorial assistance from the center, which paralleled the courses offered at the University (Kabwasa and Kaunda, 1973).

Among the Anglophone universities established in the post-independence era, the University of Zambia was amidst the few with a clear objective and vision in promoting distance education. Dr. Howard Sheath, first Director of External Studies at Australia's University of New England and a participant in the Lockwood Committee, helped greatly in the shaping of plans for the University of Zambia. The Lockwood Report of 1964 recommended that the new University of Zambia should offer degrees and other qualifications by correspondence. This programme was launched in 1966, and adopted the University of New England, Australia scheme of distance learning. With UNESCO and University of New England assistance, the programme of correspondence education was introduced in the University of Zambia, offering a limited number of courses leading to BA/BSc. degrees and a postgraduate certificate of education (Erdos, 1973).

The adoption of the University of New England program, as opposed to those of the University of London, University of Wisconsin, University of South Africa and others, was meant to ensure that correspondence students were taught and examined by the same full-time staff of the university. No stigma is attached to students studying independently on their own; they are offered the same course content, similar lectures and examinations as full time students. If they pass, they receive degrees from the University of Zambia. Furthermore, the New England plan was sufficiently flexible to permit a free flow of students from full-time to part-time status and vice-versa (Kaunda, 1973:83).

Since the early initiatives in distance learning by the University of Brazzaville and the University of Zambia, the distance education mode has proliferated to several African universities. As profiled by Roberts and Associates (1998), there are very few universities which do not include distance learning programmes. Some of the more recent tertiary open learning institutions in the world are to be found in Africa, for example, Madagascar's Centre National de T'ele-Eseignement

CNTE (established in 1992); the Open University of Tanzania (launched in 1994); the Open University of Zimbabwe (announced in 1997); and now the African Virtual University (detailed in a later section of this paper). The growth is unlikely to end there.

Characteristics of traditional distance learning programs

A topology of university distance learning programs classified as by Yesufu in 1973 (and recently reiterated by Musa in 1994) included the distance education programs found in most universities, and continues to remain valid. But before examining the classification of the nature of the programs offered, it is important to first categorize them according to their modes of operation as they apply to the institutions which carry out similar functions in other parts of the world. Distance education institutions of higher learning fall into three main categories: single mode institutions, including the open universities, dual mode institutions, which offers either conventional on-campus or distance learning to separate students, and mixed mode institutions (Mmari, 1998), which allows students to undertake some of their courses by distance and others by conventional methods. (This is described more in the chapter by Mason.)

In accordance to the classifications by Yesufu (1973) and Musa (1994), the topology of distance learning programmes are conveniently grouped into: intra-mural courses, correspondence courses, and general non-certified courses. A brief description of each of them follows below.

Intra-mural courses

These are normally directed to non-residential students living within reasonable distance of the main university campus. Such courses lead to the award of proficiency certificates, diplomas and degrees, obtained through conventional teaching and instruction, using the university library, laboratories, lecture theatres, and tutorial rooms. Almost all the Departments of Adult Education, Continuing Education and Distance Education in African universities offer such opportunities.

Correspondence courses

These are intended as "outreach" or distance education programs for a wider audience, and designed to earn degrees, certificates and diplomas. They are undertaken primarily by post, and supplemented by radio, television, and newspapers, where practicable. Supplementary contact sessions are an imperative to enable interaction between the teachers and students, between students and the universities, and amongst the students.

General non-certificate courses

They include public lectures; forums and debates; general and non-credit proficiency courses in languages, accounting, music and dance, art, drama; human resources courses for entrepreneurs, artisans, trade unionists, co-operative movement leaders and others. These are offered at most departments of Adult Education, Continuing Education, Community Development and/or Extension Services, Extra-Mural Units etc. of many African universities.

With regard to the mode of delivery in many of the distance learning programmes, the "study material" is the principal mode of transmitting content. Study material includes print, audio, video, broadcast, computer software and experimental kits. Most distance-teaching institutions provide printed texts only. Others vary in additionally providing audiocassettes, videocassettes, slides and experimental kits (Dewal, 1988). Printed study material comprises texts or other books, and/or self instructional lessons or modules. Various institutions prepare study guides which give a total view of the courses to be studied, as well as information about examination schedules, context programmes, students assignments and so forth.

Other components of study material are audiovisual materials and experimental kits, prepared as teaching materials, or as adjuncts to printed materials, to fulfill supplementary roles. Non-print material and other communication technologies can be very meaningful if the content of various media methods reinforce each other, helping the learner to achieve his/her goals. However, not many African distance education institutions have the capacity to use mixed media, as the production of multi-media study materials is a complex operation. Not surprisingly, as in many other parts of the world, a majority of institutions prefer a simple media mix such as printed text and audio-cassettes (Dewal, 1988:65).

Although there is a surge of hardware use in the technology market, such as television, radio broadcasts, videocassettes, or computers, the actual use of these technologies in African distance learning is very low and marginal. Costs are a prohibitive factor, considering that most universities in Sub-Saharan Africa are under-funded by their financially strapped governments. Furthermore, in the institutions where such technology is available, maintenance is very poor. When audio, videocassette players, or computers breakdown, repairs usually take a long time. It is also a common practice that the teacher is often not involved in the planning of how technologies, such as radio, television broadcasts, audio and/or videocassettes, or computers, are used in instruction. This type of top-down planning isolates the teacher, who ends up considering media as outsiders or intruders, contributing to the current marginal use of technology.

It has to be emphasized that it would be unwise to apply media simply because it is available. The chief criterion should not be the availability or access to media, but their instructional potential and teaching effectiveness. Some institutions use radio and television broadcasts as adjuncts to printed materials. The strength of such a media mix are that it costs little to students and can reach a large number of people in the rural areas. The weaknesses of this arrangement are that the messages are ephemeral, and students do not have any flexibility and control to refer back to the information presented. Furthermore, students receive the message passively, and there is no active interaction between students and the content of learning. Often the broadcasts are at inconvenient times, and students find it difficult to view or listen to programmes at that hour. In some countries, radio or television reception is extremely poor (Dewal, 1988).

A few countries have attempted another media mix in the use of audio and video cassettes along with print. The chief merit of audio cassettes is that they are cheap and easy to produce. They can be easily integrated with printed material. Their weakness is that they are relevant only for a limited range of topics and cannot present experience visually. Videocassettes have a great advantage over audiocassettes, and their visual reach and potential for presenting unique learning experience is wide. As compared to the production of television programs for broadcasting, videocassettes are easy to produce. The key problem is that for students in rural settings, the use of videocassette players is still a distant dream (Dewal, 1988).

In some countries, study centers constitute an important mode for the transmission of content. Depending upon financial resources, distance teaching institutions set up resources centers or resource study/centers. The former only provides learning resources, whereas the latter additionally provide facilities for individual or group tutoring, academic guidance and counseling. In most African countries, resource/study centers are conspicuously absent, or if they exist, are run ineffectively. They hardly undertake tutoring and counseling responsibilities. Even basic functions like information, provision or resource consultation are performed inadequately.

Some key challenges in distance education

Despite the progress made in distance learning at the tertiary level since the 1960s, a number of challenges continue to threaten its effectiveness. These largely center on issues of costs, financing, and policy orientation which directly affect the provision of materials, personnel, and learner support.

There is a belief that distance education is cheaper than traditional forms of education, that is, the cost per student/or per graduate is less than traditional forms of education. Since students can progress at different rates (either full-time or part-time), the cost per student is often expressed in a standard measure, such as the cost per student-hour or cost per full-time student equivalent (Rumble, 1988:91). It is now evident that this assertion cannot be taken at face value. There are examples of distance-teaching schemes where the cost per student or cost per graduate has been higher than the costs incurred through traditional systems (Perraton, 1982).

A major cost in distance learning programs is that of material production. The production of distance teaching materials is far more expensive, as has been demonstrated by the British Open University (Zahlan, 1988). This is because the cost includes the curriculum and course design, the production of materials including the author's fee, the remuneration of reviewers and assessors, and the presentation of the final product (use of graphics, language, lay-out, style etc.), and its dissemination. In addition, when television, radio and laboratory kits are utilized, the cost of preparing a course increases. Many African university distance learning programmes have tended to cut down on costs by adopting full-time (traditional) programmes with little or no supply of adequate learning materials to distant learners. A proliferation of such distance learning approaches have been exacerbated by the universities' efforts to initiate income-generating programmes following reductions in government funding. Most of these programmes are inadequately organized with insufficient human resources to manage a distance learning system. Staff are often assigned to several tasks, with distance learning not being a priority. Learners' support is weak in terms of materials. Consequently, the quality of graduates produced through these systems is mediocre and compare less favorably with graduates of conventional programs.

Economy is achieved in distance teaching where the number of students enroled per course is large. Only when these numbers are in their thousands will the cost per student be reduced to a level equivalent to that of an equivalent residential university. In many African universities, the scale of resources in the production of quality learning materials can only be justified by relatively large numbers of students who get enroled in the courses (Perraton, 1992, Makau, 1993). In both Francophone and Anglophone Africa, with the exception of Nigeria, the national populations are smaller than those other countries that have successfully established open universities. Thus, for a country to justify investment in the development of courses at the tertiary level that has led to the success of the existing open universities is difficult. Similarly, even where universities decide to establish a distance education department, they face serious difficulties in releasing adequate funds for course development, as is now commonly the practice with many units of the universities. There is therefore a strong case for an African national or regional open university, as has been considered by the Southern Africa Development Coordination Conference (SADCC), a proposal that is in accord with World Bank's concept of one or more regional centers for distance education within Africa (World Bank, 1988:111).

The issue of cost seriously undermines the provision of learning resources for small numbers of students enroled in the national distance learning programs. Students experience poor library facilities; and learners' support systems do not receive the priority they deserve. Prof. Hugh Africa's vivid description of emotional stress which learners in such programs experience illustrates the importance of providing such support. A typical indicator of the latter is the delay in the marking of assignments and examinations, which many learners remark upon. Reliable communications methods with students, be it by mail or telephone, are often lacking.

Collaboration and partnership

A number of challenges identified in the previous section could be met through partnerships and collaboration. These are a central feature of distance learning in a number of ways. For instance, collaboration of a team of content experts, course designers, material production and evaluation is critical to success in the development of distance learning materials. And partnerships facilitate the sharing of scarce financial and teaching resources, as will be demonstrated in the case of the African Virtual University. Collaborative projects could undoubtedly have a major effect on the development of tertiary distance education in Africa.

Moran and Mugridge (1993) classify partnerships into three broad categories:

- The integration of learning pathways through credentials offered by more than one institution.
- Collaboration in distance education operations such as course development credit transfer.
- The creation of new agencies of collaboration such as the African Virtual University (AVU).

Although a number of African collaborative organizations exist, by and large they do not seem to have gone through these stages. For example, Francophone countries have a strong infrastructure of collaborative organizations that include distance learning as a priority: AUPELF-UREF (Agence francophone pour l'enseigement superieur et la recherche—universite des reseaux d'expression francaise) and ACCT (Agence de la Francophonie, formally the Agence de cooperation culturelle et technique). Although not indigenous to Africa, they nonetheless play a major role in fostering the development of distance learning in Francophone Africa (Roberts and Associates, 1998).

In Anglophone Africa, various national and regional associations have been formed and have provided support to distance learning institutions: these include organizations such as the Distance Education Association of Southern Africa (DEASA)—a regional organization; and the National Association of Distance Education of South Africa (NADESA)—which is national. They have been the main vehicles for professional development, staff training, and the exchange and sharing of expertise among and between a number of institutions within Africa. Many of the institutions have carried out these activities at their own cost and with very little assistance from external sources. These associations have been instrumental in ensuring that programs are carried out and sustained in a professional manner, and that information is shared through newsletters and publications. Some of these associations have compiled their own directories containing information about their member institutions, courses, research and publications, as well as a calendar of annual activities (Roberts and Associates, 1998:25).

The Francophone Virtual University (FVU) which is planned, implemented and sponsored by AUPELF-UREF, relies on networks put in place through UREF to carry out its mission. Its purpose is to decentralize the production of knowledge, to encourage the creation and distribution of research, and to develop competencies in distance education and learner support.

In Francophone Africa, the ACCT and its partners are engaged in action research pertaining to the use of new information and communication technologies in distance learning. This acts under the auspices of the African Distance Learning Network (RESAFAD). Its aim is to develop African expertise in the design, production and management of distance learning systems. Innovative, logistical and methodological support systems have been put in place—with the support of the French University consortium—to facilitate collaborative team work, as well as the sharing of resources and products developed by different national teams in each country that rely on a local multimedia resource center linked to the Internet (Robert and Associates, 1998:27).

Indeed the list of national and regional distance learning organizations and networks is long and all the activities they undertake cannot be sufficiently discussed in one chapter. It is apparent that they are patchy and not well coordinated. This suggests that the establishment of better inter-country collaboration and partnerships. Furthermore, the strengthening of a major institution like the African Virtual University (AVU) will enable it to coordinate all the work being done in distance learning in Africa along the lines of the proposed center of distance teaching, as recommended in the World Bank's policy study on education in Sub-Saharan Africa. Other suggestions include strengthening the African Association for Distance Education, and establishing a number of broad based regional networks.

The African Virtual University (AVU)

Africa and the knowledge economy

The AVU has arisen on a stage in which tremendous changes are taking place in the global economy, involving the generation of huge quantities of knowledge and a divergence in growth and capabilities between Africa and the rest of the world. At the same time, despite forty years of development efforts, the economic situation in Sub-Saharan Africa (SSA) has gotten worse. (Baranshamaje, 1997).

The quality of life has not improved for most people in SSA as economic growth in the past decade has stagnated. Economic growth in the region has been much lower than in other developing regions. This situation is even more distressing when contrasted with the population growth in the region, which has been 3.0% during 1980 and 1992, and which is expected to grow by 2.8% from 1992 to 2000. As the inability of economic growth to keep up with population increases, the social and political well-being of people in SSA is increasingly severely affected. Entering into the 21st century, countries in SSA are confronted by poor health and a rising incidence of AIDS, a disproportionately high degree of poverty and hunger, low levels of education, increasing civil strife, and a deteriorating infrastructure base.

Fortunately, something very exciting and unique is taking place in the world, which, in one scenario, provides a reason for hope. For the past forty years, development efforts in SSA have basically tried to mimic in a linear, step-by-step manner, the industrial development of the United States and Europe. However, the emergence and combination of several powerful institutional forces; the technology revolution, the end of the Cold War, and the democratization of ideas are fundamentally changing the global economy by influencing the nature and relationship of markets, products, competition, trade, and sources of comparative advantage. For example, modern information infrastructure is creating the "Death of Distance", reducing economic isolation and allowing remote countries to diversify their volatile bases by providing an opportunity to participate in the global economic process. Non-traded goods (e.g. services) are becoming internationally tradable via information technology; previously immobile factors of production (e.g. labor) are becoming "mobile" as geographic barriers have less meaning.

The new global economy has the potential to create opportunities for developing countries to leapfrog development steps and constraints and speed up the development process, or to undermine development efforts by further accentuating the gaps between the rich and poor. For developed and developing countries alike, the ability to realize knowledge-based productivity gains depends on a country's capacity to tap the global system of generation and transmission of knowledge, generate indigenous knowledge, diffuse and transfer information, and utilize that knowledge in productive activity.

Sub-Saharan Africa's ability to actively participate in the new global economy and to solve the many social and political problems that it faces depends largely on the intellectual capacity and skills of its labor force, particularly in the fields of science, technology and business. This implies that there must exist a cadre of professionals from a broad range of disciplines who are skilled in using and adapting existing and new knowledge and information on changing local, national and international conditions. The recent economic successes of India's software industry and of the Asian Tigers can be traced to their ability to build such a cadre of professionals. There is an urgent need to develop such a cadre in SSA. On an even more basic level, Africans must have continual access to various forms of knowledge and information in a flexible and timely manner.

Unfortunately, a "knowledge gap" currently exists between the SSA and much of the rest of the world. This gap implies that the distribution of skilled professionals and of access to knowledge and information is highly skewed in favor of developed countries. For example, scientific and technological capabilities are distributed unequally in the world. Developing countries account for only 13% of the world's scientists and engineers, and only 4–5% of global spending on research and development, with a concentration in a minority of countries.

This knowledge gap currently stands in the way of Africa's ability to develop a critical mass of trained professionals in the fields of science, engineering, and business, which is crucial to its economic growth. Moreover, it places limitations on the ability of the current African workforce to quickly adapt to changes in the market and places SSA firms at a severe disadvantage in their own markets. It is imperative to the future economic and social development of SSA that swift and deliberate steps be taken to bridge this gap.

Despite these possibilities, the limitations of conventional distance learning and higher education in Africa is becoming clearer. Unfortunately, tertiary institutions in their present form, overwhelmed with problems related to access, finance, quality, internal and external efficiency, are unable to bridge the "knowledge gap". Limited space and declining budgetary levels prevent universities from servicing the growing demands for higher education. As a result, universities in SSA are affected by a number of factors: low numbers; trained faculty; limited levels of research; poor quality educational materials (e.g. African libraries have suffered immensely as collections and laboratory equipment become antiquated), and programs which do not meet modern requirements. Much of the educational methodologies are based on the model of rote memorization and do not encourage critical thinking, problem-solving and creativity—all essential skills for promoting entrepreneurship. These constraints have prevented institutions of higher education from being able to link their graduates with the needs of their countries: the distribution of graduates is inconsistent with expected labor market needs. To a large extent, many African universities have failed to remain relevant in a rapidly changing world, as a disproportionate number of students graduate in the humanities rather than in the fields of science and engineering.

Essentially, the African Virtual University (AVU) is a concept or model of distance education that seeks to address this situation through the use of satellite technology for instructional delivery. The project is funded by the World Bank and headquartered in Washington, DC It is the first of its kind—an interactive-instructional telecommunications network established to serve countries of Sub-Saharan Africa. (AVU, 1997). The mission of the AVU is to use the power of modern information technologies to increase access to educational resources throughout Sub-Saharan Africa. The objective of AVU is to build world-class degree programs that support economic development by educating and training world-class scientists, technicians, engineers, business managers, health care providers, and other professionals.

The implementation of AVU and current status

The AVU is being developed and implemented in three phases. The first phase is the prototype phase, implemented from 1997 to the end of 1998. The purpose of this phase is to establish partnerships with institutions of higher education throughout Sub-Saharan Africa, offering technology-based credit courses and non-credit seminars using digital satellite technology.

A successful prototype service will provide the foundation for the second phase of the AVU project, which includes complete undergraduate degree programs from leading universities world-wide beginning as early as November, 1999. The third phase of the AVU will follow with the development of science and technology curricula from one or more partner institutions in Sub-Saharan Africa. The full implementation of the third phase will be the sharing of resources amongst technology-based degree programs from institutions of higher education throughout Sub-Saharan Africa.

For the 1998–1999 academic year, the following 27 institutions offer AVU courses:

- Addis Ababa University, Addis Ababa, Ethiopia
- Cape Verde Higher Education Institute, Cape Verde
- Egerton University, Kenya
- Kenyatta University, Nairobi, Kenya
- Kigali Institute of Science and Technology, Rwanda
- Loko Vocational School, Ivory Coast
- Makerere University, Kampala, Uganda
- National University of Rwanda, Rwanda
- National University of Science & Technology, Bulawayo, Zimbabwe (NUST)
- Open University of Tanzania, Dar-es-Salaam, Tanzania
- Togo University, Benin
- Uganda Martyrs University, Nkosi, Uganda
- Uganda Polytechnic, Kyambogo, Kampala, Uganda
- University of Eduardo Mondlane, Mozambique
- University of Cheikh Anta Diop, Senegal
- University of Bou Moumouny of Niamey, Niger
- University of Namibia, Namibia
- University of Abidjan
- University of Benin, Benin
- University of Cape Coast, Cape Coast, Ghana
- University of Dar-es-Salaam, Tanzania
- University of Ghana, Accra, Ghana
- University of Mozambique, Mozambique
- University of Noukchott, Mauritania
- University of Ouagadougou, Bokina Faso
- University of Science and Technology, Kumasi, Ghana(UST)
- University of Zimbabwe, Harare, Zimbabwe

The AVU network

Content production

The AVU network is centered around the content producing institution, with production at the institution itself or in a nearby facility leased by the institution. Each content site includes all equipment necessary for the initialization of both live and pre-recorded instructional programs, and for the provision of these programs to either a local or remote INTELSAT-compatible uplink Earth station. The studio also includes equipment necessary to provide certain levels and types of interactivity, such as audio talkback for live programs, and to facilitate such other related functions as satellite files. The personnel on location consist of the professor or lecturer and, typically, a program technician. The broadcast can either be live or videotaped.

Satellite transmission

The signal from the content site is transmitted via a local provider, e.g. in the U.S., the Indiana State University system currently sends material to an uplink facility in Washington, DC. AVU currently contracts with COMSAT to relay the signal from Washington, DC to Africa by way of the INTELSAT satellite system. AVU currently enjoys unlimited usage with INTELSAT, which has donated the equivalent of $2,000,000 of satellite capacity to the project. The Technical Officer of AVU manages the network system.

Local Support—In Africa, each Sub-Saharan Africa (SSA) partner institution is provided with all of the equipment necessary to receive the digital satellite transmission, and a full complement of AVU programs and services. The current funding source (roughly equivalent to $54,000 per university) is the World Bank. In the future, funding for local infrastructure will be provided by the AVU franchise. The outdoor equipment provided by AVU includes the receive-only satellite antenna and necessary electronic equipment, integration hardware, and cabling. The indoor classroom equipment is designed to be "user-friendly" and to enable maximum flexibility in the use of AVU programs and services.

The indoor equipment unit includes two or more digital video receiver-decoders, one or more television monitors, one or more video cassette recorders, two multimedia personal computer systems, a push-to-talk telephone, printer, a facsimile machine, a monitor and control (M&C) unit, a custom-designed equipment cabinet, and other miscellaneous components to help insure ease of operation and uninterrupted programme reception. A comprehensive data handling and communications system has been designed to receive, store, and process incoming high-speed broadcast data channels, and to provide each receiving site with an outbound data/audio talk back link.

Classroom—The typical AVU classroom consists of between 30 and 40 students sitting at their desks watching the videotaped or live broadcast on television monitor. During live broadcasts, students have the opportunity for real-time two-way interactivity with the instructor, and the ability to collaborate with other students.

AVU pilot programs in Kenyatta University

This section provides more detail on Kenyatta University's AVU program. Kenyatta University is a comprehensive university offering undergraduate and postgraduate degrees in six faculties: Arts, Commerce, Education, Environmental Studies, Home Economics and Science. AVU courses were mainly used to enhance existing courses or provide for non-existing courses at Kenyatta University. During the Kenyatta University site's pilot phase, three semesters of courses were offered, transmitted from Universities in Canada, Europe and America as shown below in Table 10.1.

TABLE 10.1
Courses Offered at the Kenyatta University AVU Site

Course	Originating Institution	Total 1997 Enrollments (two semesters)
Calculus I	New Jersey Institute of Technology (US)	110
Calculus II	New Jersey Institute of Technology	112
Physics	Carleton University (Canada)	47
Organic Chemistry	Laurentian University (Canada)	31 (one semester)
Introduction to C++ Programming	Mount Saint Vincent University (Canada)	28 (one semester)
Introduction to Engineering	Georgia Institute of Technology (US)	20 (one semester)
Computer Organization and Architecture	Colorado State University (US)	41 (one semester)
Differential Equations	New Jersey Institute of Technology (US)	12 (one semester)
Introduction to Computing	University of Massachusetts (US)	20 (one semester)
Introduction to Internet	University of Massachusetts (US)	73 (one semester)

All the students registered for the AVU courses mentioned above were regular campus residential students, from the departments of physics, mathematics, chemistry, and appropriate technology, which are part of the Faculty of Science curriculum. However, some of the core courses—such as Introduction to Engineering, Computer Architecture and Design, Introduction to Internet—have previously not been offered in Kenyatta University.

Enrollments

Enrollments at Kenyatta AVU have increased since the first courses were offered in the 1996/97 academic year. In the first two semesters in 1997, there were a total of 167 and 299 student course enrollments respectively, totaling 494.

Gender disparity is evident with males outnumbering females in almost all the units. The evidence revealed by female participation on AVU courses does not differ from the national representation of girls in university; where under 30% are women. The picture deteriorates in faculties such as engineering and computer science. For example, in Kenyatta University, women constitute 23% in the Faculty of Science. This means that relatively few women choose science-based AVU courses in comparison to men. This imbalance does not originate at the university level, but begins from the primary level. Studies indicate that science and mathematics are generally conceived to be "masculine" subjects: the negative perception about the female ability to study sciences, and other factors related to cultural attitudes, poverty, and poor performance contribute to a low participation of girls in science subjects, at tertiary level of the education system. (Juma 1997).

Course Experiences

Some of the course experiences and impacts are as follows:

Calculus I

In the Calculus I course, originating from New Jersey Institute of Technology (NJIT) in America by Prof. Rose Dios, 28 students from the Kenyatta Mathematics department registered. Students enjoyed this course, as illustrated by the following comments:

> "Prof. Dios is a very good teacher because her lectures start with simple examples and build slowly to complex examples (i.e.) from "known to unknown". This approach enables me to understand and enjoy Mathematics" (Personal interview with first Calculus I student).

> "It has taken me two times to retake Calculus I examination without passing. I would have given up on Calculus I if it were not for AVU Calculus I course, which I managed to pass. It was possible to pass because AVU courses provide textbooks, comprehensive notes, tutorial sessions, interaction with an instructor in America and a local university lecturer who always explains concepts during tutorials. During revision, we get pre-recorded videotapes to review lectures. Therefore, all these benefits enabled me to pass" (Female student personal interview).

Internet course

This course generated a lot of enthusiasm from students, faculty members and the general public. It was oversubscribed with more than 400 students registering, but owing to limited facilities, only 40 students were admitted to the course. A loan of Kshs. 5 million (US$86,000) from the University was given to AVU to purchase 20 computers and one projector for Internet training in a separate laboratory. AVU—Kenyatta instituted income-generating measures to offset the loan and meet some of the running costs of AVU. Students paid 15,000 Kenya shillings (equivalent to US$280) per semester for the Internet course.

Students commented about the Internet course:

> "Before I joined AVU courses, I had never seen a computer nor sat before a keyboard, but from the Internet course, I have known how to use the computer, send e-mail, design websites and access courses, data etc. from the websites".
>
> "I am only a third year physics student, from very poor family background, struggling to complete fourth year fees for January 1998. Fortunately, after my Internet examination in December 1997, I managed to apply for a job with Commercial Bank of Africa, Nairobi, attended an interview, performed very well in both theory and practicals, and was employed even before completing my degree course in Physics and Mathematics. The AVU Internet course has earned me a job"!

Out of 40 first Internet class, 22 students have been employed in companies in Nairobi.

In general, the performance in all these courses/units has been good, except for the *Calculus I and Computer Architecture and Organization* course, whose average score was below the average of 50% during the summer and second semester '98. The performance was been very good, especially in *Computer-related Courses* which experienced an average of over 70% for the last two semesters.

Before the AVU was adopted at Kenyatta University, the Senate approved the integration of AVU pilot phase courses to the Kenyatta University curriculum. On this basis, therefore, semester grades from AVU courses were computed along with other grades in the respective departments. Two groups of students, one taking the traditional course and the other, an AVU course provided an opportunity for comparison in terms of participation and performance. Thus, in mathematics, while one group of students used the AVU system, the rest of the Mathematics students continued with their traditional Kenyatta University lecturer during the summer semester. Two continuous assessment tests and one final examination were given at the end of the semester. The local AVU instructor from the Mathematics Department set and marked the examination, which resulted in a pass rate of 80% compared to a mere 40% for students studying via the traditional mode of delivery (which consists of lectures and a few demonstrations as pointed out by the teacher who taught both courses). It is important to mention that in all courses except for the Internet and Introduction to Computing courses, interested students from related departments at the university register for AVU courses.

Other AVU courses

Short computer courses

Apart from AVU semester programs, which targeted on-campus students, there are brief computer courses lasting from two weeks to one or two months, depending on the nature of the course. The courses are open to the public for professional upgrading and continuous learning. The target groups are civil servants, people working for organizations such as banks, ministries, parastatals, universities, schools leavers, graduates, and doctors etc.

The schedule for short computer courses runs from seven in the morning to eleven o'clock in the evening. The following short computer courses were taught: Introduction to Microcomputers and MS-DOS, Windows '95, MS Word '97, MS Access '97, MS Excel '97, Using Internet, Advanced Internet, C++ Programming, and Visual Basic Programming. Enrollment in the three phases of these courses in 1998 totalled 276, 462 and 251 respectively.

Statistics reveal that enrollment in the AVU Short Computer courses has been very positive. The most popular of these courses being the Introduction to Microcomputers and MS-DOS course, which enrolled more than 100 students in the last two phases. Enrollment has been higher for males than for females but this discrepancy is decreasing with time. The performance in these computer courses has been very good, with a mean score of over 70% in all courses.

The computer certificate courses are extremely popular and have generated a lot of income during the last six months. Although AVU's pilot phase is drawing to an end, short computer courses will continue to be delivered to interested people.

Pre-university courses

A program for a pre-university course has been put in place to prepare students for university level study. The pre-university course takes three semesters. After the advertisement, selection and enrolment, 105 students registered for the pre-university course. Subjects taken include Introduction to Computer Science, Mathematics—Calculus I, Chemistry, and Physics. The pre-university course prepares students to undertake the AVU degree course, which is expected to begin in October 1999. A total of 105 students enrolled for the pre-university course but 4 students dropped out. In this program, students in the chemistry course had the highest mean score while those in computer science had the lowest, falling below average (50%).

Seminars

The marketing of AVU seminars to the business community in Nairobi resulted in an overwhelming response: 40 executive managers attended a seminar for top executives, which focussed on "Purchasing Policies and Practices," transmitted "live" from Virginia Tech in the U.S.A., from 14th to 21st August 1997. The participants included individuals from Barclays Bank, Standard Chartered Bank, Central Bank, Kenya Commercial Bank, Commercial Bank of Africa, Kenya Breweries etc. During the live seminar, participants interacted with Professor Murphy from Virginia Tech and their counterparts from Uganda, Zimbabwe, and Ethiopia. The virtual seminar was extremely exciting as expressed by managers' comments:

> "It is unbelievable that a virtual session typical of situations in America is taking a place in Kenya. This is the first time in my life to participate in a seminar of this nature" (seminar participant).

> "The seminar was very educative, informative, and enriched what I already knew on purchasing".
>
> "The case study approach gave solutions to practical problems in an organization".
>
> "The use of modern technology enhanced proper learning and sustained our span of concentration".

This was a successful seminar, indicating the potential for collaboration between the business community and the university. Such support augurs well for the sustainability of the project, as the business community can be expected to attend and support such courses in the future.

The Economic Development Institute (EDI) of the World Bank, Washington, DC also organized a series of seminars for the AVU. Seminars for 30 journalists from Kenya ran for 14 weeks. The journalists assembled at the AVU premises for live transmission via satellite. During the first transmission, one female journalist could not conceal her excitement when she enthusiastically exclaimed:

> "I feel like I am in Washington, DC attending EDI seminar, we can ask questions easily and interact clearly with course instructors as if the delivery was face to face. It is 'real' that I cannot believe we are in Kenya". (Female journalist participant March 31, 1998).

This is an income-generating venture whereby the EDI paid AVU—Kenyatta University a total of US$5000 for utilizing the AVU infrastructure to transmit courses to journalists. Several other important seminars have been held at AVU—Kenyatta, such as seminars on Y2K, Balanced Score Card, Strategy and Innovation, and recently, on influence and leadership.

The impact of the AVU on the whole of Kenyatta University

During the last two years that the African Virtual University (AVU) has existed in Kenya, it has had a considerable impact on students, faculty members and the entire community. The following achievements have been experienced:

- *Provision of Educational Resources*—Two hundred computers for the AVU enable students to participate effectively in computer courses. Prior to the AVU, the University had a significant shortage of computers: for instance, the Computer Centre possessed four and the Mathematics Department had three, which was undoubtedly inadequate for any meaningful course. Students who enrol for AVU courses are also provided with textbooks and notes. The trickle effects of AVU resources have permeated the entire university science students' community. Internet facilities provided by the AVU are utilized by everybody interested in accessing e-mail and conducting Web searches over the Internet. Thus, the electronic mail facility has modernized the university environment.

- *Introduction of New Courses*—Computer science as a discipline has never existed in Kenyatta University but with the emergence of the AVU, many computer-based courses have become very popular. The Internet course, for instance, is a new development and students find that it provides lucrative skills that heightens job-prospects. As a consequence of these new information technologies, the University, with the assistance of AVU students, has created an AVU website.
- *Capacity Building*—A culture of continuing education has been perpetuated through AVU computer-related courses, particularly among professors, lecturers, students and the public. Seminars and workshops have attracted the executive cadre of the business community, who find relevant courses via satellite very "educative and refreshing". The use of teleconferencing and interactive technology empowers them to communicate with their counterparts in Africa and Europe. The enhancement of local capacities to participate in activities within the country and Africa offers an opportunity for the sharing of expertise within the region. For instance, qualified engineers, namely, the technical coordinator of AVU and the campus coordinator, have effectively assisted in setting up satellite technology in other African countries. This means that Kenyatta University experts assist other countries and enhance the sustainability and maintenance of AVU within Africa.

AVU will target courses that provide training for secondary school leavers. This will undoubtedly have a significant influence on the expansion of opportunities for them. Currently, of the estimated 160,000 students who complete secondary school annually, around 7–8% are admitted into universities, another 15–20% into other tertiary institutions, leaving over 30% of the qualified students without places.

AVU prospects, income generation and financial sustainability

A weakness observed in African university programs is their emphasis on the production of lengthy programs, as opposed to short duration courses, which serves as a way of continuing education for the first degree. For example, training in business administration has been focused on full-time MBA type postgraduate education. Not only is this formula bound to be limited to a tiny minority, but more importantly, small African firms which form the backbone of African economies cannot afford to enrol their employees in long duration programs. What these organizations require are short duration courses in a well-focussed domain, e.g. marketing purchasing and supplies, computer systems and applications, programming, leading to certificate and diploma accreditation. In this respect AVU—Kenyatta has designed short certificate courses in computer applications and systems, Internet access, and others. Through high quality content delivery, affordability through better technologies for delivery, and a market-driven program, these courses are very popular and generate reasonable income. To illustrate, within seven months, short computer-based courses generated six million Kenya shillings, (US$100,000). Additionally, pre-university courses have become very popular; in two semesters (July–December, 1998) income of 4 million Kenya Shillings (about US$66,600) has been generated. Of this amount, US$66,000 was channeled to meet the running costs of AVU, marketing and payment of resource staff. AVU technologies such as Internet, e-mail services, fax services, cyber cafes, also play a part in generating income. As a matter of fact, AVU—Kenyatta University has been self-sustaining since last year, when limited income generating measures were put in place.

Indeed, the virtual model promises viable prospects for private institutions by reducing the investment requirement and operating costs of higher education institutions. In addition, the new technologies allow improved student achievement, and the introduction of better pedagogical techniques. For example, students are able to pursue more hands-on-learning opportunities, and conduct more Internet search and research. During the brief period the AVU has existed in Kenyatta University, the new technology has contributed to the improvement of internal efficiency within some departments (the Mathematics Department has reduced drop-out rates), and increased pass rates (as students receive unlimited access to instructional material, and more effective interaction with faculty and fellow students). In essence, the AVU is creating a favorable environment for the introduction of cost-recovery through student's fees.

Problems and Challenges

There are pertinent challenges impinging on the operations of the African Virtual University (AVU) in Africa. The following challenges are significant, particularly to Kenya:

- *Electricity interruptions*—Power fluctuations in Africa tend to affect the satellite-receiving terminals, particularly computers. The power "black outs" or "change of power" in relation to generators, results in damage to the hard disk and sometimes the system board of computers.
- *Lack of a clear national communication policy in Kenya*—Communication services are strategic infrastructure which are core to the country, and therefore private ownership of Kenya's telecommunications services is limited. Privatization and liberalization go hand in hand with ownership, and multi-nationals have not been given that kind of ownership in Kenya. Consequently, communication services are very expensive due to the monopoly of services by Kenya Posts and Telecommunications. For instance, the cost of Internet is very high in Kenya. This is because Kenya does not own an Internet "hub" as yet and so local Internet providers have hubs in the USA, Canada and Europe. The cost of international links are expensive, hence Internet services are very costly for local consumers. The African Virtual University (AVU) relies on Internet, e-mail services etc. which immediately places it in a disadvantageous expense bearing position.

- *Poor Internet connectivity at AVU – Kenyatta*—Internet access is very slow because connection is still made through dial-up telephone lines.
  Slow Internet access result from reasons such as the following:
- The bandwidth for transmitting signals over the telephone lines in Kenya is normally slow with the highest rate of transfer averaging 19.2 kbps.
- The telephone line signal is also of poor quality, resulting in average signal loss and errors. This slows access in two ways: it cuts connection, and finally a loss of the link. Hence, the sending and reception of mail is frustrating.
- Congestion of Internet accessibility by users from western regions may create "traffic jams", resulting in a slowdown within the Internet network. This occurs during the day interrupting Internet courses at the AVU.
- Frequent "link" breakdowns, either as a consequence of natural or physical calamities.
- A breakdown at the Kenya Posts and Telecommunications (KPTC), which may last one day: cutting off all forms of telecommunications between the whole country and the outside world. This can often be the result of flooding amongst the cables or other technical problems.
- Poor maintenance of KPTC telecommunications equipment, resulting at times, in loss of communication for one or two days.
- Telephone problems due to a breakdown at the university telephone exchange, blocking external phone calls for an extensive number of hours.
- There have been a few cases where the Internet service provider (ISP) pilot telephone line goes out of order, causing frustrations and anxiety as dialing and connection becomes very difficult.

Despite the problems the challenges identified above pose, they do not undermine the basic soundness of the AVU in Kenya, or its abilities to generally meet its objectives. Experiences and lessons learnt have taught AVU Kenyatta University to solve and overcome obstacles. At the national level, a Parliamentary Bill on Kenya's Posts Telecommunications policy is soon to be tabled for discussion. It is hoped that with a policy framework, postal and communications services will be liberalized and better services ensured.

Kenya Posts and Telecommunications Cooperation is expected to launch a national hub or backbone by the end of December, 1998. With this facility, the Internet access is likely to be fast and affordable by institutions of higher learning.

The African Virtual University (AVU): prospects for distance education in Africa

Since AVU is a distance education model using modern information technologies, it complements the traditional mode of distance education delivery systems: namely, through print, correspondence, radio, cassettes, television etc. which have existed in Africa for a long time. For instance, dual mode institutions like Nairobi University,

University of Zambia etc. have had external degree programs and extra-mural centers, providing degree, diploma and certificate courses via distance learning.

Unfortunately, distance education programs in Africa cannot effectively meet the demand. Therefore, most programs have primarily focused on the liberal arts, while AVU tends to supplement existing programs by providing access to scientific and technological opportunities for people who are unable to attend school for a variety of reasons.

Most traditional-distance modes of delivery lack an interactive–instructional network. Unlike traditional modes of delivery in distance education, AVU has an instructive-instructional telecommunications network, which enhances participatory learning throughout Africa. AVU functions as a single network that enhances North–South cooperation and South–South cooperation in terms of delivery in content and technology.

Conclusions

Prospects for AVU in Africa are promising, where the revitalization of university education is concerned. Furthermore, the skills and knowledge imparted by AVU are relevant to the job-market, hence contributing to the formation of a knowledge-based economy. In this way, the AVU can be seen as a complement to traditional forms of distance learning in Africa.

References

AVU, (1997) *The Campus Coordinator Manual for 1997 Prototype Phase.*

AVU, (1997) *The Course Coordinator Manual for the Prototype Phase.*

AVU, (1997) *Inaugural Training Workshop Manual,* Addis-Ababa 16–21 February.

Baranshamaje, E *The African Virtual University, Concept paper*, Washington.

Delors, J. et al (1996) *Learning: The Treasure Within; Report to UNESCO of the International Commission on Education for the Twenty-first Century,* Paris, UNESCO Publishing.

Dewal, O. S. (1988) Pedagogical Issues in Distance Education, *Prospects, UNESCO Quarterly Review of Education* 1988; **18**(1):65.

Erdos, R. (1973) UNESCO and Correspondence Education for Teacher Training. In: Kabwasa A. and Kaunda M. M. (eds.) *Correspondence Education in Africa,* London, Routledge and Kegan Paul.

Juma, M. N. (1997) *AVU Progress Reports*, Kenyatta University.

Harries, H., (1996) *The African Virtual University Library*, Washington.

Kabwasa, A. & Kaunda, M. M. (eds.), (1973) *Correspondence Education in Africa,* London, Routledge and Kegan Paul (1973).

Kaunda, M. M. Post-secondary Education by Correspondence: An African Experience. In: A. Kabwasa & Kaunda M. M. (eds.): *Correspondence Education in Africa,* London, Routledge and Kegan Paul (1973).

Kaye, A. (1988) Distance Education: The State of the Art. In: *Prospects, UNESCO Questerly Review of Education* Vol. XVIII No. **1** 1988 (65).

Mackenzie, N., Postgate, R. & Scupham, J. (eds.) (1975) *Open Learning: Systems and Problems in Post-Secondary Education,* Paris UNESCO.

Makau, B., (1993) The External Degree Programme at the University of Nairobi. In: Perraton H. (ed.) *Distance Learning Education for Teacher Training,* London, Routledge.

Mmari, G. (1998) Increasing Access to Higher Education: The Experience of the Open University of Tanzania, in UNESCO Regional Office, *Higher Education in Africa: Achievements, Challenges and Prospects,* Dakara.

Musa, M. B., (1994) Extension Education and the Role of University Extension Department *International Review of Education* Vol. **40** No. 2 1994.
Perraton, H., (1982) *The Cost of Distance Education,* Cambridge, International Extension College.
_______________, (1992) A Review of Distance Education in P. Murphy & A. Zhiri (eds.) *Distance Education in Anglophone Africa: Experience with Secondary Education and Teacher Training,* Washington, DC, The World Bank.
Roberts, J. M., Keuagh, E. M. & Howard J, (1997) *Diffusion of Appropriate Educational Technology in Open and Distance Learning in Developing Commonwealth Countries,* Vancouver, BC: The Commonwealth of Learning.
Roberts & Associates, (1998) *Tertiary Distance Learning in Sub-Saharan Africa: Overview and Directory to Programmes,* Toronto, ADEA Working Group in Higher Education.
Rumble, G, (1998) The Economics of Mass Distance Education *Prospects, UNESCO, Quarterly Review of Education* 1988; **18**(1):65.
United Nations Educational, Cultural and Scientific Organization, 1963 *The Development of Higher Education in Africa: Report of Tananarive Conference 1962*, Paris, UNESCO.
World Bank, (1988) *Education in Sub-Saharan Africa: Policies for Adjustment, Revitalization and Expansion*, Washington DC The World Bank.
Yesufu, T. M. (ed.), (1973) *Creating an African University: Emerging Issues in the 1970s,* Ibadan, Oxford University Press.
Zahlan, A. B., (1988) Issues of Quality and Relevance in Distance-Teaching Materials, *Prospects, UNESCO Quarterly Review of Education* 1988; **18**(1):65.

11

Graduate Studies at a Distance: The Construction of a Brazilian model

RICARDO BARCIA, JOÃO VIANNEY, ROSÂNGELA RODRIGUES, MARIALICE DE MORAES, FERNANDO SPANHOL and DÊNIA DE BITTENCOURT

Introduction

The Laboratory of Distance Education (LED) of the Graduate Program in Production Engineering (PPGEP) at the Federal University of Santa Catarina (UFSC) is constructing models for graduate programs through distance education. This also includes the development of pedagogic strategies and the use of advanced digital communication resources. They will provide the grounds for innovative inter-university programs and company-university interactions.

It is possible through these programs to increase the numbers of students, institutions and business organizations involved, so as to promote a more democratic access to graduate education. Meanwhile, follow-up and evaluative research on distance education in a digital communication environment is being carried out. These bring together results which further question the whole set of procedures and concepts firmly held and institutionalized by the scientific community and national educational bureaucracy.

Such procedures and concepts were established by discussions in the 1970s to determine quality standards for higher teaching institutions that wished to obtain credentials for research and graduate programs from the Education Ministry. These have remained practically untouched in three decades, despite profound scientific changes, a better understanding of learning theories and research, developments in the notion of the creative individuality of human beings, a technological revolution including IT tools, and resources making possible instantaneous and low-cost communication. In particular, the new technologies have made "face-to-face" communication viable among people all over the world, and have provided local or remote access to databases and information processing units, causing an exponential growth in the production capacity and dissemination of knowledge (Miller, 1992).

Within the same scenario, the international economy is undergoing a process of intensified competitiveness and technological innovation. Renovation of knowledge and products takes place on a scale that makes former models of graduate programs progressively more obsolete, since a large number of workers have to leave their workplace in order to pursue graduate studies. This scenario demands continuous participation of workers/students in new models of continuous education that meet the labor market's demands.

With methodological support and technological management in education carried out by the Distance Education Laboratory, the Graduate Program in Production Engineering of UFSC created the first master's and doctorate courses by the distance education modality, utilizing the instructional design shaped by the new media, already available in Brazil. One doctoral program, eight master's programs and three on the specialization level were offered from September 1996 to June 1998. This was undertaken from the city of Florianópolis and provided access to a total of 240 students and 44 cities from nine Brazilian states. Through university qualification and extension programs at a distance, the LED has structured about 20 courses with a total of 130,000 registrations from June 1995 to July 1998 (Table 11.1).

TABLE 11.1
Graduate Distance Programs Offered from '96–'99 by PPGEP/LED

Course	Co-participating Institutions	Number of Students-Distribution	Media Used
Doctorate in Production Engineering	UNOESC, UNISUL,UNIVALI, FURB, ETEFESC	8 students in the cities of Blumenau , Itajaí, Chapecó, Florianópolis and Tubarão in SC	Videoconference, Internet, Obligatory and supplementary bibliography Meetings with students present
Certificate Management in Technical Education Institutions	SENAI I	50 students, in 31 cities in 6 states	Internet Printed material Meetings with students present
Certificate Management in Technical Education Institutions	SENAI II	60 students, in 29 cities in 8 states	Internet Printed material Meetings with students present
Certificate Management in Technical Education Institutions	SENAI III	60 students, in 27 cities in 9 states	Internet Printed material Meetings with students present

Course	Co-participating Institutions	Number of Students-Distribution	Media Used
Master's Degree in Logistics	PETROBRAS I	21 students in the cities of Salvador (BA), Natal (RN), Belém (PA), Rio de Janeiro e Macaé (RJ)	Videoconference, Internet, Obligatory and supplementary bibliography
Master's Degree in Logistics	PETROBRAS II	27 students in the cities of Natal (RN), Belém (PA), Rio de Janeiro e Campos (RJ), Itajaí (SC), Aracaju (SE), Manaus (AM), Porto Alegre (RS), S. Mateus (ES), Santos e S. José dos Campos (SP).	Videoconference, Internet, Obligatory and supplementary bibliography
Certificate in Ergonomics	SIEMENS	8 students in the city of Curitiba (PR)	Videoconference, Internet Obligatory and supplementary bibliography Meetings with students present
Certificate in Production Engineering	ALUMAR	25 students in the city of São Luiz do Maranhão	Videoconference, Internet, Obligatory and supplementary bibliography
Master's in Production Engineering	SIEMENS	33 students in the city of Curitiba (PR)	Videoconference Internet Obligatory and supplementary bibliography Meetings with students present
Master's in Media and Knowledge-Major in Information Systems Engineering	UNOESCFURBFE-DAVI	16 students in the cities of Blumenau, Chapecó, and Rio do Sul (SC)	Videoconference Internet Obligatory and supplementary bibliography Meetings with students present
Master's in Applied Intelligence	UNOESCFURBFE-BEETEFESC-UNISUL	16 students in the cities of Blumenau, Chapecó, Florianópolis, Tubarão and Brusque (SC)	Videoconference, Internet, Obligatory and supplementary bibliography Meetings with students present
Master's in Technological Evaluation and Innovation	UNOESCUNIVAL-IUNISUL	16 students in the cities of Chapecó, Florianópolis and Tubarão (SC)	Videoconference Internet Obligatory and supplementary bibliography Softwares/Enterprises Games Meetings with students present

Course	Co-participating Institutions	Number of Students-Distribution	Media Used
Master's in Productivity and Quality Management	UNOESCETEFES-CUNISUL	16 students in the cities of Chapecó, Florianópolis and Tubarão (SC)	Videoconference, Internet, Obligatory and supplementary bibliography Meetings with students present
Master's in Environmental Quality Management-Major in Agribusiness	UNOESCETEFESC	16 students in the cities of Chapecó and Tubarão (SC)	Videoconference, Internet, Obligatory and supplementary bibliography Meetings with students present
Master's in Environmental Management	UNOESCETEFES-CUNISUL	16 students in the cities of Chapecó, Florianópolis and Tubarão (SC)	Videoconference, Internet, Obligatory and supplementary bibliography Meetings with students present
Master's in Media and Knowledge	FEPESMIG	27 students in Varginha (MG)	Videoconference, Internet, Obligatory and supplementary bibliography
Master's in Media and Knowledge	CEFET/RN	30 students in Natal (RN)	Videoconference, Internet, Obligatory and supplementary bibliography
Master's in Media and Knowledge	INSTITUTO ISABELA HENDRIX	26 students in Belo Horizonte (MG)	Videoconference, Internet, Obligatory and supplementary bibliography
Master's in Production Engineering	TECPAR I	29 students in the city of Curitiba (PR)	Videoconference Internet Obligatory and supplementary bibliography
Master's in Production Engineering	TECPAR II	31 students in the city of Curitiba (PR)	Videoconference Internet Obligatory and supplementary bibliography
Master's in Production Engineering	TECPAR III	32 students in the city of Curitiba (PR)	Videoconference Internet Obligatory and supplementary bibliography
Master's in Production Engineering	TECPAR IV	29 students in the city of Curitiba (PR)	Videoconference Internet Obligatory and supplementary bibliography
Master's in Production Engineering	TECPAR V	30 students in the city of Curitiba (PR)	Videoconference Internet Obligatory and supplementary bibliography

Course	Co-participating Institutions	Number of Students-Distribution	Media Used
Master's in Media and Knowledge	TECPAR VI/CEFET I	30 students in Curitiba (PR)	Videoconference, Internet, Obligatory and supplementary bibliography
Master's in Media and Knowledge	TECPAR VII/CEFET II	30 students in Curitiba (PR)	Videoconference, Internet, Obligatory and supplementary bibliography
Master's in Informatics	UNEB	27 students in the city of Brasília (DF)	Videoconference, Internet, Obligatory and supplementary bibliography

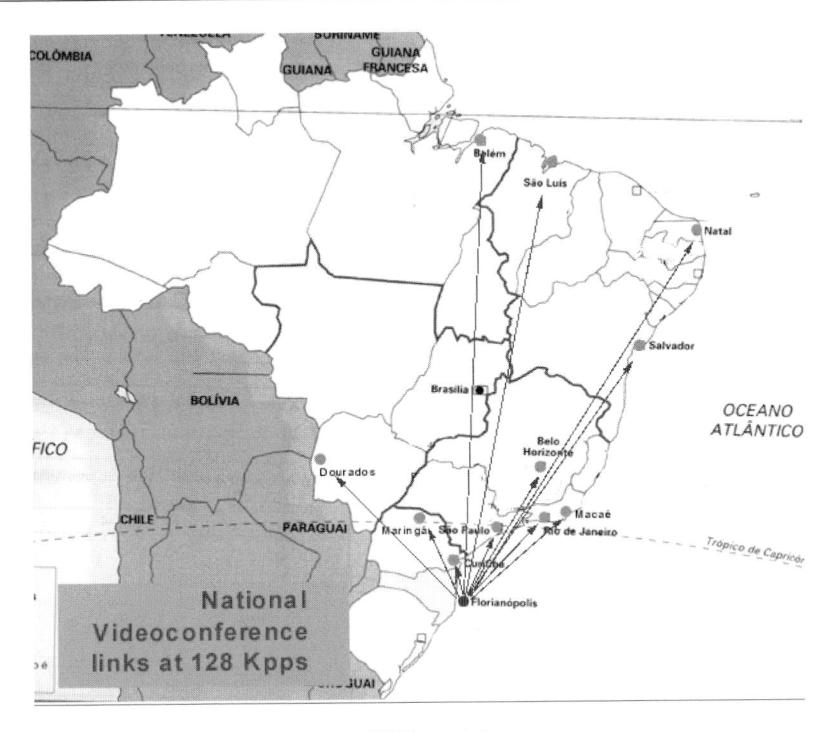

FIGURE 11.1.
USFC Videoconference Links

Creating a Distance Education Culture

Apart from planning and organizing, the development of distance educational systems involves the analysis of client profiles, accessibility of technology, selection of contents according to learning objectives, and the acquisition of knowledge and skills. In addition, academic requirements have to be maintained, and considering the available basic supplementary resources. Beyond this, the main strategic thrust of the Distance Education Laboratory was the designing and implementing of systems that generated a culture of distance education for undergraduate and graduate teaching in Brazil.

Scanning the historical records of Brazilian university experiences with distance education, one discovers considerable variety in the use of DE to reach students who are geographically or economically disadvantaged, considering the demands for education in a continental and developing country (Landim, 1997; Niskier, 1993; Nunes, 1994).

However, these academic experiences were often failures, leaving disappointing imprints in the history of national experiences in distance education. There have been very few attempts to encourage reflection on the causes of the failure in these experiments. Most of them merely contained a record of the administrative discontinuities and the inadequacy of the technology and methodology utilized (Mattelart, 1994).

Memory of DE in Brazil

In 1936, the "Instituto Rádio Técnico Monitor" appeared, with programs on electronics, followed in 1941 by the "Instituto Universal Brasileiro", aimed at professional training on the elementary and secondary levels and utilizing printed material (Alves, 1994).

The Diocese in Natal, Rio Grande do Norte created radio broadcast schools that gave rise to the Basic Education Movement (MEB) in 1959—which Alves (1994), Nunes (1992) and Pimentel (1995) considered to be outstanding educational experiences. Their basic concern was "to achieve literacy and to support the first steps in the education of thousands of young people and adults, mainly in the north and northeast of Brazil. The project was dismantled after the post-1964 government" (Nunes, 1992).

In 1970 the Minerva Project arose, with Junior and Senior High School level courses for adults, produced by the "Fundação Padre Landell de Moura"—FEPLAM—and by the "Fundação Padre Anchieta". According to Alonso (1996) this program was implemented as a "short-term solution to the country's problems of economic, social and political development. It developed in a period of economic growth known as the "Brazilian miracle", when it was assumed that the purpose of education was to prepare a labor force for this development and for international competition" (Mattelart, 1994). This project continued until the early 1980s, in spite of severe criticism of its low rate of completion (77% of those registered were unable to obtain a certificate).

In the beginning of the 1970s, the number of illiterates in Brazil was an obstacle to the country's modernization, mostly in the north and northeast. The government then opted to provide the first experiences of satellite education, based on the Advanced System for Communications and Education for National Development

(ASCEND) as elaborated by Stanford University. The objective was to create an efficient "total system" prototype for audiovisual use, aimed at elementary education.

In 1974 there arose The "Saci" Project, which, in a serial television format, covered the first four years of elementary school. The project was interrupted in 1977–1978, under the "official pretext that it would be too expensive to buy another satellite; showing clearly the contradictions in the different instances of the Brazilian State, in terms of telecommunications, education and scientific policy" (Mattelart, 1994, p.190).

Another unsuccessful initiative was a project developed by the University of Brasília (UnB) (Nunes, 1992) in the mid-1970s, when, influenced by the success of the British Open University, it acquired the rights of translation and publication and then began to produce some of its own courses.

Nunes attributed the failure of this project to inappropriate discourse on the part of the directors, who presented distance education as a substitute for conventional teaching with joint teacher-student presence, diverging policies, and lack of administrative competence. Now the UnB has its own Center of Open, Continued Education at a Distance (CEAD), linking the Administration to University Extension, and it has already produced several different programs in print, video and diskettes.

There were various initiatives that did not achieve success or continuity. Nunes believes that the "most significant problems that held up the progress and mass adoption of the distance education modality were:

- Organization of pilot projects with no suitable preparation for their continuation;
- Lack of evaluation criteria for design programs;
- Non-existence of systematized memory of programs developed and of the evaluations carried out (when these evaluations existed);
- Discontinuation of programs without accounting to society or even to the government or financing organizations;
- Non-existence of institutionalized structures for managing projects seeing that the objectives are carried out;
- Programs only slightly linked to the country's real needs and organized with no specific links to government programs;
- A deeply rooted administrative and political view that took no account of the potential and needs concerning distance education, causing this area to be permanently administered by personnel who lacked the necessary technical and professional qualifications;
- Organization of pilot projects only to test methodologies.

Some projects may be considered exceptions to Nunes' observation on the apparent failure of projects in Distance Teaching. One outstanding example is "Fundação Teleducação de Ceará" (FUNTELC), also known as Educational

Television (TVE) of Ceará, which since 1974 has been developing a regular teaching program for the 5th–8th grades and in 1993 had a total enrollment of 102,170 students in 150 municipalities.

Several different initiatives for open education—professional training, support for traditional teaching, adult education and college preparation courses—promoted by private organizations, such as "Instituto Universal Brasileiro", "Instituto Monitor" and "Instituto Padre Reus", are examples of success. Although they are viewed with some degree of prejudice in academic circles, they meet the needs of the students registered in their various courses.

In 1978, the "Fundação Padre Anchieta" (TV Cultura) and the "Fundação Roberto Marinho" inaugurated the "Telecurso 2º Grau" (High School Telecourse), which is still being broadcast, utilizing TV programs and printed material sold at magazine stands, to prepare the students for the conclusion exam. In 1995 "Telecurso 2000" was inaugurated based on the same model (Preti, 1996).

Considering this description of distance education, it is not at all surprising that the scientific-academic community views distance education initiatives with incredulity and resistance. This is compounded by a lack of serious evaluation and the fact that the involvement of universities and government is subject to political oscillations every time there is a change of administration.

Distance Education in Brazil in the 1990s

Recent examples of formal schooling at a distance which do not exhibit the above problems of discontinuity and inappropriateness can also be found in all the literature. These include the "Project Acesso" on elementary and secondary levels of supplementary adult education, for PETROBRÁS employees, the certificate course in distance education at the University of Brasília, and in the late 1990s, the graduate courses created by the Federal University of Santa Catarina.

This lack of academic structure makes possible the analysis and creation of distance education models to effectively meet needs. At the core of most of the previous approaches, one can find the source of failure not in the DE model itself, but in the conceptual and behavioral processes that create deeply rooted practices in basic Brazilian formal education and Brazilian educational culture. Therefore, the matter goes beyond mere organization and management of DE systems, and reaches the culture of Brazilian education. How do we work with distance education in a country so used to rules that made the educational system like a notary public office; overly concerned by the need for certification; where very few have had the habit of self-education; historically dominated by an educational fundamentals which still bear the mark of its Jesuit heritage? (Gadotti, 1997).

The Brazilian debate on the possibilities of distance education as an alternative for graduate programs, instead of offering lines of research on the limits and applications of pedagogical innovations, is marked by the attacks and pressures of the Brazilian Society for the Progress in Science (SBPC). This organization held studies and discussions for the elaboration of Decree 2494 on February 11th, 1998, which established the rules for distance education, together with restrictions for graduate programs. Via the bulletins of this institution, the entire Brazilian academic community followed this discussion from October 1997 to April 1998. The preliminary version of that decree provided for the possibility of offering master's and doctorate programs.

Because of the Education Ministry's restrictive position, the SBPC has further managed to prevent the development of competence in universities by offering and managing distance graduate programs on their own. This has turned the Brazilian university market inside-out, resulting in the promotion of international traditional DE Universities to Brazilian students while at the same time resulting in rigid rules for the national ones. More about this debate may be found in the *Jornal da Ciência* (SBPC, 1998).

Even though international examples show successful experiences in dealing with thousands of students (see Table 11.2), the value of DE is still doubted, as an educational mode able to meet a clearly identified demand on the most diverse levels (e.g. formal and continued education, qualifying for a profession, etc.). (Rodrigues, 1998)

TABLE 11.2
Universities at Distance

University	Country	Students per year	Began in	Courses	Media
Athabasca	CA	12,500	1985	41	Printed material, teleconferences, www, audio, video, tutoring
Wisconsin-Extension	USA	12,000	1958	350	Printed material, radio and TV programs, kits, video & audio conference, www, tutoring
Penn State	USA	20,000	1892	300	Printed material, video & audio tapes, teleconference, www.
FernUniversität	GE	55,000	1974	7*	Printed material, audio & video tapes, CBT (computer-based training), www, tutoring
UK Open University	UK	150,000	1971	116*	Printed material, kits, www, workshops
Netherlands Open University	NL	22,700	1984	300	Printed material, audio & video tapes, CAI, tutoring

University	Country	Students per year	Began in	Courses	Media
Indira Gandhi OU	IN	95,000	1987	487	Printed material, audio & video tapes, tutoring
Radio & TV Universities	CHINA CN	530,000	1979	350	Printed material, Radio & TV programs, tutoring

(*) Considering only undergraduate and graduate courses.

With the growing demand for part-time advanced graduate programs, Brazilian students are looking abroad for institutions which might offer such features. Thus, British, Canadian, Spanish and North American universities are registering Brazilian students in graduate courses through distance education.

Furthermore, universities largely dedicated to teaching students in the on-campus tradition, such as Stanford, Johns Hopkins, Caltech, Michigan, Cambridge, Harriot-Watt and Oxford, have also incorporated distance education into their activities in the 1980s and 1990s, and begun competing in a market which had previously belonged to the universities created exclusively for distance education, such as the Open University in England, and the "Universidad Nacional de Educación a Distância (UNED)" in Spain, and the open universities of Israel and Portugal (Moore and Kearsley, 1996). For instance, in the first trimester of 1998, the University of Michigan set up a videoconference room in the city of São Paulo to reach students at distance, helping create a model for a transnational university.

Thus, the determination of the SBPC to restrict and rule postgraduate distance programs in Brazil is aiding the opening of the Brazilian market to internationalized universities from overseas. At the same time, it can be argued that the Education Ministry does not recognize degrees obtained from distance education offered by foreign institutions. In actual practice, it is only the market for hiring teachers for graduate courses that is subordinated to the Education Ministry's consent; namely, the restriction for hiring postgraduates who have held academic positions.

On the other hand, for enterprises which choose to be competitive and wish to survive in the market, their main concern for hiring professionals will always be competence, not titles. It is obvious that the present-day context points to the need for increasingly more qualified human resources, prepared to face a growing sophisticated labor market.

According to Alvin Toffler (1990), an outstanding scholar of social transformation through the conquest and implementation of technologies, education has become a constant concern to advanced sectors of the business world, and the latter's leaders increasingly recognize the interrelation between education and competence.

However, according to what has already been observed in Brazil, what remains is a formal teaching system with little efficiency for qualifying its human resources to meet the emerging needs of the market. For this reason, the transition to more sophisticated technologies, one of the requisites for facing competition in a globalized world, becomes difficult and traumatic. Experience shows that the whole process of technological change involves the formation and continual retraining of large contingents of the labor force, as it has been in Europe, United States and Japan (FGV, 1993).

At the moment, there is a great expectation in Brazil regarding new human resource policies, which endeavor to train and develop large contingents of the labor force both easier and faster. Distance education appears to be the road to follow, as it has the potential for serving a large part of the population.

Thus, distance education becomes a necessary tool for this new scenario, involving not only matters of creating a technological culture, but also that of the growing need to prepare individuals for continual lifetime learning. Intellectual capacity has become the main investment and the main product of the new economy based on knowledge (Miller, 1992; Bates, 1995; and Visser 1997), and this intellectual capacity should follow what Levy (1993) calls the "*Informatic-Mediatic Pole*", where speed is the determining factor and the returns are immediate. It is supposed that the faster one learns, the greater are one's chances in a competitive world (FGV, 1993).

Despite the need for education in Brazil, the conservative positions assumed by SBPC and academic leaders may not be easily reversed. Within SBPC itself, six months after criticism and pressure against its ruling about graduate studies by DE, the Cyberspace Research group again proposed the creation of a national network of communication to coordinate graduate distance programs, expressing one condition: that the programs must be managed by SBPC (SBPC Report, 1998).

The Brazilian education scenario, even considering the currents for a liberating educational practice, is still dominated by students subordinated to authoritarian teachers who consider themselves as holders of all knowledge, (re)producing dependent students.

The tutelage of such teachers will perpetuate practices like pass or fail grades, and the testing of students' memory, their abilities to do predefined calculations, and abilities to pass university entrance examinations as the main goal for education; instead of knowledge construction, and creative and critical thinking.

The UFSC Distance Education Laboratory (LED)

To institutions seeking to create reform, such as the LED, distance education means breaking with the atavistic educational scenario, and establishing instructional modeling structures based on cooperative and shared teaching–learning environments. Above all, it means breaking with the Brazilian tradition of distance education, organized from traditional education models. These models often rely on simple instructional materials printed and formatted for self-correction with promises of easy solutions to complex problems, and are often poorly committed to the students' learning and the impact of its performance on society.

Although it is clear that there is a need to break with the basic premises concerning distance education in Brazil, it is necessary to meet the requirements imposed by the on-campus Graduate Program in Production Engineering at the UFSC, which was the first UFSC graduate program to be delivered at a distance. The requirements stipulate that distance courses must maintain the program's standards, and that the quality of the diplomas granted should be unquestionable, reversing the present image of discredit, discontinuity and superficiality.

In LED we have incorporated, parallel to the training process, pedagogical innovations that seek to overcome the difficulties of previous DE experiences in Brazil. This has resulted in a distance mediation process based on off-line and on-line monitoring, with tutoring supported by systems that diagnose and provide for feedback on students' learning difficulties, including questions about the inappropriateness of course contents, instructional materials and methodologies.

The customized approach to the contents taught in each course, and an instructional design suitable to the students' profile and learning environment offered to students by institutions to UFSC, allows the students to work individually, with a firm commitment and realistic objectives, breaking with the history of failures and goals unsuited to students' needs.

The term of the present general coordinator of the distance education program, Ricardo Barcia, began during the third generation of distance education, with the intense use of communication networks in education. During this phase, the experiences of the open universities such as the successful British Open University, starting in the 1970s, provides a background for the work of the Virtual Universities emerging in the 1990s, including Simon Fraser, Canada, New Brunswick, Canada, Open University of Catalunia, Spain, Polytechnic University of Wellington, New Zealand. These institutions are examples of a non-institution-specific model of a university in a globalized and integrated world. Observe in the following table the DE generations, identified by Moore and Kearsley (1996).

TABLE 11.3
The Distance Education Generations

Generation	Beginning	Characteristics
1st	Until 1970	Studies by correspondence, in which the principal media of communication are printed materials, generally a study guide, with written essays or other assignments sent by mail.
2nd	1970	The first open universities emerge, with systematized design and implementation of distance courses, using, in addition to printed materials, transmission through open television, radio and audio/ videotapes.
3rd	1990	This generation is based on conference networks by computer and multimedia workstations.

In the LED, to orient the work groups, we have developed the concept *of Integrated Media for DE,* using new technologies in consonance with learning objectives. To adapt to the characteristics of Brazilian society, the Integrated Media concept incorporates, rather than excludes, the use of first and second generation DE media.

Based on communication theories, LED has developed research on the educational use of the Internet, and on the use of such tools as videoconference, teleconference, video-classes, computer-based training, printed materials, fax and telephone media. These stimulate in students the responsibility for learning, be it through interaction of student–media or student–content, student–teaching institution, student–teacher–tutor, or student–student, so as to establish a practice and culture of cooperative learning.

The idea was that "besides generating products of communication for use in education, the Distance Education Laboratory is also a center for producing knowledge, research and academic reports on the theme, involving the generation of new opportunities for action and continuous improvement, of the aesthetic and pedagogical qualities of the products that both generate and apply" (Barcia, 1996). The search for reference material to form the group has represented a challenge, since scientific research in Brazil on the area is scarce and what has existed does not focus on integrated media concepts, especially those of the 3rd generation.

As an alternative, we have concentrated our research on international practices. Knowledge was acquired in various ways:

- Intense research over the Internet looking for institutions that have worked with DE and its "modus operandi" and the articles placed on the World Wide Web.
- The acquisition of books, journals and proceedings of national and international congresses, seeking to start a library on this theme.
- Invitations to leaders like Prof. Michael Moore (Penn State University), Wolfram Laaser (FernUniversitat-Hagen) and Luiz Lobo (Iowa Public Television) in this area to administer workshops and teach classes in Florianópolis.
- The sending of students linked to the program to study in leading institutions in their area, who keep in permanent contact with Brazil, bringing further sources and references.
- The participation of students linked to LED in national and international congresses.

Student attitude

For the Distance Education Laboratory, both the traditional on-campus and the distance student are understood to be agents in building his/her own knowledge. Namely, not merely one who is self-educated and who seeks his/her own way, but as a participant, aware of an organized and systemic process, in which his/her teaching institution offers him/her the tools of mediation, tutorial resources, access to contents, and to manage the route to arrive at that required knowledge.

LED understands the distant student as a "Giddenian being" (Giddens, 1989), an agent capable of reading the scenarios in which s/he is immersed, of setting goals, of designing strategies, and developing procedures to reach the desired results. This agent cannot be understood as a subject determined by the environment, but as an individual who can overcome environmental limitations and conditions to "undetermine" any process. This meaning of a student as an agent, an empowered student, surely is emerging—if not incipient—in the traditional education scenario in Brazil, whether amongst traditional on-campus or distance learning students.

In the process of doing this, the LED has had to break with canons established over centuries of reproduction in the Brazilian schools, to form a new culture of distance education.

Empowered students and autonomous students

Paralleling the creation of LED, the PPGEP developed the teaching and research area of Media and Knowledge, in which master's and doctorate students in the lab improved the DE programs by their research. This included theses, dissertations and articles. The development of professionals linked to LED breaks with past practices by creating a new research practice, where courses and products are developed and accompanied during their implantation and implementation time, serving as case studies in academic research. All this led to shorter research–application–evaluation cycles.

TABLE 11.4
Masters Dissertations Using LED's Courses as a Study Case

Author	Title	Date	Description
Bolzan, Regina		1998	
Rodrigues, Rosângela	Evaluation Model for Distance Education	1998	Construction of a evaluation model, defining the criteria and tested in a real course
Paas, Leslie			
Bittencourt, Dênia			
Spanhol, Fernando	Videoconference in Distance Education	1999	Technological and environmental structure for videoconference systems in distance education. Study of LED/UFSC case.

The LED Model

The LED model is structured as a generative process, that is, no other model has been adopted as a reference. Despite an international benchmarking operation to identify the causes of success or failure in diverse experiences, the strategy adopted was to create our own models from the Laboratory's competencies and adjusted to the clientele's characteristics.

From this situation, the following stages were planned and implemented: Technological Acquisition, Pedagogical Innovation, Content Modeling and Distance Mediation[1]. In all the stages implemented between June 1995 and July 1998, the greatest changes in academic culture involved:

- pedagogic preparation of teachers and of scripts for performing and interacting on offline and on-line media;
- preparation of content writers, instructional designers and script writers for modeling contents and activities;
- the dissemination of the concept of the distance student as an agent capable of overcoming the *habitus* (Bourdieu, 1989) of the Brazilian educational system, centered around the need for teacher–student presence.

The change of paradigms for these three actors: teachers, technicians and students resulted from the nature of the Distance Education Laboratory. The LED started from the Production Engineering Graduate Program (PPGEP) of the Federal University of Santa Catarina. The latter's Production Engineering masters and doctorate programs alone account for 47% of the national scientific production in this area, when compared to the other nine Production Engineering graduate programs in Brazil (according to data from CAPES, 1996). Even though UFSC and PPGEP set distance education as a new frontier of knowledge and teaching service capacity, a new vision was necessary to transform teachers and technicians—who had the habit of personal teaching—into content suppliers, pedagogical strategy designers and technology managers for distance education.

A short report on the activities performed between 1995 and 1998 describes the Technological Acquisition stage—the first stage that deals with installation and trial operation of the media used. At this stage, the appropriate communication networks via videoconference, teleconference and Internet were developed; and competency was developed for producing storage and retrieval information systems in CD-ROM, video, audio and printed material directed to distance teaching.

LED technological infrastructure

The main technology used at the LED for distance graduate teaching is videoconferencing. This technology was defined by Cruz (1996) as one that transports digitized video and audio signals, properly compressed by software and multiplexed into a single signal. A virtual meeting can take place between two sites (i.e.

point to point) or across several sites (i.e. multi-point), depending on the capacity of the multi-point managing equipment.

The interactive system was considered to be the most successful in integrating the traditional classroom with multimedia resources, according to Novaes (1994). Videoconferencing was the medium chosen as the basic technology for the PPGEP/UFSC Virtual University, because it allows a gradual change from the present classroom to distance teaching.

An information channel of two megabits speed has been established to serve as the Internet provider for all Higher Education Institutions in the State of Santa Catarina. Also, a multi-point videoconference system has been set up, with virtual classrooms installed in the seven main universities from Santa Catarina, with the headquarters at the Federal University of Santa Catarina (UFSC).

The implementation and operation of the distance education network required the hiring and specialized training of telecommunication technicians, for the LED as well as the state telecommunications operators, in order to deal with videoconference signal traffic (in the 384 Kbps fixed range), and to manage the multi-point network.

Videoconference telecommunication links with speeds varying between 128 Kbps and 3840 Kbps where installed. These were dedicated or dialed between UFSC and: the University of Port (Portugal), Fernüniversität (Germany), University of Wisconsin and University of South Florida (USA), United Nations University/ Institute of Advanced Studies (Japan), Ministry of Education (Italy), European Common Market Committee (Brussels) and, in Brazil, between UFSC and the enterprises/institutions EQUITEL (Curitiba), TELECEARA (Ceará), PETROBRAS (Rio de Janeiro), ALUMAR (Maranhão), TECPAR (Paraná), ELETROBRÁS (Rio de Janeiro), TELEMIG, Fepesmig, and Izabela Hendrix Methodist Foundation (Minas Gerais), Federal Center for Technological Education (CEFET—Rio Grande do Norte), UNEB (Brasília), and Getúlio Vargas Foundation (São Paulo).

The accumulated know-how on videoconference operation has allowed the LED to become a partner of institutions that want to use technology with a minimum expenditure of time and money. The FINEP/RECOPE project setup is an example of this. It foresees the acquisition and assembling of video post-production equipment, a Prism multi-point and four videoconference systems, to develop a network integrating USP (Carlos Vanzolini Foundation) and FINEP in São Paulo, UFRJ (COPPE) in Rio de Janeiro with UFSC.

The goals of this project are to transmit the experiences and methodologies developed at the Distance Education Laboratory, to develop high-performance multimedia products for teacher competency development, and to strengthen university–private sector connections through university activities. It uses new communication technologies for technological education, meeting the objectives of the Program for the Development of Engineering Courses (PRODENGE). From the media used to generate products for distance education, the LED started the production of video-classes.[2]

In following the new trends and integrating the Fourth generation of DE, the LED developed an integrated model, based on the acquisition of equipment, site editing and management, and www servers. The LED has also formed a research and technical team to evaluate and develop tools for Internet use in distance education.

Pedagogical innovation and content modeling

The stages of Pedagogical Innovation and Content Modeling occurred simultaneously. The former involves the development of support teams for the planning and executing of teaching materials. This was done for various media, including CD-ROMs, video-classes, printed material and Web pages. Classes were prepared and transmitted by electronic media such as teleconference, videoconference and Internet.

With the help of communication grids and electronic media, a nucleus for the generation and research of knowledge on distance education has been created in the laboratory. This group involves professionals from the areas of psychology, education, communications and engineering. The group was qualified through development seminars on distance education from Penn State University and from Fernüniversität, as well as by the worldwide benchmarking of distance education. The laboratory nucleus has sought to develop strategies and products for customized distance education. In other words, the courses offered should be both pedagogically and technologically developed, according to users' profiles, with special consideration towards:

- The identification of learning needs
- The identification of the cultural profile of students and of metaphors and language
- The verification of conditions of technological access
- The assessment of the need to implement or expand the communication and information systems to students, as a requisite to start or continue the courses offered
- Users' needs to acquire skills for using the media and teaching materials for educational purposes in each course
- Course planning, taking into account the goals of knowledge acquisition and the improvement of individual and corporate performance (as when assisting in the process of qualifying and training programs), or improving skilled work for enterprises and institutions
- The need to customize the pedagogical approach, instructional planning, materials projects and the use of interactive media

The initial nucleus of the Distance Education Laboratory also fulfilled two roles: *tutoring* in teacher–training programs on the use of multimedia in distance interactive teaching-learning systems; and *advising* on content created by the writer teams, TV script writers, Web designers, programmers, text and image editors, illustrators and other professionals in the production of electronic and printed media for distance education.

This stage has been identified as the Content Modeling stage, comprising both the formatting of teaching material and the preparation of scripts and techniques for live activities by teleconferencing, videoconferencing, or Internet systems.

It is necessary to point out that the LED always works as a support to teachers, and does not control the teachers' content. The team's role is to find appropriate means for communication (whenever the teacher does not know the target audience), as well as to format classes and teaching materials. The only instance where the LED develops content is the subject of distance education.

At this stage, the models for distance education relied on prior references. This was achieved through systems which combined the use of printed resources, video, computer–based systems, corporate nets, with teleconference and videoconference systems, and the Internet.

The Distance Education Internet Laboratory

The Distance Education Internet Laboratory evolved from the need to formulate a strategy for the educational use of Internet which would overcome the limitations of e-mail, remote file access, discussion lists (e.g. forums or chat lists), or demonstration Web pages. This viewed the development of an integrated environment devoted to teaching learning as a necessary advancement.

The Distance Education Internet Laboratory *site* was modeled after the actual functions and cognitive processes seen in the daily life of an actual university campus.

In the course of developing virtual learning environments, teachers ought to have adequate tools to select, indicate or even present the contents the students are to work with. The students, as a consequence, ought to have objective means of accessing the needed contents, as well as knowledge of the criteria used by teachers. This should also involve exercises and activities, including the use of simulation, experimentation and exploration by students.

Learning and evaluation activities carried out by the students, such as tests, workshops, reports, reviews, etc. should be automatically handed in to the teacher, who should maintain control over the student's work. Students and teachers should have time scheduled for meetings, including formal ones such as study groups, and informal ones such as unsupervised communications. Moreover, teachers and students should be able to access each other without trouble. To achieve these goals, tools and learning systems were developed, such as the Virtual Library, Paper Delivery (production room), Study Groups, Advising and Tutorial, Discussion Room, Meeting Room, News and Mailbox, all of them integrated together into one Internet learning environment.

Distance mediation

The application to custom-built courses, with contents and pedagogical planning adapted to specific kinds of students, has created opportunities for learning in each project.

The opportunity for continuous interaction with students allows advance diagnosis so as to identify areas of content which cause the greatest learning difficulty amongst students. This enables the teaching and technical team to offer the students supplementary assistance in approaching the content, such as metaphors to reinforce understanding, as well as permitting the team to provide supportive exercises or learning activities for each stage of the intended learning. We call this strategy Mediation at a Distance, a flexible and dynamic tutoring mode for the planning and execution of new teaching materials or of new approaches to live transmissions.

With the concept of media integrated with distance education, the activities of tutoring and research orientation take place on multipoint videoconference networks and Internet-based virtual learning environments. These are supported and complemented by printed mediating instruments and periodic evaluations with the student, as well as by the use of first and second generation DE systems, reinforced by telephone contacts, printed material, etc.

LED's products also involve student interaction or feedback, such as instruments to "remove doubts". These are made part of the pre-produced instructional material, and lead to the means for instantaneous interaction with the electronic media. For example, in a course on Company Accounting with 2,800 students, distributed throughout the whole country, the textbook delivered to the students during the first week of July 1997, contained an additional interactive tool called "Remove Doubts". On this form, the students were to point out, as early as July, the chief difficulties they encountered in the understanding of the content and the procedures indicated. Once the course coordinators received the "Remove Doubts" forms, they planned a synchronized tele-education stage, held in August 1997 through the teleconference system, with a reception organized for students in 527 cities across Brazil. This consisted of 12 class hours via satellite, in six sessions (two a week), structured according to the principal learning needs revealed by the students in "Remove Doubts".

During the teleconference the students had further opportunities to remove their doubts or they could simply posit their inquiries through a system of free direct dialing into the television broadcasting studio or to the team for pedagogical support offering on-line tutoring. In the concluding teleconference stage, all the "Remove Doubts" and questions received by DDG were answered individually. In this way, the systematization of the most basic needs provided an opportunity for the mediation and elaboration of a whole contingent of students, and the students" needs for having their questions cleared up were met promptly. Only after those steps were concluded did conclusive learning evaluation instruments get applied.

On the other hand, in another course offered, that of the Master's in Logistics, the Distance Mediation strategies utilized were based on the Internet and videoconferencing. These were customized for a selected group of engineers and administrators at Petrobrás, with live classes conducted via network videoconference from Florianópolis to the cities of Natal (RN), Macaé (RJ), Salvador (BA), Belém (PA), and Rio de Janeiro (RJ).

The videoconference meets the need for presence, where the students see and hear one another in real time. While relations are established in only two dimensions (by means of the screen), the instantaneous contacts and the possibility of interaction overcome the "filtering" effect of the media.

Through the Internet Distance Education Laboratory software, students have access to the contents of the Virtual Library, do exercises in the Case Bank, turn in their assignments to the teacher in the Production Room, form study groups in the Discussion Room, and promptly send questions off-line through the Mailbox. Live questions can be asked through videoconference during class time. Thus, the modeling of the contents of the next units, and even of the units being offered, can be altered in a dynamic way so as to achieve the established goals.

The group's usage of videoconference and Internet were originally stipulated by legislation concerning the control of the learning environment, and that of creating legitimate master's programs in Brazil by means of DE. This was carried out as an extension of the UFSC campus to companies and institutions that are parties to an agreement with UFSC. In practice the videoconference takes the teacher to the students' classroom, where they can become fully integrated, and the Internet brings a wide range of existent services to the university campus.

Even while complying with legislation, this model has brought to the student a closer level of interaction, which has proved to be very rewarding. During the classes via videoconference, there is a synchronized contact with the site, and the students have flexibility of access literally 24 hours a day, seven days a week.

To meet client needs, in 1998 the Distance Education Laboratory at UFSC developed a Specialization Course in Management of Technical Teaching Institutions for the directors and technicians at SENAI (National Apprenticeship Service). It is geographically distributed across seven states and 31 cities in Brazil, and utilizes distance mediation in an Internet environment for learning activities. Apart from following the integrated media strategy, it also supplies students with printed material and a study guide as a complementary support.

This is a pioneering project in Brazil, being the first university to offer a specialization course based on the WWW interface to students and service organizations. Along with opportunities for interactive studies with the university, this course also uses the generative model directed towards client needs. The modeled design aims to meet the needs of both the students and the institution: bearing in mind the student profile; aptitudes and abilities for using new tools; the cooperative learning practice; the use of networks and virtual environments; the conditions of

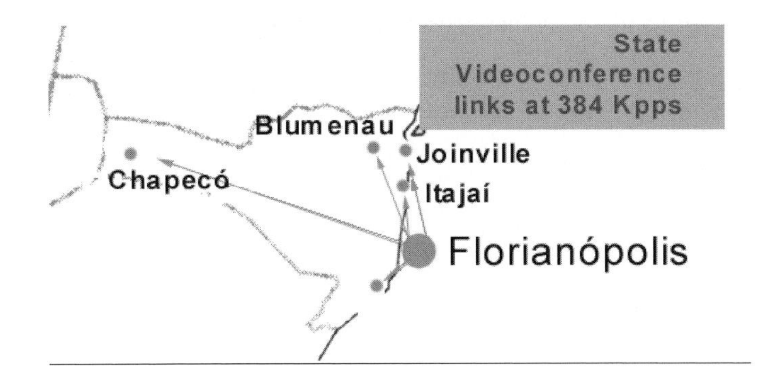

FIGURE 11.2

technological access available in work stations and residences; and the company's strategic goals.

To meet all these needs, the Distance Education Laboratory has had to invest in research and development of pedagogies and technologies in order to establish within the WWW environment a motivation for learning and assimilation of expected contents.

Researchers involved in the project are constantly evaluating student satisfaction (e.g. in percentage terms) for each of the components of the course. The process of adapting learning theories, methods, and strategies to online teaching requires increasingly greater flexibility, in order for the product to meet client satisfaction. On a scale of one to five (five being the best), the majority of students rated the general aspects of the program as either a "four" or "five". The "general aspects" include: achievement of the goals proposed; adaptation of the content to the aims; and applicability of the content, among other evaluation items.

The students' satisfaction at being aided continue to be evaluated with respect to material access (study guide, chat, course site), and with regards to the ease of communication with the team—from webmaster, monitoring, teachers to UFSC/SENAI coordinating staff. The teaching materials and tools developed are also assessed both for their functionality, applicability and visual graphic appearance.

After concluding the first module of subjects, the results of the course evaluation have proven DE research in WWW environments to be highly satisfactory, as illustrated by student statements such as the following:

> "The strongest point in the course is the innovation it shows. We are navigating in the right direction. Let's continue straight ahead!" "This kind of study is a pleasant experience. It is worth the sacrifice, for I am learning and, whenever possible, applying what I learn in my daily life—both personal and professional".

For the Distance Education Laboratory, statements of this kind are indications that we are on the right track and offering a stimulus to research, thus leading to a constant improvement of the pedagogical tools offered by its programs.

A New Teaching Institution

The posture of a teaching institution is fundamental for changing from a traditional educational process to a modernized process. The former is characterized by a teaching process that places the teacher at the center of the stage, whereby the teacher is the master and "lord" of the arts. Now the student takes his/her place in the center of the action, as an agent of a process. In this, teachers still have an important part to play, but they are now oriented to supporting student needs, and have to respect the learning strategies and resources the student is developing. In this, a new culture is slowly establishing itself.

As a believer in the multiple forms of competence and intelligence in individuals, Howard Gardner (1994) maintains that, in a process of cultural formation,

> "If a particular kind of behavior is considered important by a culture, if considerable resources are dedicated to it, if the individual him/herself is motivated to operate in this area and appropriate means for crystallization of learning are made available, almost every normal individual can achieve an impressive degree of competence in a symbolic intellectual domain. On the contrary, and perhaps more obviously, even the most innately intelligent individual will come to nothing without a positive supporting environment."

In Gardner's quotation, we can see the importance of a distance education process that considers the culture of the student as a determining variable for the learning environment. It even considers the need to establish a culture for the process. For instance, in the specialization program for technical school managers developed using the distance education modality and based on the WWW environment, the evaluation of results reveals the students" discovery of a work culture in a network—a sub-product of the course. Scattered over 31 cities in the southern, southeastern and northeastern regions, students could only develop the activities requested by the course programming through the intensive use of the Internet. An example of this social transformation can be found in the statement by Fernando Jefferson (1995) quoted below:

> "What matters most is not technology in itself, but its impact on people and organizations. The world is changing very fast. Technology helps to speed up the necessary and deep changes in order to enable companies to survive in an increasingly competitive and aggressive market."

Companies undergo changes, not only in their structures, but also mainly in their way of working and in their relationship to the market. Working together in groups (workgroups) enables the employees to work as a team, inasmuch as each one collaborates in carrying out the tasks as well as possible."

Conclusions

The strategy we have developed in the Distance Education Laboratory at UFSC sets out to assimilate communication and information technologies and develop applications for their use in distance education, specifically for the Brazilian context. The results have produced in Brazil a new scenario with characteristics of a process, with educational consequences that breaks through the vicious circle of theorizing and procedures. This overcomes the institutional and cultural barriers that subordinated graduate education to educational planning, thus limiting universities' capacity to offer graduate courses, and preventing them from developing competence and capacity for international competition in the provision of distance type graduate courses. Assuming that there are no boundaries to knowledge, the possibility of improved quality comes from increasing competition in the world market. This is augmented and made feasible by the multiplication and international distribution of telecommunication equipment, lower telecommunication rates, and the expansion of the Internet.

Through the various programs that assist students, located in practically all Brazilian states, LED has stimulated the formation of inter-state groups. This has created the possibility of integrating all the different regions in Brazil, transcending geographic and political barriers, and bringing equality of access to knowledge, plus a first class technological education. For instance, the LED's programs have enabled students far from major centers of intellectual and academic production (located largely in the metropolitan areas of the south and southeast) to continue their studies while remaining in the labor market, with the almost immediate applicability of their new knowledge to their jobs.

Notes

1 This model for implementing a distance education system has been developed by João Vianney as his Production Engineering doctorate thesis research subject—PPGEP/UFSC—under way.

2 To this end, LED acquired a SONY video–recording and editing system, BETACAM SP format, involving external recording units, studio units, lighting equipment linear and non-linear post-production facilities, and graphic effect producer stations. The generation of class transmission via satellite (teleconferences) was built up in another DE modality, with video-programming generation in standard studio broadcast with an EMBRATEL microwave link in Florianópolis and Brasilsat B2 up-link, with B and "C" for open reception throughout the whole Brazilian territory.

References

Alonso K. M. (1996) A Educação a Distância no Brazil: a busca de identidade. In: Preti, O. (ed.) (1996) *Educação a distância: inícios e indícios de um percurso. (Distance education: beginnings and indications)* Cuiabá: NEAD/IE–UFMT, pp. 127–154.

Alves, J. R. M. (1944) *A Educação a distância no Brasil: síntese histórica e perspectivas.(Distance Education in Brazil: historic synthesis and perspectives)* Rio de Janeiro: Instituto de Pesquisas Avançadas em Educação.

Barcia, R. & Cruz, D. & Vianney, J. Bolzan, R. & Rodrigues, R. S. (1996) *A experiência da UFSC em programas de requalificação, capacitação, treinamento e formação a distância de mão de obra no cenário da economia globalizada.* In: International Symposium On Continuing Engineering Education For Technology Development, Rio De Janeiro.

Bates, T. (1995) *The future of learning.* URL: http://www.bates.estudies.ubc.ca.Bourdieu, P. (1989) *O Poder Simbólico.* Rio de Janeiro: Difel.

Cruz, D. & Barcia, R. M.. (1996) *A Videoconferência na Educação Continuada em Engenharia: A Experiência de Santa Catarina. (The Videoconference in Extended Education in Engineering: the*

Experience of Santa Catarina). International Symposium on Extended Education in Engineering for the Development of Technology, Rio de Janeiro.

Fundação Getúlio Vargas (1993) *Os recursos humanos para a ciência e tecnologia » (Human resources for science & technology)*—a paper prepared for the project « O Estado Atual e o papel Futuro da Ciência e Tecnologia no Brasil » (The State at Present and the Future Role of Science and Technology in Brazil.

Gardner, H. (1994) *Estruturas da Mente: A Teoria das Inteligências Múltiplas*. Porto Alegre: Artes Médias, p. 243.

Giddens, (1989) *A Constituição da Sociedade (The Constitution of Society*) Martins Fontes, São Paulo.

Jefferson, F. (1995) *Entrevista* Revista Nacional de Telecomunicações – RNT p. 35.

Landim, C. (1997) Educação a Distância: algumas considerações. (Distance Education: some considerations) Rio de Janeiro.

Lévy, P. (1993) Tecnologias da inteligência: o futuro do pensamento na era da informática Trad. Carlos Irineu da Costa. Rio de Janeiro: Ed. 34.

Mattelart, A. (1994) *Comunicação Mundo: história das idéias e das estratégias. (Communication in the World: history of ideas and strategies*) Petropolis: Vozes.

Miller, G. (1992) *Long-Term Trends in Distance Education* DEOSNEWS. V.**2** No.23.

Moore, M. & Kearsley, G. (1996) *Distance Education: a systems view* U.S.A.: Wadsworth,

Niskier, A. (1993) *Tecnologia Educacional: uma visão política. (Educational Technology: a political view)* Petrópolis: Vozes.

Novaes, A. G. (1994) *Ensino a Distância na Engenharia: contornos e perspectivas. (Distance Education in Engineering: contours and perspectives*). In Gestão e Produção, v.**1**, n.3, pp.250–271.

Nunes, I. B. (1994) *Noções de Educação a Distância. (Fundamentals of Distance Education)* Revista Educação a Distância. Vols. **3**, **4** & **5**. Brasília: INED.

Pimentel, N. (1995) *O Ensino a distância na formação de professores*. Revista Perspectiva, Florianópolis, n.**24** pp. 96–128.

Preti, O. (1996) Educação a Distância: uma prática educativa mediadora e mediatizada (Distance Education: a mediating and mediated educational practice). In: Preti, O. (ed.) (1996) *Educação a distância: inícios e indícios de um percurso. (Distance education: beginnings and indications)* Cuiabá: NEAD/IE – UFMT, pp. 15–56.

"Pós-graduação à distância divide opiniões da comunidade", *Jornal da Ciência (SBPC.net)*, número 387, de 17 de abril de 1998. URL: http://www.sbpcnet.org.br/jc/jc387.htm#7.

Rodrigues, R. (1998) *Modelo de Avaliação para Cursos no Ensino a Distância: estrutura, aplicação e avaliação. (Evaluation Model for Distance Education Courses: structure, application and assessment)* Master´s Dissertation, PPGEP/UFSC.

Toffler, A. (1990) *El cambio del poder. (The Road to Power)* Barcelona. Ed. Plaza & Janes.

Visser, J. (1997) Learning without frontiers: beyonds open and distance learning. In: *1997 World ICDE Conference, 18 Th. Proceedings*. Pennsylvania: Pennsylvania State University.

12

The ITESM Virtual University: Towards a Transformation of Higher Education

MARISA MARTÍN PÉREZ

The Monterrey Institute of Technology and Higher Education (ITESM)

The ITESM is a national education system with an international scope that opened 55 years ago, and today has 27 campuses in 26 Mexican cities, covering a large part of the Mexican territory, and extending its services into other Latin-American countries. All 27 ITESM campuses have been accredited by the University Commission of the Southern Association of Colleges and Schools (SACS) in the United States to award bachelor's, master's and doctoral degrees.

The student body of 75,000, taught by 5,800 lecturers, is concentrated in the fields of engineering, computer sciences and administration. The system currently offers 31 undergraduate, 37 master's and nine doctoral degree programs. The entire system is interconnected via computer networks, and the use of telecommunications for academic activities is strongly encouraged for both lecturers and students. Each campus has a computer center which provides one computer per eight students. Students and lecturers are being encouraged to obtain their own computer, since an increasing number of courses each semester use computer technology in the learning process.

In 1998, 14,000 laptop computers were distributed to students, and this same rate of distribution is expected to continue over the following years. Beginning the next semester all new students will be required to have their own laptop. Since 1989 new technologies have been explored, giving rise to the Interactive Educational System by Satellite (SEIS). In 1996 the Virtual University collected and organized the experience generated on distance education and the use of technologies for interactive learning.

Introduction

Not long ago, as we lecturers began using new communication and information technologies as teaching supports, we asked ourselves some very challenging questions: Will our students in Mexico soon have "classmates" from universities on other continents? Will young Mexican students be able to share a learning space with young Japanese, Muscovite, American, Canadian or Colombian students? Will our students have access to experts willing to share their experience with us? Can we achieve and surpass Simón Bolívar's dream of a united America or, to be more global, a culturally united world?

In our Institute we already have experiences in networked courses where this dream has been realized. Students and lecturers from various institutions, including the University of Paris, Pontificial Catholic University of Chile, Carnegie-Mellon University, Thunderbird (American Graduate School of International Management), University of British Columbia, and others, now exchange ideas in virtual forums with ITESM students. Each week we receive an average of three video-link conferences from lecturers and experts from any geographical point of the globe, who share their experiences and knowledge, opening our students' minds to a global reality. Thanks to the communication and information advantages offered to us by the unprecedented qualitative development of these new technologies, collaborative and inter-cultural working groups can be formed to carry out assignments and projects, and solve problems in virtual spaces using the electronic networks.

The ITESM has been using these new educational technologies in Mexico for over a decade. The purpose of this chapter is to share our experiences, problems and achievements, as well as the technological and educational decisions, trends and questions. We will also look at the implications at both an organizational level and on the lecturers and students themselves. Over these ten years we have built an extremely valuable knowledge base on using these new virtual spaces. With this knowledge, the shape of the VU has become increasingly defined, likened to a sculptor's workshop opening out to a widely-varied public, and integrating the principles and trends of education into the technological options. We can characterize this experience as a true learning laboratory and workshop.

The VU is a constantly changing institution which is continuously evolving, which has incorporated innovation and creativity as a work culture, and is always adapting to the benefits of new technologies. The institution opens itself to national and international educational needs and presents diverse options in order to provide a top quality education for the future. It thus fulfills its social commitment to contribute to the improvement of its own country, and from a national platform, to the construction of a technologically united world.

The new social and work context

We have learned that technology alone does not automatically facilitate the educational change we are seeking, nor is it advisable for those who are unaware of the direction of change. Wit and creativity are required to integrate technology into the educational process, and these qualities are developed when we know where we are going and how we should get there. Only then can we choose the ship that will take us into port. Therefore, the key questions are: What is having such a strong impact on peoples' lives and work in the real world? What challenges does this situation present to higher education? What role does technology play in all this?

According to the Committee Report to UNESCO on Education in the 21st Century (1996), the scientific and technological advances in the last quarter of the century have produced great cultural and human changes, transforming people's way of life, the economies and the very nature of work. Humanity today must undertake the enormous task of understanding and harmonizing with these changes.

The most recent research shows that training and education are the two main forces behind a country's development and growth. This situation demonstrates the role human capital plays in a nation's productivity, and therefore the importance higher education will have in the future. Universities must begin by perceiving society's demands and responding to current needs. This has increased the necessity of thinking about how the educational process should occur in the future.

Information technologies have transformed communication and knowledge, shrinking the world and changing peoples' global view. They have also liberalized economic and financial borders, allowing humanity to move on from the basic community to a world society, thus achieving an interdependence in problems and an interest in solving them collaboratively. In the future, "obtuse nationalisms should give way to universalism, tolerance, understanding and pluralism. From a divided world we should become a technologically united world" (UNESCO, 1996).

Current trends in the world are having a strong impact on work:

Technology becomes a necessity and distance communication a way of life. Over the next few years, an increasing number of employees will work from home. Work organizations are more focused on group decisions, departments are giving way to collective work on mutual interests where group conversations and actions are centered on common objectives. These work groups are based on cultural diversity and cross disciplines, organizing themselves, distributing responsibilities and establishing their own norms.

People need to communicate effectively and make decisions in an atmosphere of democracy, solidarity and respect for the different cultures they represent. Organizational management will be less directed, more horizontal and will rely more on employees' decisions. Since we are aware that people's competencies are important to the success of services and products, employees will be expected to take responsibility for and commit themselves to lifelong learning and a continuous improvement process (Oblinger and Rush, 1997).

Developing countries in the face of globalization and change

For developing countries this situation involves facing the difficulty of participating in the international technology competition and of access to world markets. Developing countries cannot neglect the indispensibility of their entry into the universe of science and technology. Companies try to penetrate transnational markets with their services and products, implementing global strategies and increasing their level of proficiency (UNESCO, 1996).

Mexico, like other countries, is undergoing a profound transformation. Globalization, technological advancements, and alliances on a universal level are urgently needed to ensure international competitiveness of Mexican products, to promote well-paid employment, and to improve the population's standard of living. Corporate improvement is a Mexican priority, as this will also make the national workforce more competitive. Lifelong education and training are becoming indispensable. A company cannot survive this environment if it does not have sufficient human resources and does not prepare for change. If education is so important, "more" and "better" training become imperative.

Challenges for universities

Higher education must change if it wishes to meet the educational needs of society (Dolence and Norris, 1995). The changing working environment demands new functions and new ways of understanding and implementing education. The gap between industrialized and developing countries must be narrowed in the field of scientific and technological teaching, and these educational opportunities must be made available to everyone.

Teaching must go beyond a narrow national focus to broaden students' horizons and sensitize them to other cultures. Peace, economic advancement and global harmony depend on strengthening the links between people from different cultures. International trade, energy resources and foreign markets, diplomacy and cross-cultural interaction require a deep understanding of how other people live. Without doubt, learning to live together is one of education's most important undertakings. (UNESCO, 1996).

Another great goal that universities face today, particularly in developing countries, is how to make their people citizens of the world without losing their roots, and to continue participating actively in the life of their nation and communities. A permanently changing world will require permanent education. This notion must be integrated into the working world, while offering flexibility and accessibility in time and space.

On the other hand, the following will have to be inculcated into basic and professional education: a liking for and pleasure in learning, as well as intellectual curiosity and "learning how to learn." This will make people capable of evolving, of adapting to a rapidly changing world and of mastering change, applying knowledge to solve problems and participating in the improvement of reality.

Living in the information age requires putting all possibilities contained in the new information and communication technologies at the service of education and training, and integrating them into educational spaces to enrich learning and provide equal opportunities, acquiring, in turn, a wide range of knowledge and skills to work in institutional networks and connections.

In such a vast world, with such a profusion of information, by the end of their education all students should have the tools necessary for an autonomous and critical understanding of the world around them. In an interdependent society, a global conscience representing the entire world must be developed.

In order to develop realistic perceptions, our students must understand the most significant problems of human society. The study of isolated disciplines does not foster to this purpose; we must instead develop proposals for integrated curricula. Young men and women with a more cross-disciplinary education are better trained to detect, analyze and solve new problems. "The cross-discipline of today will be the discipline of tomorrow" (Torres, 1996).

Cross-cultural communication and teamwork must be encouraged. It is important to see diversity as an asset in which students can (a) consider other points of view, (b) encourage debate on situations which demand ethical decisions and (c) make decisions as a group.

Reflecting on this situation, many of us believe there is no relationship between what today's education offers, what society currently needs, and what graduates will require in the future. The traditional structure of the university serves young, full-time students who live on campus. It is subjected to rigid and fragmented schedules with one expert lecturer in a limited space (the classroom) for each discipline, supplemented by local resources. Will such institutions, designed to serve primarily a homogeneous group of classmates, nearly all of whom share the same culture, be able to adapt to internationalization, a wide range of cultures, abundant points of view and develop a global conscience? Will this model provide an answer to the need for continuing education in corporate life and in society in general?

We believe universities should diversify their ideas and knowledge, extending their scope of education and capacity for action. It will be important, for example, to incorporate into career training a way to develop personal qualities which are acquired during the training period, whereby students get to know themselves and others better, thereby improving interpersonal relationships. The context in which learning experiences develop is fundamental to achieving these objectives. On the other hand, the educational proposal will not be limited to the classroom only, but it will also offer services to corporations. Educational institutions have the opportunity to improve corporate competitiveness and to achieve social benefits. "According to the American Society for Training and Development, by the year 2000, 75% of the current workforce will be retrained just to keep up" (Oblinger & Rush, 1997).

Many institutions of higher education have initiated processes to adapt their mission and actions to meet these requirements. ITESM has also recognized the

TRADITIONAL EDUCATION	WORKPLACE REQUIREMENTS
• Facts • Individual effort • Passing a test • Achieving a grade • Individual courses • Receiving Information • Technology separate from learning	• Problem solving • Team skills • Learning how to learn • Continuous improvement • Interdisciplinary knowledge • Interacting and processing information • Technology integrated with learning

FIGURE 12.1
Characteristics of Traditional Education and Workplace Needs (Forman, 1995)

need for change in order to fulfill its commitment to offering a quality education for the coming years.

The ITESM Virtual University: Approaching the Future

Every ten years the ITESM reviews its mission in order to better serve the country's and society's needs. In 1995 the ITESM undertook a wide-ranging consultation to define the institution's course for the next ten years. The board, non-profit organizations, employers, students and faculty participated in this process.

The result of this was the Mission 2005 statement, which defines the Institute as a university system charged with forming people committed to the development of their community, to improve it socially, economically and politically, and who are internationally competitive in their field of knowledge. The mission includes conducting research and continuing education relevant to the sustainable development of Mexico. This mission is founded on the perceptual context of Mexico's needs, and on the principles and values that guide world-wide educational thinking for the coming century.

In order to accomplish these goals, several strategies were established, four of which integrate the new technologies and virtual spaces as tools:

- Fully implementing the Virtual University
- Reorganizing the teaching–learning process
- Orienting research and continuing education towards the development of Mexico
- Promoting internationalization

History of the Virtual University

The ITESM has always recognized the importance of telecommunications, computer networks and multimedia resources to support learning, as well as to develop new educational models and proposals in both distance education and traditional classrooms. Since the early 1990s these technologies have been applied to educational processes, with recognition for and economic incentives from the Institute for faculties who develop projects.

Although these were generally isolated cases, the results were publicized to all faculties, from the different campuses in annual forums for exchanging experiences. The ITESM created the Educational Multimedia Project Design and Development Fund, and the Educational Technology Research Centre (CIETE) to promote these projects. Still in operation, CIETE offers an advisory service for lecturers who wish to use informatics and multimedia technology both inside and outside the classroom. At present we have many examples of virtual laboratories, Web pages and multimedia.

The following figure represents the evolution of the ITESM's VU in relation to the technologies it has used.

The VU began in 1989 as an Education by Satellite Project (PES), responding to the Institute's need to connect its different campuses, and to be able to increase faculty expertise by offering master's programmes. The satellite made it easier to offer distance courses with a conventional teaching plan, based on lectures transmitted from the Monterrey campus. Monterrey broadcast lectures and other information to the system's other campuses.

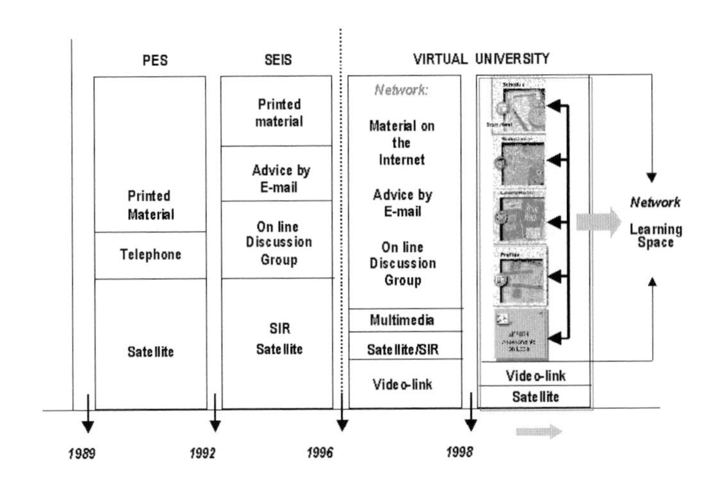

FIGURE 12.2
The Evolution of the ITESM VU

Satellite broadcasts were sent to an office equipped with a reception monitor, as well as a telephone and fax machine used to interact with lecturers during the sessions. During these sessions, students seemed very unmotivated and unhappy with the passive, fairly impersonal nature of instruction, and seemed to have difficulty interacting with instructors.

In 1992, PES became the Interactive Educational System by Satellite (SEIS). Computers were installed in the receiving classrooms with a remote interactive system known as SIR online. This allowed students to contribute to the course by sending replies, asking questions, etc., during each satellite session. Using this arrangement, the teacher could establish a truly simultaneous interactive process between the students in the different receiving classrooms. These technologies, when properly used by instructors, make it possible—apart from simultaneous interaction—to more fully communicate knowledge using videos, images, graphics and animations. However, the majority of students still perceived this experience as rather unsatisfactory, passive and without meaningful interaction. The ongoing problem with this system has to do with student anonymity and impersonal interaction. Only a few lecturers possessed a special flair for handling the technological resources well enough to promote a rich educational experience in the virtual classroom.

The courses offered by SEIS were centered on the professor and his/her lecture, individual homework, review and evaluation by the lecturer, followed by feedback or advice at the request of the student. That is, they were based on traditional educational models which were more familiar to lecturers, where the student was dependent on the lecturer, there was little interaction, and the individual learning process was centered on the receptive acquisition of knowledge.

To accomplish this, lecturers gave printed materials (normally a few texts), and assigned written homework, which students completed and sent to instructors via express mail. Instructors then returned the homework, with corrections, back to students. The process required several days, thus delaying feedback. Telephone consultations were also available at pre-arranged office hours. Students generally lived in a state of constant anxiety, never knowing "where they stood," and this led to a high dropout rate.

This stage in distance education by satellite was highly instructive for ITESM. We needed to find solutions to the multiple problems and explore alternative "technologies" to overcome obstacles and provide solutions. We were provided with ample knowledge of the features of distance education, while at the same time, new technologies were developing which were able to serve learning requirements.

The Virtual University, based on SEIS' distance learning experiences, was created in 1996. It was defined as "a higher education institution that offers distance courses with an international scope, based on an interactive teaching-learning system that operates through a wide variety of telecommunications and electronic network technologies." Its mission is to support the Institute in fulfilling its commitment to improving the competitiveness of large and medium-size companies, participating in democratization and educational improvement processes, and helping the private and public sectors to plan and execute decisions. In addition, it offers high-quality programs on an international level, and develops innovative educational models with cutting-edge technology that lecturers can apply to classroom teaching, and gradually incorporate into the new educational models. The VU is a diversified educational supplier, transmitter and receiver, as well as a place for collaboration and creation of knowledge, models and experiences, and a research space for students (Lopez del Puerto, 1997).

Globalization of the virtual university

The Virtual University currently focuses on three main objectives: graduate studies, faculty development, and education at work. In doing so, it serves people from widely dispersed geographical points. In general, the programs offered by the VU are developed through agreements with other universities. These agreements are of different types and with different objectives, depending on the purpose of the program (we are currently working with 50 agreements).

All VU programs are supported by satellite and teleconference technology, which forms a system of transmitting and receiving with associated sites across the American continent. We currently have 17 transmitting sites that can use two satellites (Monterrey ITESM and Mexico City ITESM) and 15 teleconferences, 10 on the different ITESM campuses and five distributed in different countries. The receiving sites are constantly expanding. At the end of this semester (May, 1998) we had 873 sites, the majority in Mexico, with 17 in Latin America. These sites receive master's and bachelor's programs, as well as faculty development and business training programs. The programs for the latter two groups are held at corporate workplaces.

In addition to a monitor in the signal receiving room, these sites have a learning center which allows students to interact with their lecturers via computer, access the relevant information for course work, receive advice and interact with classmates in group discussions using the Internet, e-mail, the WWW, home pages and groupware. Educational activities are carried out individually and in groups, and interaction is generally asynchronous. The satellite sessions and video teleconferencing both allow nationally or internationally recognized experts in specific fields to communicate their experiences.

Programs offered by the virtual university

The new technologies offer opportunities to break the barriers of distance and to facilitate access from any place and at any time. It is now possible to offer various learning options which were once impossible. The extremely positive response

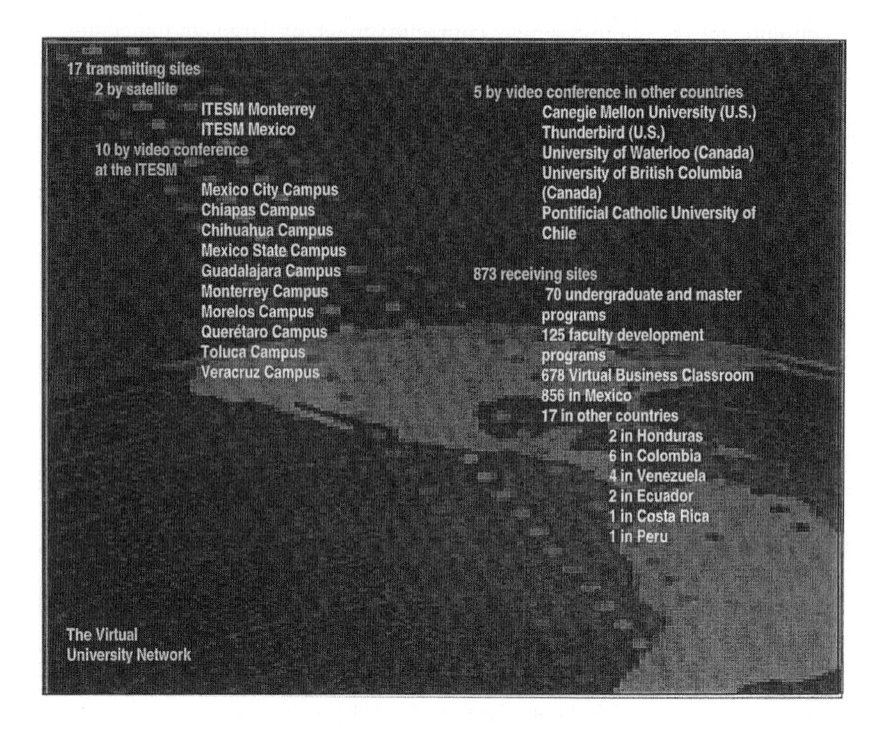

FIGURE 12.3
The ITESM Virtual University Network

which the Virtual University proposals received in the marketplace indicates that we are offering a necessary service which will grow exponentially in the future, and challenges us to provide top-quality service.

Over the past two years the VU has developed the following programs:

- Virtual Business Classroom

A communication company offered its advantages in technological experience and satellite infrastructure for providing educational and training services. This project was developed through an alliance between Monterrey Tech, and the commercial firm, MVS Communications. It is an innovative learning system aimed at raising business competitiveness and internationalization by making Monterrey Tech's educational services available at the workplace using business education models. In the coming years, plans are for these continuing business education programs to be expanded across the American continent.

For Latin-American companies, especially small and medium-size businesses, the concept of virtual enterprises offers a new way to improve competitive position against international firms. In this project a framework for conducting global business in a virtual form has been developed and used to operate a Mexican virtual enterprise. The framework, which consists of three entities, Virtual Industry Clusters, Virtual Enterprise Broker and Virtual Enterprises, was first developed to better understand the relationships among the entities and the conditions necessary for successful business in the new global economy. The Virtual Mexican Company (VMC) in operation is a leading manufacturer with a virtual production line, which means a line of specialized, exclusive and qualified workshops that give a flexible and efficient focus. The main specialized areas include silk screening, metal components, textile operations, plastic injection and ceramic decoration. The activities performed by the VMC include product marketing and sales, product development, process design, logistics, client servicing, and inventory control. (Lopez del Puerto, 1998).

BOX 12.1
An Example of an ITESM Project: Virtual Enterprises

- Virtual Bachelor's Degree and Continuing Education Programs

Statistics show that a large number of Mexico citizens have not finished secondary school, and very few hold bachelors' degrees. The Institute is taking advantage of its alliance with a Mexican partner—Galaxi for Latin America (MVS):

- To offer virtual degree courses and continuing education programs "at home" for Mexican and Latin-American adults. This reaches the most underprivileged and low-income areas and the places least culturally-recognized (such as our native Mexican communities). Our aim is to offer flexible educational alternatives, to educate a growing number people every day and improve the cultural level and standard of living of our communities
- Educational Support Programs

In countries with low-quality secondary education, the national scientific capacity is critical and requires urgent attention. In order to compete with the rest of the world, we need highly qualified professional teachers. This is a long-term investment that will have a strong impact on the development of the Latin-American community in the future. At present, the Teaching Skills Updating Program is being developed for teachers from different levels of education in Latin America. At present, over 3000 teachers from 17 Mexican regions and four Latin-American countries are participating in this program, which expects to train 13,000 teachers next year.

- Support Programs for the Private and Public Sectors

The objective is to address the most immediate needs of this group of officials, whose activities affect all Mexicans, by offering courses for improving strategic planning, developing management styles that encourages participation, commitment and responsibility in workers, and improving government attitudes and values.

- Academic Degree Programs for Professionals

These programs, directed at master's and doctoral degrees, are built on agreements made with foreign universities that are recognized world-wide in their fields of knowledge. Although the strategy used in each program is different, they all take advantage of satellite and video-link resources and establish agreements with leading universities in a specific field. This allows the exchange of ideas between students from different geographical points in real time, enriches the course with videoconferences by well-known experts, encourages team teaching with faculties from other universities, and allows the offering of degrees common to both universities.

Eleven master's degree courses and one doctoral program are currently being offered in the areas of Education, Educational Technology, Engineering and Technology, and Business Administration, some in cooperation with the other international universities.

The VU is looking for partners in Latin America for these programs, and has an office in Venezuela for this purpose. The trend is to promote team teaching and teamwork, with more international students and faculties, with common objectives and in more democratic settings which will form real on-line collaborative, cross-cultural learning communities. Experiences with these characteristics, until now, have only occurred in isolated cases and in classroom teaching. An example is the course "Trade between Mexico and France" from the master's degree in Marketing. This same course (with the same characteristics) took place last year with students from the Universidad Católica in Chile, and the results were very gratifying, but greater effort and commitment are required from both parties. A similar case was a joint research agreement established between Monterrey Tech's Integrated Manufacturing Systems Centres and Techniker, a Spanish Basque advanced technical company.

- Support for ITESM undergraduate courses

From the beginning of the SEIS, the VU has offered a great deal of support to academic needs which arise on many ITESM campuses regarding undergraduate degree courses. Quite often, a large number of students from the smaller campuses must transfer to larger campuses to finish their studies with sufficiently high academic standards. The Institute has always seen the VU as a way to provide all the Institute's students with equal access to the best teachers, resources and educational experiences possible, without having to leave their home communities. To accomplish this the VU is offering 16 terminal courses for the most densely-populated on-campus courses. Another significant advantage, particularly for campuses with fewer resources, is that VU courses cost 90% less than the same courses offered in a traditional classroom setting. This allows the Institute to develop top quality courses that can be distributed to a large number of people.

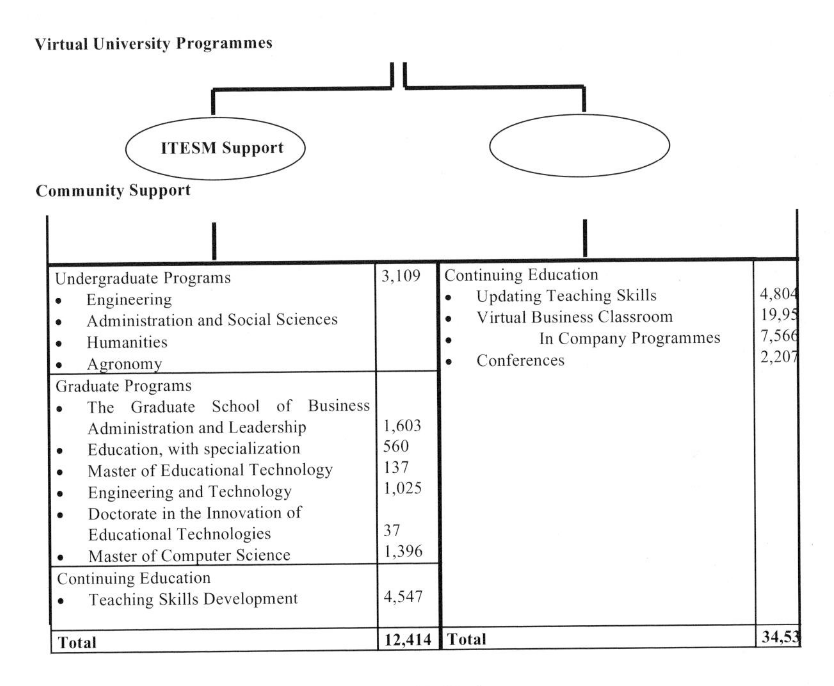

Undergraduate Programs • Engineering • Administration and Social Sciences • Humanities • Agronomy	3,109	Continuing Education	
Graduate Programs		• Updating Teaching Skills	4,804
• The Graduate School of Business Administration and Leadership	1,603	• Virtual Business Classroom	19,95
• Education, with specialization	560	• In Company Programmes	7,566
• Master of Educational Technology	137	• Conferences	2,207
• Engineering and Technology	1,025		
• Doctorate in the Innovation of Educational Technologies	37		
• Master of Computer Science	1,396		
Continuing Education • Teaching Skills Development	4,547		
Total	**12,414**	**Total**	**34,53**

FIGURE 12.4
Numbers of Students by Program

In the future, the VU will evolve to offer courses which will be of a higher educational quality, backed by research, the experience of the VU itself, and the technological and didactic support by faculty members. The programs will become more diverse as more needs are identified, and will serve a more heterogeneous and geographically dispersed public as the internationalization process advances. As part of its vocation to service, the VU, rather than merely offering ready-made academic packages, will continuously develop new projects to meet new needs. For example, next semester, the Master's in Educational Technology Management will be offered in response to a request from Venezuela. To accomplish this VU developed a new approach based on distance learning that is more oriented toward adults, responds to working needs or self-improvement interests, and is integrated into workplaces.

The Emerging Educational Model

The technological resources used by the VU to offer its teaching–learning process are multifarious and have different educational features and possibilities. Their use is not based on their supposed modernity, but on the educational intention that connects them and gives them meaning and measure. Our experience tells us that this is the most critical aspect, as well as the "soul", of the VU. The question is: how can we make the use of technology a pedagogical experience? Technology is of no use to us without some criteria for helping the lecturer to make the appropriate choice and application, and extend the possibility for him/her to continuously redirect learning. The key component at the VU is the educational model and the support technologies.

The education ITESM seeks to provide will fulfill its mission of developing people who are both internationally competitive and committed to their country. The following provides a profile of the "new ITESM student".

Attitudes and Values	Skills
• Responsibility and honesty • Leadership and innovation • Spirit of personal • Working culture • Clear awareness of the their country's and region's needs • Commitment to the sustainable development of the country and their communities • Committed to acting as agents of change • Respect for people's dignity and their inherent obligations and rights • Appreciation for culture • Vision of the international milieu	• Self-teaching • Analysis, synthesis and evaluation • Critical thinking • Creativity • Identify and solve problems • Decision making • Teamwork • High capacity for work • Quality Culture • Efficient use of informatics and telecommunications • Command of the English language • Good oral and written communication

FIGURE 12.5
The New Characteristics Sought for an ITESM Student

The responsibility now lies with the teacher. How can students acquire an attitude of respect for people's dignity as a basis for coexistence? How can their permanent education and an efficient performance in the new society of knowledge be guaranteed? How can we prepare them to make decisions in problem-solving and improvement projects collaboratively? How can we help students to become aware of other cultures and to communicate effectively? How can we make them competitive on an international level by using informatics and telecommunications efficiently?

Teachers are being presented with a highly exigent panorama. The academic learning that forms part of the curricula must be incorporated into an education that has more to do with how to act and how to relate to others—both their fellow students and the world community at large.

This education is related both to *what* is learned and *how* it is learned. The context in which education is provided, the manner and climate of working, holds an important position. The faculty's most important role is to facilitate experiences so that learning will occur. "The faculty must create a virtual environment in which students succeed, and in which they develop, along with a healthy self-interest, an understanding of the sociological and cultural necessities for preserving the community. It is not only for their individual good that students must achieve" (Oblinger, 1997).

The effort to put this learning into practice is moving us away from the traditional model, in which exposition accounts for 80% of teaching. The main characteristics of the traditional model is defined by the faculty making all the decisions on the "what" and "how" of the course, and a student passively dependent on the lecturer. This discourages individualism, and instead involves learning by rote, is reproductive and uncritical, and reduces interaction to merely asking questions related to the content of the teacher's class.

This is expressed in a 1996 VU study on the quality of interaction in SIR during satellite sessions. Another study on the effectiveness of this model (Oblinger, 1997) concludes it is not the best method for teaching students to solve complex problems, to think critically, to collaborate in common objectives and to retain what they have learned.

Beginning in 1997, the VU proposed an educational model for all courses that transforms the lecturer-centered process to one of group learning, with emphasis on the following principles:

- Students should develop the ability to learn on their own.
- They should learn from their classmates through dialogue and discussion.
- They should participate in solving problems as a group.

The VU builds the whole learning process on two basic technologies: The Learning Network and the Satellite. By networking we mean an educational infrastructure which makes it possible to hold course activities on-line, and which integrates a series of tools such as computers equipped with information resources, and communication possibilities that are highly valuable to the faculty and students. Learning networks accomplish the application of electronic technologies to the learning process. This integrated system supports all the above learning processes. The following is a description and analysis of the applications of these technologies and the different educational possibilities they offer.

Internet

The value of the Internet lies in the almost limitless amount of information resources it offers, far more than any individual workplace, their community or their local library. ITESM engineering students and faculty members frequently use many of the Internet's resources as developed by internationally renowned experts, which greatly enriches the course content. This resource provides much desired educational improvement. Through the Internet, students can browse, select, analyze, compare and evaluate a variety of relevant information, thus expanding the research and discovery process. In the near future, VU students will have access to a digital library, with various connections apart from the Internet, which can offer them a whole world of pre-selected and pre-classified scientific information.

E-mail

E-mail is used at ITSEM to facilitate basic communication, but also for students to mail homework, and for teachers to provide feedback and monitor individual students' progress. E-mail is widely used at ITESM among faculty and students, and is totally integrated into their working environment. Students also send a large number of daily e-mail contributions to the Continuous Improvement System (CIS), which allows them to give their opinion on the real impact of the introduction of the new educational model and technology in the redesigned courses. Students' input to CIS is turning out to be an excellent source of information to discover needs and guide future decisions.

Computer conferencing

This application, also known as groupware, is a restricted virtual space that creates a particular social system where a group of people with a common objective can use computers to fulfill the commitments they establish among themselves. The communication software allows groups to co-ordinate and organize the material in a appropriate manner for their objectives and to progress according to the requirements of the work itself. This tool complements e-mail and provides the truly interactive space that students and faculty need to accomplish a learning process.

The conference is a database of a discussion held by a group in an easily accessible format. All the participants can access the others' data. The program also maintains a list of group members, permitting identification of participants. Through this interaction system, changes can be made to some of the contributions, facilitating progress towards the objective. The computer conferencing systems offer two kinds of activities: group discussions and group work, and facilitates the activities necessary for each type of task. Because the Internet offers access to all group members regardless of physical location, electronic group discussions enable students from different campuses, schools or business training centers, either from the same country or different countries, to work together on common projects.

For example, the on-line courses offered by the University of British Columbia in collaboration with ITESM in the VU's Masters in Educational Technology, unites students from all over the world in group working sessions, to form a highly cross-cultural learning community together with their instructors. The course is organized so that knowledge is gradually built up through readings and contributions to group discussions with everyone involved. The group discussions provide multiple perspectives on each topic and cross-cultural communication.

The possibilities for collaborative group work are enormous: seminars, projects, subject discussion, problem solving, etc. Every task has its own requirements, established by the objectives and the components of the group.

Some distance courses generate an effective "socialization" group for "faculty only" or "students only". These groups have often resulted in a participant wanting to find the places and people they have been in contact with, thus promoting social cohesion, solidarity and broadmindedness.

Computer conferencing offers the faculty many educational and course management possibilities. For example, the teacher could pose a question in a group discussion and the computer can stop students from seeing the others' answers, until they themselves have answered. This offers a range of possibilities to adapt the process to the group and to the requirements of each task.

When properly facilitated by a teacher, all these information and communication possibilities offered by Internet technologies, computer conferencing and e-mail can simply and economically deliver a course totally on-line using the learning requirements we have thus far established.

Learning network advantages

Our experience with this learning network demonstrates its enormous potential for providing significant educational benefits to both students and faculties, as well as to the educational process itself. Some of its many advantages are set out below.

The learning network favors active, student-centered learning. Students who participate in a network course are constantly involved in activities. Students only go on-line when they need to make a comment or contribution. But active learning is more than just pressing keys on a computer keyboard. Contributing to discussions, answering fellow students questions and sharing ideas is a cognitive and social commitment. Active participation also strengthens learning. Putting ideas into writing in a coherent manner requires a mental effort. Formulating and articulating an affirmation or declaration is an especially rich cognitive action if these contributions are followed by an argument.

The learning network also offers students equal opportunities to participate by making comments when required. The fact that on-line learning is offered 24 hours a day allows students to reflect more on the ideas and to think them out at their own rate, compared with exposition which assumes that all students will advance at the same time and with the same background.

The learning network changes the faculty–student relationship completely, breaking with hierarchies, and the lecturer becomes more of a facilitator than an authority figure. The faculty normally provides a reference for an objective, some texts or other instructions that could be useful in guiding students' work. Students carry this out collaboratively, interacting among themselves, co-operating and evaluating the process. The teacher constantly monitors what is going on, providing him/her with the means to continuously evaluate the process while remaining totally integrated with it.

Network learning communities are formed, resulting in the personal and formative enrichment of all the members of the group. The communication that flows leads to friendship, intellectual stimulation and personal satisfaction. The members of the community share interests, know each other and are interested in each other. When students who have experienced group work are asked who they would go to if they had a problem, the most common answer is not the faculty, but their fellow group members. The data we have on the faculty and student opinions regarding communication show that it becomes more frequent, more profound and more intimate.

Finally, learning networks promote responsibility. For students to be successful in an on-line course, they must take the responsibility for their own learning, as well as help their fellow students to take responsibility for their own work. They must read and reply to other student's comments, reflect on their opinion on matters that arise, and seek additional information, if necessary, to reply.

Students are more committed to the process. They establish the norms that will guide the group's work and assess themselves continuously, reflecting on how they are doing, what they have achieved and what needs to be improved. This produces a culture of continuous improvement in their work and prepares them for lifelong education. An example of the norms a discussion group might establish would be: to answer messages promptly, contribute with well-founded ideas, personalize contributions, avoid hostile or aggressive comments, and promote participation by helping and supporting others, etc.

This facilitates the internationalization of higher education. Through these experiences, both faculty and students achieve an understanding of a more global world. With this on-line work, students acquire a wide range of knowledge and skills for working on networks and institutional connections. We note the exchange of experiences with students from other countries and take the richness of cultural diversity to the "virtual classroom." (Bates, T., 1995)

Learning network evaluation

Evaluation is one of the most critical aspects of an on-line course, according to faculty and student experiences. At the same time, an on-line course provides many advantages for effecting an evaluation process. In order to pass a course, for example, not only the results of assignments are evaluated, but also students' skills, and their attitudes towards work and others. Teachers often ask how they should evaluate participation and individual student work in group projects.

One way of evaluating participation is by taking into account the quality and quantity of that participation by determining the degree of relevance and the opportunity for contribution to the common good. Group work requires the constant participation of all the members of the group. Activities should be promptly distributed and deadlines set for contributions to each activity, keeping all the members of the group constantly busy. Deadlines and requirements for assignments should be set by the lecturer at the beginning of the course. This allows students to organize their time properly.

The lecturer should also indicate, from the beginning, the weighting that participation will have in the final grade, as well as which other activities will be taken into consideration and how they will be evaluated. Students want to earn good grades and feel frustrated and anxious if they do not know what is expected of them. If this is made clear from the beginning of the course, students' attitudes becomes much more relaxed, flexible and positive, and a climate of trust is generated.

The members of the team should evaluate group participation. This is easier to do if criteria are first established as a reference framework. This evaluation is highly desirable as it develops critical thinking and commitment to the process. As in any learning process, classroom or distance, there are many options for evaluating group work. An on-line course has advantages over classroom courses; having access to a database with all the contributions recorded and with a personal reference, as well as the date of when they took place, makes it possible to create very rich evaluations. This does not halt when the course has finished, since formative evaluations can take place outside the course, aimed at learning how the different elements behave, and identifying areas of opportunity for improvement.

This is one of the main functions of an on-line course lecturer. In this analysis of participation in the group, instructors can also identify who made a special effort in contributing to the group's work. On the other hand, if one of the participants has not contributed according to the pre-established guidelines, this student can be evaluated individually. Finally an on-line course encourages students to evaluate themselves, as well as each other, as members of a group, and permits a formative evaluation as an integral part of the process. This allows the lecturer to monitor each student's situation at all times, offers a pertinent feedback and makes the evaluation of the group easier. (O'Banium, T.)

Satellite

What role does the satellite play in this context of networking in VU courses? The satellite as a learning technology in the new educational model has its individual advantages, but cannot support the entire learning process. This is why we have redefined its application over the past two years. The satellite makes it possible to extend messages to unseen limits. Now we have coverage throughout the American continent, and soon we will be able to reach all continents through the development of agreements with telecommunications companies.

The satellite has its advantages. It helps to enrich a course, for example, with conferences offered by experts from anywhere in the world, providing students with the results of an investigation, or the explanation of a theory by the expert that is relevant to the students in relation to their course. In addition, different experts from different places, who represent contrasting trends, visions or experiences, can offer conferences that help students to acquire a more real and global understanding of the world in which they live, allowing them to place themselves in a more international context.

The satellite also offers the ideal method for team teaching with expert faculty from renowned universities, with whom students can interact through the SIR system. The use of video links enriches interaction, since students and faculties from another institution can discuss relevant topics. Satellite use is complemented by other technological applications to make rational use of the diverse technologies integrated into the VU courses. The time that a student spends in a satellite-image receiving room has been reduced to only 10% of the total course time. Its use is being optimized and frees the channels through which images are transmitted. Many more courses can be offered and other activities can be incorporated that enrich people's lives. Last year, for example, 16 leading world figures were invited to give conferences on politics, the economy, literature, business, etc.

Learning space

This is ITESM's focus for the immediate future: The tendency to use the Internet instead of the satellite is increasing. This will help to reduce costs by applying more integrated technologies and standardizing them throughout the Institute, as in the case of Learning Space which makes it easier for the lecturer to integrate technologies. Apart from lowering costs, using the Internet instead of satellite makes it possible to provide a quality education. The long-term financial yield of this system is difficult to measure. The most recent decision made on the use of technology was to standardize the platform on which a course operates by using the Learning Space operating platform with Lotus Notes software. Standardizing the program means standardizing the operating system on AIX. The AIX operating system provides for applications in knowledge management. The disadvantages are its being limited to one supplier. The advantages are that it is more economical, and we only need to specialize in one technology. With the variety available on the market, it would be very difficult to continually incorporate new software. Training would be very expensive as we would have different computer languages, requiring many types of software and training experts.

However, all this does not mean that we will always use the same system. The Institute is constantly surveying markets. This investigation is so important that a Vice-President's Office for Technology was created to provide support for the system and to control the decisions in this field that has a promising future in our institution. Initially, standardization meant a heavy investment, owing to the problems involved in standardizing the speed and capacity of a network. From just two Lotus servers, we now have 57 on all the campuses, and this investment will grow exponentially. Medium- to high-capacity personal equipment is required, as well as a large investment in infrastructure. Consistency is expected of the technology being used. Continual upgrades are necessary if we are to remain on the cutting edge of technological advancement.

Apart from these economical and administrative advantages, Learning Space (LS) offers considerable educational possibilities for faculty members who are trying to introduce the Institute's new educational model. Like other multimedia and distributed learning tools, LS offers the course designer the possibility of structuring information, presenting content and supporting asynchronous collaboration among students and instructors.

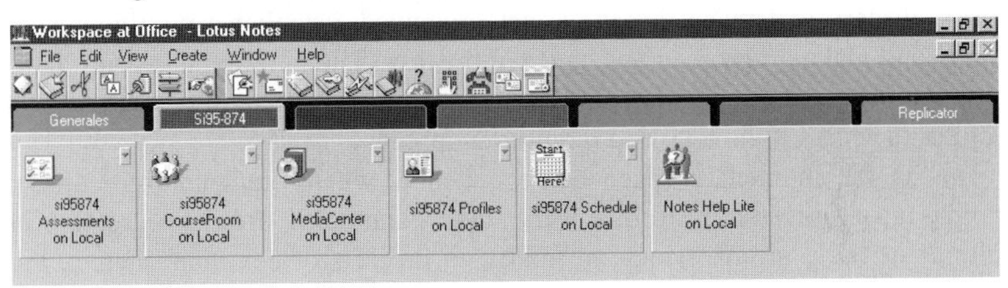

FIGURE 12.6
Front Door of the Learning Space Platform

Learning Space contains five integrated databases that users can easily browse through according to their interests and needs. The first database, and the door to the other four, is Schedule, where the instructor records the entire course curriculum in a highly organized and structured way.

The program presents options the instructor has to choose from, including the manner of organizing the course content (modules, sessions, stages, etc.), the type of activity to be affected, the space where it will be performed, to whom it should be sent and the preparation dates. The course introduction, general objectives, table of contents, instructor's expectations for the students, global methodology or strategy to be followed, evaluation system and criteria, form of communication, instructors' functions and guidance on how to use the different technological applications are also recorded in this database. In general, it includes everything that the lecturer considers important for the success of the course.

This space is prepared before the course begins, but changes can be made during the process if necessary. For example, new activities can be included, dates changed, strategies modified, etc. and will be saved in the "change history" section, which students should always visit. The possibilities offered by this program facilitate flexibility and openness in instructional design, which is particularly necessary for distance education. It also makes possible continuing negotiation between the students and the lecturer throughout the process.

The Media Center is the database wherein lies the course content. It is a sort of student library where documents, texts, CD-ROMs and general presentations are available in a wide variety of multimedia formats. It can also be linked to external sources, such as the World Wide Web or electronic libraries from other universities. This space is where students process and interact with information by analyzing, contrasting, and synthesizing it, amongst other things. This activity is normally carried out individually, allowing the student to create his/her own file and personal notes.

The Course Room records the data from the student discussion groups, as well as the results of individual activities or group projects. It has two sub-spaces: Discussion and Work, which facilitate collaborative work, permit private and collaborative discussions. This allows students to exchange ideas, discuss relevant issues, discern and make decisions, get organized, establish commitments as a group, and make progress in the stages of the task. The program classifies data by participating students, by date and by working teams, making it easier for the lecturer to access, follow up and monitor the process.

This is the interactive space par excellence, where students from different campuses or universities from a very wide geographic area can create cross-cultural working groups which are united in projects of common interest. From this database the instructor can send private comments to an individual student, to a specific group, or to all the groups in general. Projects can be made available to the instructor alone or to all the students, so that they will have the opportunity to learn from others, make comments, and even evaluate each other if desired. This database holds many educational options for a creative lecturer.

Another database is Profiles, where students record their photograph and personal information. In this way students can get to know each other, ask for help when needed and when required, as well as comment on other topics. In general, these data favor the group's socialization and the formation of a community of learners. Students often say that they know their on-line classmates better than their fellow students from traditional courses, and that they prefer to seek help from on-line classmates rather than from their instructor.

The final component of Learning Space is the Assessment Manager, which is used to design evaluations and exams. Through this space the lecturer can prepare questions to evaluate students' educational needs and achievements, as well as chart individual progress. All students can access this document to monitor their progress. The lecturer can also design quick tests, polls and self-evaluations for students. This helps make students aware of their needs and accomplishments, and thus achieves educational autonomy and responsibility.

Although the Institute is currently using Learning Space (LS) as the "official platform" for the VU courses as a support for the Institution, its use has not yet been generalized, and is only being applied to 20% of courses, while the rest continue to use the electronic page. Both media are complemented by the satellite and video link.

The computer equipment required for working with LS should have a large memory capacity, be fast, and be loaded with Lotus Notes software. A 20-hour training course is needed for both students and faculty to handle the program efficiently. Users have found this program easy to operate and very useful. Both students and faculty need their own personal computer for these courses and the Institute managed to obtain very affordable prices with easy financing systems from distributors. It is the System's policy not to increase the cost of fees with the use of technology.

Some Problems and Solutions

We have faced all kinds of problems with the application of new technologies. One is related to infrastructure. Although we have greatly increased the number of servers, we do not know, due to a lack of experience, the amount of simultaneous workloads the system can bear, and quite often the server fails at certain times. To solve this problem the Institute continues to invest resources and offers a control service so that the server can be put back into working order as soon as it fails. In this way we can determine server capacity and purchase equipment in a more rational manner. Short-term solutions do not solve the problem entirely, as both faculty and students often have to interrupt their work.

Another problem is the constant internal failures of the program itself, which appear as operational errors when transferring data from one base to another. This has been very frustrating for students and faculty alike. We confronted this problem by providing alternatives. For example, by placing all data on an electronic page so that the course will be on two platforms: LS and the web. A group of Lotus and ITESM researchers are continually creating improved versions which are better adapted to our educational needs.

Another very serious situation is generated by logistics problems. To provide students with timely access to the databases, an on-line course requires distributed work that is well coordinated among those responsible for technology and the academics. Although all the campuses have a coordinator for each of these areas of the virtual courses, this operation is far from being the norm. These are new functions in a highly structured system that is difficult to change. Access to databases is still a complex process and this problem is far more notorious in LS than on the web. Students often look to the lecturers for a solution, who are presented with all kinds of problems that they cannot solve. However, lecturers should pay close attention to this matter during the first days of the course, and follow up and dedicate course time to solving these problems. We believe these critical situations in the virtual courses will be solved as students and faculties gain more experience in courses of this type, and as infrastructure becomes more ad hoc.

Learner-Centered Learning: The Changing Roles of Faculty and Students

The role of the faculty

An educational model centered on learning has an effect on new roles both for the teacher and the student. Many of the activities which used to be the faculty's responsibility are now shared with the student. From being an expert in knowledge transmitted to students through expositions—the faculty's dominant role for many years and the quickest and easiest way of offering students knowledge—the faculty has now taken on the role of the learning-facilitator and student guide. To make a comparison, in this new educational approach the students have become the players and the teacher the coach. Students therefore play a very active role and the teacher is expected to give constant attention to what is happening during the process, in order that they are able to guide students properly.

What do facilitator-teachers do?

Facilitators-teachers provide learning opportunities for students using adequate teaching support methodologies and techniques, as illustrated below:

- They offer the necessary resources and create the maximum facilities for quality learning; they prepare or define appropriate working environments. In group work, for example, they create a relaxed atmosphere of mutual respect so that students will feel at ease when giving their opinions.
- They facilitate effective communication between participants, listen carefully to discover whether and to what extent the agreed objectives have been met, and they continue to refocus the process as necessary.
- They help students to engage further within their own learning process. They invite students to make decisions, such as to select information relevant to the course, propose strategies to meet objectives, reflect and evaluate their own process and make proposals for improvement.
- They attempt to be always up-to-date on each student's activities and quality of work. They help to create and develop virtual collaborative groups from national and international institutions.
- They constantly update and evaluate the information to be offered to students, which require constant research and updating of their continually expanding area of knowledge.
- They establish and clarify the criteria and the levels of command of knowledge, skills, competencies and results they wish their students to achieve, helping them to overcome obstacles.
- They continually guide students, assisting them in their individual needs and organizing new options for new needs.
- They motivate students, challenging them to put their ideas into practice, questioning their values through discussions and debates, and encouraging them in their explorations and projects.

As students become more autonomous in their professional and personal development, they will gradually be able to carry out these functions together, making the group a true collaborative learning community.

The teacher takes on the role of programmer, process administrator, guide and controller. In this new relationship with the student, the faculty moves from "soloist" to companion, and becomes not so much the one who imparts knowledge and who evaluates the student's capacity to reproduce it, but rather, one who teaches students to construct their own learning experience.

Students in this educational model require certain cognitive skills, motivation and computer culture. The VU has a very diverse public, so we can talk about courses at different levels: basic, advanced and very advanced. The focus changes from presence and orientation on the part of the lecturer during the entire process, to more autonomy and self-regulation from the student.

The role of the student

In a learning process with these characteristics, the role of the student radically transforms. Students form part of a community of learners who participate in making decisions on course strategy, and establish work commitments to achieve this learning. The student, normally as an individual, seeks, selects and processes information to obtain results that he/she will share with the group in discussion and debate sessions. In these sessions students consolidate what they learned, eliminate any errors or prejudices formed, in the process they extend their vision and contribute to enriching their fellow students' results. This situation allows students to become aware of the importance of "others" as a valuable human resource on which they can count on, to do things better, thus departing from individualism, promoting social relations and making this a gratifying experience.

We have a great deal of information on the opinions of students who have experienced the VU process. In general, what they like best is the teamwork, declaring that it extends their individual vision and that they learn more.

Students' work in this educational model does not end with these activities; they continue with a learning process in which they have to apply what they have learned to reality, in order that they understand it better and more scientifically. This allows them to analyze ethical or scientific problems presented by society or companies operating in their field of interest. This process is performed by working in virtual groups with diverse strategies, possibly in cooperation with a company that presents a need, and the students contribute by developing a project to solve this need.

Alternatively, the students themselves identify an area requiring improvement or innovation through a process of investigation or field studies, that rigorously applies scientific research methods. In this case, the students have total responsibility for the project: they define their objectives; determine their strategies; delegate responsibilities; establish commitments and assess themselves through a permanent process of reflection on what they achieve and how they achieve it; and thus creating a portfolio of all the documents in which they collect evidence for the entire process.

A Teaching and Learning Network

In the previous paragraphs we discussed the educational model which will guide the VU's decisions and actions over the next few years. A basic variation resulting from this educational change is that the satellite will no longer be the primary technology around which all the distance courses academic activity revolves, but instead the focus will switch to the workstation and network learning. What makes network learning so thought-provoking is that in such a short time at our Institute, it has become not only the center of distance activity, but is also now included in all aspects of classroom education. Part of the allure of network learning lies in its capacity for asynchronous interaction regardless of location, which offers a wide range of possibilities for distance education. These advantages have also been taken into consideration for ITESM classroom education, to the extent that the educational change in the entire Institute, both classroom and virtual, must be supported by network learning.

The line between classroom education and the VU is becoming less clear. Some VU programs such as the Master's in Educational Technology with UBC have eliminated the satellite and are offering totally on-line courses. Some classroom lecturers use the network for up to 80% of the whole course. The application of technology per se does not constitute an educational process, but depends on the use the lecturer makes of it, which in turn is related to the agreed educational targets or goals for his/her course. Its applications can be as varied as the didactic processes themselves. The new educational model presented previously includes multiple processes, many of which not only become more efficient with the use of technologies, but also more enriched and extensive. Students who work on the network interact with their computer (desktop or workstation) in a variety of ways, determined by the nature of the task and the learning objective. Therefore, the use of technology presents several challenges and the questions that arise from its use are, according to Harasim, (1996):

- Which network technologies are appropriate?
- How can we integrate networking into the virtual course?
- How can we best teach and learn using a networking system?
- How can we transform the network into an effective educational environment?

What is the role of faculty in an on-line course?

The role of the lecturer in an on-line course is fundamental. In some courses in our experience, using the same technology but offered by different lecturers, student opinions varies from "I'll never take a course like that again" to "this course has been the best thing that has happened to me in my student life". An on-line course has fundamental requirements which must all be strictly observed if the course is to succeed.

The lecturer is fundamental in accomplishing these requirements. There are two very different stages in an on-line course during which the lecturer carries out different activities. In the first stage the lecturer designs and creates the work, his/her educational project. In the second the project is introduced and the lecturer supports and guides students, constantly monitoring the progress in light of the project's mission, insuring that what he/she wants the students to achieve is fulfilled in the best possible way.

When the faculty prepare course curricula, they clarify their educational intentions and objectives, select course content, determine the plan of action to be followed in activities and assignments, identify the technologies they will use in each of the activities, estimate the time and workload required for the student, and establish an evaluation system. They also explain to students why they have chosen some strategies over others, commenting on the benefits they will reap from these strategies. These comments have a very important motivating effect and a course with these characteristics lightens the student's workload.

All decisions the faculty make regarding the course are documented in an orderly, structured and integrated manner, on an electronic page that provides students a complete overview and lets them know where they stand. The faculty should choose a technological platform that allows them to establish links to other technological applications so students can carry out all the activities on-line. At the VU, we are using two types of platforms: the Internet and the Learning Space. In the former, the lecturer develops his/her own Web page and in the latter he/she only transfers the information into the Schedule space using software specified for this purpose.

Apart from the aforementioned data, the faculty should clearly express to students their expectations for the course, the technological equipment required, as well as most common problems students will face, along with strategies for overcoming them. It is advisable to provide addresses of people or areas that can help them in these cases, always trying to offer a reliable alternative if a solution cannot be found. This electronic page would be the obligatory entrance to any of the course's spaces.

Some faculty members also create a video to welcome students to the course, or digitize a photograph to personalize communication and generate a more personal atmosphere. This is a rich space in which the faculty can develop creatively in order that students can feel at ease and acquire confidence. It is also a space for the faculty to transmit an attitude of constant availability. From this electronic page, students can visit all the spaces they need to perform activities, including Web information, discussion or work groups, access to previous students' work, and classmates' and faculty profiles (to provide familiarity).

The possibilities for creating valuable spaces for students are vast. Designing an on-line course takes up a lot of the faculty's time, particularly time for reflection. It is not easy for a new faculty member to design a course with these characteristics, but it must be done. The success of the course depends greatly on this design.

Lecturers devote at least one semester to designing a course and they are provided with all the support, both technological and didactic, they require.

During the development of an on-line course, the role of the lecturer is very different from his/her role in traditional education. In an on-line course, the faculty must have the skills and attitudes required for their role as facilitators. They must organize interaction and pay close attention to everything that happens during the process. They must know how to guide a group without dominating it, not intervene with their opinion unless it becomes necessary, know how to create a framework that will guide students in their tasks without confusing them by indicating the directions step by step, always respond to and encourage contributions, not overload information with comments, frequently read the comments and send personal messages to the students that do not participate. In addition, they have to divide the class into small groups and know how to manage them, encourage reflection among students, make suggestions to improve the experience, help students to realize what works in the process and what can be improved, foster collegiality between students with close communication to maintain motivation. This requires an open and flexible attitude towards the group's contributions and suggestions, and know how to accept new points of view and options in several topics. They must also be able to apply a variety of learning models that help them to structure and sequence the process.

In summary, we can illustrate the various roles and activities of faculty and students by the following diagram.

	Technologies	Aplications	Teacher activities	Student activities
	Satellite SIR	• Updating • New research	• Experts in the discipline	• Sets the discipline in a national and international context
	Internet	• Develop cognitive skills • Analysis • Synthesis • Critical thinking	• Guide • Helps the information process	• Surfs • Investigates • Processes information
	Computer Conferencing	• Exchanging information • Professional development • Actions with common objectives • Problem solving • Social development	• Controls • Guides • Monitors • Evaluates	• Contributes to carrying out tasks jointly • Group decision taking • Evaluates results as a group
Networking	E-mail	• Personal Interaction • Advise • Personal Comunication • Feedback	• Advises • Evaluates • Guides	• Documents results • Seeks solutions • Interacts with the faculty and classmates

FIGURE 12.7
Different Technologies Used at ITESM and Faculty and Student Activities

Support for the ITESM Lecturer Facing Change

The VU is also committed to supporting the Institute in its own teachers' training needs. Just as we began this chapter by emphasizing the crucial role that higher education will play in the development of nations in the years to come, we will now finish by stressing the vital role played by the lecturer in this change. "The contribution of teachers and lecturers is essential to prepare young people, not only so that they can approach the future confidently, but also so that they can build the future themselves in a resolved and responsible manner. Education, beginning in primary and secondary school, should concentrate on facing these new challenges: contributing to development, and helping everyone to understand and master to a certain extent the phenomenon of globalization and favor social cohesion. Teachers are the necessary conditions for the success of formal teaching and permanent education" (UNESCO, 1996). This is a challenging situation we are going through at ITESM, as we make a transition from a traditional educational model to the new paradigm of learning centered on the student, aiming at forming excellent professionals and people with the skills and attitudes needed to successfully meet the changes created by the current global context. This requires lecturers to totally redefine their teaching practices.

Yet again many questions arose: What do our faculty need to change? Do they themselves know how to use the informatics and telecommunication media efficiently? How do they put this new role into practice? The learning processes that change implies and their administration through technology, is based on the development of the faculty. In order to facilitate this process, the Institute offers the Skills for Change Development Program to its faculty, focusing on technological and didactic skills, as well as emphasizing their integration. The axis of the program is the development of an educational project by the lecturer, which we have called "Course Redesigning". By redesign we understand the didactic, learning and work-culture changes that lecturers must make in their individual teaching approach, as a result of having incorporated this new educational paradigm we have been analyzing throughout this chapter.

The program consists of four stages that correspond to the stages of the preparation of the project. The first is aimed at all the Institute's lecturers and its purpose is to generate a positive attitude and commitment to change within the faculty community. This stage involves participating in group analysis, reflecting on social needs and their impact on education, which are facilitated by workshops and on-line courses by the VU.

After this stage, lecturers who are interested in initiating change formally register on their campus and continue with their training, which involves the next three stages: planning of a redesigned course, introduction and evaluation of a redesigned course, and transference and continuous improvement. This program is based on the same practice the lecturer will use to bring about change with the support of a facilitator acting as a critical companion in each of the stages. As the lecturer advances in the change process, he/she identifies diverse needs. Some are technological, such as knowing how to handle a specific tool or mastering a program for on-line documentation. Needs are probably more of the didactic kind, such as: organizing activities based on the problem solving technique, or preparing a network evaluation plan based on collaborative learning.

In order to respond to the faculty's multifarious needs, the program offers a series of workshops and courses related to the new technologies, as well as to the new ways of teaching and learning. All the workshops and courses are offered on-line and have a similar methodology: a theoretical basis, a technical direction and a practical application, and they are also based on reflection and collaborative work. When faculty members finish each of these activities, courses and workshops, they must prepare a document that expresses the impact of the workshop on their teaching in such a way that by the end of the program they will have a portfolio, providing evidence of their changes and the process followed. The ultimate objective is for the lecturer to include innovation and continuous improvement in his/her teaching as a work culture.

Other applications

The feedback from those who have participated in this educational change experience is that incorporating technology into learning and integrating them properly is making it easier for them to carry out the educational model ITESM has developed for the coming years. It would be difficult for the Institute to fulfill its mission without using these powerful electronic spaces. Technology is also having a strong impact on the Institute's relationships dynamics, facilitating the creation of a culture of participation and collaboration among all the faculty members on the different campuses.

Academies have been created throughout the Institute, organized in working teams united by electronic networks in which directors, faculty members and student affairs personnel from all campuses participate. These groups hold synchronous and asynchronous virtual meetings, contribute to discussions and reaching agreements directed at the continuous improvement of teaching methods and the curricula.

The entire ITESM community is invited to attend satellite meetings, where the President reports directly on the Institute's position, course innovations and future projects, and the faculty can clarify and express their opinions. The forums lead to a clear understanding of where we are going as a group, and generate an attitude of responsibility and commitment to the Institution. Hans Aebli (1991) says, "the light of clear thinking helps people to understand each other and work to be more fertile; in the darkness of the lack of clarity all kinds of prejudices and insincerity flourish".

We also have a database on the Web which is available to all faculty members, so that they can learn about the innovative educational projects developed by their colleagues. The e-mail addresses of the authors of these projects are recorded on this database, so that those interested can contact them directly.

Project information is complemented by annual satellite forums aimed at the exchange of experiences. The best examples of that year's educational, didactic changes and innovations are shared. These settings not only bring about change but also accelerate the process. We expect all classroom and VU courses to be redesigned by the year 2002.

Comments and Conclusions

The most visible problems in on-line learning are technological. Students complain that the servers frequently fail, that there are insufficient machines, the network is too slow and traffic is saturated. The most difficult challenges, however, are still educational. We are concerned by students' opinions such as 'this course had no academic relevance for me,' 'the work is rather impersonal' or 'my doubts were not settled.' The technological problems are easier to solve. Fifty-seven servers were made available to the campuses this year, compared with the two we had last year. Bandwidth, and therefore network speed, was increased. Students and faculty were able, through special loans and agreements with IBM, to acquire laptop computers. The didactic problems will take longer to solve but students' opinions, such as the following, encourage us:

> "Working on-line has made me change my study habits. I have something to do every day and you can't neglect your work."
>
> "It makes you responsible and prepares you for studying by yourself."
>
> "Assignments take a long time because you have to use your own initiative and apply your knowledge."
>
> "You have to do your homework to be able to participate in the discussion dynamics."
>
> "At first it's a lot of work, but then you get used to a constant rhythm."
>
> "We can publish our comments and questions to the whole group, getting rapid feedback."
>
> "I can participate in a space where I give my opinions and where I learn about my classmates' opinions and points of view on specific topics."
>
> "it makes it easier to interact with the other students and the lecturer."
>
> "It gives me access to highly valuable information on the Internet that I would not otherwise have had."
>
> "It broadens your knowledge and teaches you more about a subject."

We conducted an opinion poll of 2460 students throughout the system, in order to gauge the real impact of the use of technology on the teaching–learning process, and obtained the following data: an average of 60% of students thought that technology let them work in collaboration with their classmates to carry out course activities; an average of 60% also thought that they had been able to interact with the lecturer to obtain feedback on assignments and contributions; an average of 50% said that it gave access to information that would not have been available to them otherwise; and an average of 60% thought that working with technology was a rewarding experience. Nevertheless, teachers still face the following challenges:

- How to increase the lecturer's convening capacity and enhance his/her relationship with students.
- How to motivate and help students to think about and study a topic more deeply.
- How to adapt a model to the different levels and expectations of such a widely varied public.
- How to encourage socialization to avoid student isolation.
- How to integrate technologies to enrich and extend learning.
- How to encourage students to use the Internet to acquire information and knowledge out of a genuine intellectual interest—a desire to know and understand.

The ITESM's priority projects for this year will to continue supporting the VU, and to redesign the teaching–learning process. In spite of Mexico's current financial crisis, the President has just confirmed that the Institute will continue giving these projects priority and that the heavy investment will be aimed at infrastructure for creating virtual spaces without increasing costs for students.

References

Aebli, H. (1991) *Factores de la Enseñanza que Favorecen el Aprendizaje Autónomo*. Narcea.

Bates. A., (1995) *Technology, Open Learning and Distance Education*. Routledges.

Dolence, M. G., & Norris, D. M. (1995) *Transforming higher education: A vision for learning in the 21st century*. Society for College and University Planning.

Forman , D. C. (1995) "The use of multimedia technology for training in business and industry". *Multimedia Monitor,* **13**(7), pp. 22–27.

Harasim, L. & Colbrs. (1996). *Learning Networks*. The MIT Press.

López del Puerto, P. & Reyes, A., (1998) *Education for technology transfer in Latin América Countries. The case of the ITESM.México.*

O'Banion, T., (1997) *A Learning College for the 21st Century*. ACE.ORIX Press.

Oblinger, D. & Rush, S. (1997) *The Learning Revolution*. Anker Publishing Company, Inc.

Sartori, G. (1979) *Homo videns. La Sociedad Teledirigida*. Taurus.

Torres, J. (1996) *Globalización e Interdisciplinariedad: el Curriculum integrado*. Ediciones Morata.

UNESCO. Informe (1996) *La Educación Encierra un Tesoro*. Santillana & UNESCO.

Related sites:

- UBC.
- Homepage of Monterrey Institute of Technology
  http://www.mty.itesm.mx/

- Homepage of the ITESM Virtual University
  http://www.ruv.itesm.mx/
- Homepage for the Redesigning the Teacher-Learning Act Program
  http://cc.viti.itesm.mx/rediseno/rediseno.nsf
- Homepage for Learning Space Description
  http://www.mty.itesm.mx/dinf/dit/si/public.htm
- Homepage for Export Marketing to Mexico and France
  http://www.ruv.itesm.mx/cursos/pgade/sep98/mt289/

13

Distance Higher Education and a New Trend of Virtual Universities in Asia

AYA YOSHIDA

Introduction

In this chapter we will examine the current situation of the so-called "virtual universities" in a number of Asian countries. Although there are many definitions of the term "virtual university", I will define it very broadly here as computer-based instruction for distance higher education, or computer-based distance higher education which employs information technology for course delivery, including two-way virtual communication.

We will examine the following four points:

- An overview of the current state of distance higher education in Asia as a background for understanding the developing concept of virtual universities. We'll look at the major distance higher education institutions in Asian countries and examine them in terms of governance, size, cost, educational programs, student profile, and instructional technology used.
  1. An examination of some recent efforts to introduce technology into the institutions mentioned above, and to examine the purpose and effect of introducing new technology.
  2. An examination of the governance, cost, educational programs, student profiles, instructional technology, etc. of some virtual universities and projects.
  3. A discussion of some new trends in virtual universities in some Asian countries. I do not evaluate these trends because they are so new that it is almost impossible to predict their development in the near future. Although the situation is in a state of flux, I will attempt to determine how things might develop in the near future. Asian virtual universities share some

characteristics with both current distance higher education and conventional institutions of higher education.

Virtual universities within Asia will be compared with each other, but not with similar institutions elsewhere. Useful data from other regions is easily available, and, using my information, readers will be able to compare Asian institutions to those in other parts of the world.

Distance Higher Education in Asia

Open universities

Distance higher education has emerged as one of the most feasible modes of instruction. It has developed rapidly, especially in Asia, where it plays an important role in increasing access to higher education. Higher education opportunities have always been limited in developing Asian countries, while the demand has increased rapidly within the last two decades due to economic development and democratization. Distance higher education in Asia is emerging to bridge this gap.

TABLE 13.1
The List of Major Distance Higher Education Institutions in Asia

	Country	Name of Institution	Year-established	Governance
1.	Bangladesh	Bangladesh Open University	1985	National
2.	China	China Central Radio and TV University	1979	National
3.	Hong Kong	Open University of Hong Kong	1989	Private[1]
4.	India	Indira Gandhi National Open University	1985	National
5.	Indonesia	Universitas Terubka	1984	National
6.	Japan	University of the Air	1983	The University of the Air Foundation[2]
7.	Korea	Korea National Open University	1972[3]	National
8.	Malaysia	The Centre for Off Campus Studies, Universiti Sains Malaysia[4]	1971	National
9.	Myanmer	University of Distance Education	1975[5]	National
10.	Pakistan	Allama Iqubal Open University	1974	National
11.	Philippines	University of the Philippines Open University[6]	1995	National

12.	Singapore	Singapore Open University	1992	National
13.	Sri Lanka	Open University of Sri Lanka	1980	National
14.	Taiwan	National Open University	1986	National
15.	Thailand	Sukhothai Thammathirat Open University	1978	National
16.	Vietnam	Vietnam National Institute of Open Learning	1968[7]	National

Sources: ICDL database (1998) and NIME and UNESCO (1994)

Notes:

1. Since 1993/4 it has become a self-financing institution.
2. The Foundation administers and runs the University which is mainly financed by the government
3. The Korea Air and Correspondence University was established this year as a branch school of Seoul National University, offering 2-year junior college courses. In 1981, it had grown to a 5-year university
4. It is one unit of the University Sains Malaysia
5. The University Correspondence Course was established this year and in 1992 it was reorganized as the University of Distance Education
6. It is one of six autonomous units of the University of the Philippines System.
7. In 1968 an institution in Hanoi employing correspondence techniques was established, and it was converted into the Vietnam Institute of Open Learning

What are the characteristics of Asian distance higher education? Based on NIME's survey (NIME, 1992, 1994) and the UK Open University's ICDL database (ICDL, 1998)—shown in Table 13.1—I will identify the major distance higher education institutions and examine their governance, size, cost, educational programs, student profile, and instructional technology for instruction.

We see three major characteristics. First, these institutions are relatively young, with most having been founded in the 1980s or 1990s. Although some were established in the 1970s, they were not necessarily independent institutions at that time, but became autonomous in the 1980s. An institution in Hanoi, Vietnam, for example, began as a correspondence program in 1968 and later became the Vietnam National Institute of Open Learning. In another case, the Korea Air and Correspondence University, offering two-year junior college program, was established in 1972 as a branch of Seoul National University, offering a two-year junior college program. It became a full, degree-granting university in 1981.

Secondly, many of these institutions are known as "open universities," and most seem to have been modeled on the Open University in the United Kingdom. They reflect not only the name but also the mission of this UK institution. "Open" implies that all applicants to the university are admitted. Some institutions, however, do require applicants to pass an entrance exam or to possess certain academic qualifications. China's TV Universities, for example, requires applicants to pass an entrance examination. The Open University Degree Program in Singapore is open only to people who have a minimum of two "A" Levels.

Nevertheless, open universities in Asia were intended to increase access to higher education, and were considered a feasible method for institutions to meet the increasing demands for higher education. The important point is that open universities are autonomous, independent institutions, which provide higher education for a large population at a relatively low cost. I will examine this point in detail later.

Governance

Third, almost all institutions have been established, administered, and financed by national governments. Governments throughout Asia have traditionally played a strong role in maintaining the country's higher education system. Many Asian governments are likely to use distance education as a tool for developing educational opportunity, as well as controlling the quality of education.

These two roles are clearly stated in the case of Universitas Terubka in Indonesia. According to the institution's stated purpose, it is "not only responsible for expanding educational opportunity, but also has the responsibility to strengthen the government's commitment to improving the quality of education and to make education more relevant to national development needs (ICDL, 1988)."

Since open universities have been financially supported by their respective governments, it has been possible for them to grow quickly, although they are subject to government regulation.

Cost-effectiveness

One of most important effects has been the expansion of students' educational opportunities in inexpensive ways. Distance higher education has been very cost-effective compared with conventional on-campus universities.

As shown in Table 13.2, the cost per student of some Asian institutions is low. At the Korea National Open University, for example, the cost per student is only 5% of that of the average on-campus universities in Korea. Although this might be an extreme example, other cases show a range from 15% to 40% savings.

TABLE 13.2
Cost of Institutions and Student Profile

	Name of Institution	Unit cost*	Graduation rate**	Enroll-ment	Age distribution −21	21–30	31–40	41–
1.	Bangladesh Open University			1,380,002				
2.	China Central Radio and TV University	40	84	5,300,002				
3.	Open University of Hong Kong			150,001)	12.0	66.0	26.0	2.0
4.	Indira Gandhi National Open University	35	10.2	4308,328)				
5.	Universitas Terubka	15	25.5	3530,005)	3.0	41.2	41.2	13.7
6.	University of the Air			680,006)	0.0	6.7	29.4	41.1
7.	Korea National Open University	5	11.0	2,157,884	7.7	59.4	27.0	5.6
8.	Universiti Sains Malaysia			55,006)	0.1	29.0	70.0	0.0
9.	University of Distance Education			1,980,005				
10.	Allama Iqubal Open University			790,671)	7.0	15.0	25.0	53.0
11.	University of the Phil-ippines Open University			14,396)				
12.	Singapore Open University							
13.	Open University of Sri Lanka			200,006)	2.1	50.0	30.0	18.0
14.	National Open University							
15.	Sukhothai Thammath-irat Open University	30	12.2	2,168,003	8.0	61.9	20.4	9.7
16.	Vietnam National Institute of Open Learning			484,805)				

Sources: ICDL database (1998), NIME and UNESCO (1994), and Daniel (1996)
Notes:
* Unit cost per student as percentage of average for other universities in the country (Daniel, 1996, p.31)
** numbers of graduates per year/ numbers of annual intake
Enrollment: 1) 1993, 2) 1994, 3) 1995, 4) 1996, 5) 1997, 6) 1998

Asian open universities, therefore, can educate two or more students at the same cost as educating one student at conventional universities. This form of distance education does not require campus buildings and other facilities for classroom instruction, and this is the primary reason costs are so low. Also, because physical space is not an issue these universities can accept a large number of students, which again lowers the cost per student.

One indication of any educational institution's effectiveness is its graduation rate. It is often pointed out that dropout ratio is high in distance education because, for most students, learning is a solitary activity. As shown in Table 13.2, the graduation ratio (i.e. the number of graduates per year/number of annual admissions) is, in most cases, approximately 10%. China is the exception because it requires a strict entrance exam to ensure high-quality students, and most students are on leave from their jobs and are required to graduate within a specified period of time.

Most distance institutions strive to keep their degree requirements equivalent to those of traditional on-campus universities. Indonesia's Universitas Terubuka states that "an UT degree is academically equal to one of any Indonesian State university. (ICDL, 1998.)" As a national body, Indira Gandhi National Open University in India has a mandate for the promotion, coordination and maintenance of standards in distance teaching in all Indian universities. Nevertheless, it is difficult to investigate this point. Degrees awarded by open universities may be seen as less valuable than those from conventional universities, even though they are considered to be equivalent. Cases in Thailand and China illustrate this situation.

However, there could be another way to measure the value of degrees earned through distance education, and this is through perceived ability in the workplace. Working adults who have earned degrees, even in open universities, may be promoted because of the self-confidence have developed while pursuing their degree (Yoshida 1991, Yoshida, et al., 1995).

Student profile

As Table 13.2 shows, open universities have huge enrollments. Some institutions with over 100,000 students in degree programs are called "mega-universities" (Daniel, 1996, p. 30) According to Daniel, the characteristics of "mega-universities" is not their size, but that they are primarily new institutions which have become the largest universities in their national higher education system. This is one of characteristics of open universities in Asia.

Most students are in their twenties or thirties (Table 13.2). Although some institutions (i.e. Singapore, Taiwan and Thailand) require applicants to meet age requirements or have work experience, most require only that applicants have completed secondary school to enter regular courses. This means open universities admit college high school graduates as well as working adults, and Asian open universities are likely to admit significant numbers of high school graduates (Daniel, 1996, p. 37). In Table 13.2 we see this is the case in Hong Kong, Korea, Pakistan and Thailand.

The profiles of these students' differ completely from those in conventional universities. Open universities not only offer working adults a second chance to earn a degree, but they also provide access to lifelong educational opportunities. In developing Asian countries the opportunity for traditional higher education has generally been limited to especially promising youth, or the children of the elite. For those who missed the opportunity to enter traditional higher education, open universities offer another chance to earn a degree. For working adults wishing to develop their knowledge and skills, open universities have been the place for recurrent, lifelong learning. These characteristics are closely related to educational programs offered.

Educational programs

As shown in Table 13.3, the types of curricula and certificates offered make open universities distinctive.

TABLE 13.3
Educational Program

		Educational Program				Curriculum	
	Name of Institution	Diploma/ certificate	First degree	Bach- elor	Post- graduate	Liberal Arts	Voca- tional
1.	Bangladesh Open University	*		*			*
2.	China Central Radio and TV University	*	*	*		*	*
3.	Open University of Hong Kong	*	*	*	*	*	*
4.	Indira Gandhi National Open University	*		*	*	*	*
5.	Universitas Terubka	*		*		*	*
6.	University of the Air			*		*	
7.	Korea National Open University	*	*	*		*	*
8.	Universiti Sains Malaysia			*		*	*
9.	University of Distance Education			*		*	
10.	Allama Iqubal Open University	*		*	*		

	Name of Institution	Educational Program				Curriculum	
		Diploma/ certificate	First degree	Bach- elor	Post- graduate	Liberal Arts	Voca- tional
11.	University of the Philip- pines Open University	*	*	*	*		*
12.	Singapore Open University	*		*		*	*
13.	Open University of Sri Lanka	*		*	*		*
14.	National Open University	*				*	*
15.	Sukhothai Thammathirat Open University	*		*	*	*	*
16.	Vietnam National Insti- tute of Open Learning	*				*	*

Sources: ICDL database (1998) and NIME and UNESCO (1994)

Most institutions offer both degree programs and continuing education programs. Degree programs, an alternative to conventional higher education, offer a wide range of diplomas and certificates, including postgraduate degrees. The continuing education program attempts to address the social and economic demands for skills-oriented education. Open universities provide many kinds of educational programs—from liberal arts to vocational training—in a single institution.

Most institutions offer both liberal arts and vocational/technical education curricula. Here, the various programs are classified using the Carnegie Classification. Accordingly, "liberal arts" are courses of study which do not relate directly to a specific occupation. In open universities, economics and language are popular liberal arts programs, and courses as a whole tend to be of a practical or applied nature—a trend which seems to be increasing. Since 1990 the IGNOU, for example, has modified its curricula away from liberal arts and toward applied subjects (Daniel, 1996).

Course delivery technology

As shown in Table 13.4, most institutions in Asia use printed materials, broadcasting, and audio/visual cassettes as the primary methods of delivering course material.

TABLE 13.4
Technology of Course Delivery

		Technology					
	Name of Institution	Text	Radio/ TV	AV cassette	Satellite	Tele-confer-ence	Computer tele-communi-cation
1.	Bangladesh Open University	*	*	*			
2.	China Central Radio and TV University	*	*				
3.	Open University of Hong Kong	*	*				
4.	Indira Gandhi National Open University	*	*	*	*		
5.	Universitas Terubka	*	*				
6.	University of the Air	*	*	*	*		
7.	Korea National Open University	*	*	*			*
8.	Univrsiti Sains Malaysia	*		*		*	
9.	University of Distance Education	*	*				
10.	Allama Iqubal Open University	*	*				
11.	University of the Philip-pines Open University	*	*	*			*(under-way)
12.	Singapore Open University	*		*			
13.	Open University of Sri Lanka	*		*			
14.	National Open University	*	*				
15.	Sukhothai Thammath-irat Open University	*	*	*			
16.	Vietnam National Insti-tute of Open Learning	*	*	*			

Sources: ICDL database (1998) and NIME and UNESCO (1994)

Distance education in Asia has traditionally relied on printed texts for delivery of course instructional material because courses can be delivered inexpensively via the national postal systems. Broadcasting is useful to disseminate courses to many people in audio and visual form.

In terms of equity, these two technologies are also effective in Asia. Students entering open universities are often from lower-income groups, and these technologies enable the general population to access higher education. A major drawback, however, is that there is only one-way instructor/student communication. Due technological limitations, it takes time for students to receive individual responses from tutors.

Open universities usually offer local face-to-face tutorials and /or counseling services to foster two-way communication, and this improves the education students receive.

New technologies in Asian open universities

Some open universities have recently introduced new technologies to foster two-way communications. As shown in Table 13.4, both Japan and India use satellites. Japan's University of the Air has used a telecommunications satellite since 1998 for digital broadcast, which supplements standard radio and TV broadcasts. This system, however, does not offer two-way communication.

In India, satellites are used differently. Indira Gandhi National Open University and University Grants Commission use a joint initiative called INTEND-OPENET Program for employing information technology in higher education. The technical system consists of a national hub at the IGNOU headquarters that can control and monitor the use of about 10,000 VSATs of various types. Under this system two types of two-way communications could be carried out: 1) interactive television, and 2) computer communications. This system has just started and is not used for daily instruction. Although the current delivery system depends on printed material, postal and telephone services, this will change in the near future when the new system establishing Internet-based virtual classrooms is introduced (IGNOU, 1998a, b, c).

Beginning in 1987 the Universiti Sains Malaysia has offered tele-tutorials through audio conferencing, and the system was expanded in 1989 to include audiographic facilities. A full-motion video conferencing system was introduced in 1995 to replace the audiographic teleconferencing system. This system enables two-way communication between instructors based at the Universiti Sains Malaysia campus and distance education students at their respective regional centers.

The Korea National Open University was the first university in Korea to develop CD-ROM titles or Internet courseware for degree-seeking students, although the primary instructional method is the broadcast lecture. These new technologies are employed as a supplements to the courses "Introduction to the Computer," "Korean Customs," and "English Phonetics." KNOU has also introduced a video conferencing system, which links regional study centers and is used for course instruction, special lectures and student orientation. This system makes two-way communication between students, faculty and staff possible.

The Open University of Hong Kong also uses some computer technology for course delivery. It develops a CD-ROM, and 24% of the university courses have this material. Currently, pilot studies on using the Internet for teaching are being conducted.

The University of the Philippines' Open University also plans to introduce computer networks for distance education.

Complications with using technology

Information technology is not yet used widely for instruction or course delivery in Asian open universities. Although many institutions are trying to introduce and employ IT for delivering courses, these efforts are currently in the experimental or supplementary stages. A major reason for this lag is that most countries still do not have a nationwide IT infrastructure, and the cost of setting up a network for educational purposes only is prohibitive for the institutions.

Even where an IT infrastructure is established, many potential students have no access to computers at home, and few can use their workplace computers for personal use. Since it is almost impossible for many students to access computers needed for higher education at an open university, the number able to participate may actually be limited. This contradicts with the mission of "open" universities to increase educational opportunities.

Information technology makes easy two-way communication through computer networks possible. However, can this technology replace the face-to-face communication of a traditional classroom or tutorial at local study centers, and can it serve the huge number of students at "mega-universities?"

Even with their large student bodies, open universities have tried to maintain academic standards through schooling or tutorials. Whether open universities can continue to survive through the use of "computer-networked education" remains to be seen.

Because of this, the effects of introducing information technology into open universities remain unknown. It is less likely that today's open universities will be transformed into virtual universities. In Asian countries, virtual universities have emerged as a distinctly different type of distance education than open universities.

Virtual Universities in Asia

Virtual universities offer degrees through online instruction, and these differ from conventional universities, which offer online courses as part of a more traditional approach to education. The former has begun emerging as new types of open universities.

As far as educational programs and governing structure, virtual universities in Asia can be classified into four types:

1 The Government Initiative Model operates under the auspices of a national government which has the resources to develop integrated infrastructures of information technology, and upon which virtual universities are established. These are most commonly seen in Malaysia, Singapore and Korea.
2 The "Joint Venture Model" virtual universities are established by organizations in collaboration with overseas universities. These are most common in China and Hong Kong.
3 The "International Cooperation Model", third type is that an international cooperation sponsors a project to establish a virtual university. The APEC Project is an example of this model.
4 The "Satellite Model" uses satellites to connect campuses and is the case in Japan.

Government initiative model – Malaysia

Multimedia super corridor

The Malaysia government is establishing an integrated environment of new information technology called Multimedia Super Corridor. It has a new physical infrastructure and is 15 kilometers wide and 50 kilometers long, located between Kuala Lumpur and the new international airport. Its main sites are Cyberjaya and Putrajaya. Cyberjaya is becoming an "intelligent city" with multimedia industries, R&D centers, a Multimedia University, etc. Putrayaya is the new government and administrative capital.

The government decided that the following seven clusters were to be developed via IT: electronic government, multipurpose card, smart schools, telemedicine, research and development cluster, worldwide manufacturing webs, borderless marketing. The government recruited business and commercial companies to invest some of these clusters.

In the research and development cluster, the establishment of a multimedia university in Cyberjaya is involvement. Based on this MSC project, businesses and commercial concerns are planning the following virtual universities.

Universiti Telekom (multimedia university)

Established by Telekom Malaysia, the country's government-owned telecommunications provider, Universiti Telekom is Malaysia's first government-approved private university. Malaysia's Ministry of Education granted Telekom Malaysia permission to set up the university in 1996. This institution was previously known as Institut Telekommunikasi dan Teknologi Maklumat, organized in 1993. In 1997 Telekom Malaysia was given the task of setting up Universiti Multimedia, which will be located at Cyberjaya in the MSC. It will commence operation in the MSC in September, 1998, focusing on telecommunications, multimedia, computers, creative digital art, information technology, software and management, and the like. The university currently is organized into the Faculty of Engineering, the Faculty of Information Technology, the Faculty of Management, and the Faculty of Media Arts and Sciences, all of which offer three-year programs leading to degrees in their respective fields.

Universiti Telekom also offers a distance education program leading to Bachelor of Management with Honors. Students can pursue a degree without needing to attend traditional classes. Instead, they use state-of-the-art technology—including the Internet, the World Wide Web, CD-ROM, and video conferencing. The program consists of three main components:

1 On-line access to educational material and assignments
2 Electronic discussions or Tele-tutorials, whereby students interact with instructors.
3 Electronic discussions whereby students interact and hold discussions with other classmates.

A minimum of three years is necessary to complete this distance-education program, although students may take up to six years.

Universiti Tun Abdul Razak (UNITAR)

UNITAR, Malaysia's first virtual university, opened in May 1998. It was established as a private institution by KUB Malaysia Berhad in response to a government request. UNITAR's educational program focuses on business administration and information technology—awarding certificates, and bachelor's and MBA diplomas in each field.

The educational process at UNITAR involves course work which utilizes multimedia technology, and interaction with tutors. UNITAR also plans study centers linked to UNITARnet, and an advanced networking system centrally administered from Cyberjaya's Multimedia Super Corridor. Learning resources include independent study, online interaction and virtual classrooms, homework assignments, and study teams. These are conducted in virtual environments including CD-ROM, online discussions, e-mail, bulletin board systems, etc. Students will be required to participate in university-sponsored and community activities such as sports, integrated groups, and the arts. These real-life activities supplement the virtual education.

The university will accept 14,000 students and has 12 study centers, each with the ability to serve 300 students per semester. Every student will spend the first two semesters (eight months) at one of these study centers.

Universiti Putra Malaysia and Universiti Teknologi Malaysia

These are large Malaysian conventional universities and leaders in science and technology research and development. The majority of their students are in on-campus degree programs. It is planned to link them with the MSC to develop the information technology environment.

Universiti Putra Malaysia's Cyber Creative Lab is charged with multimedia training and production. It is operated on a core multi-disciplinary approach involving biology, engineering, computer science, economics, human ecology and biotechnology.

Univeristi Teknologi Malaysia has Cyber Campus Project is intended to link to the MSC in order to develop a multimedia environment campus-wide. Malaysia's Multimedia Super Corridor is an epoch-making initiative of the Malaysia government for establishing a nationwide infrastructure of information technology.

Along with these efforts, all educational experiences at virtual universities can be accomplished in a virtual environment—students are able to earn degrees without any face-to-face communication. Although these are currently at project levels, today's conventional teaching model may change in a few years. Curricula of these universities focus on management and computer science, and these features differ from those of open universities.

Singapore

The IT2000 Vision

The *IT2000 – A Vision of an Intelligent Island* project was begun in 1991 by the National Computer Board in partnership with more than 200 senior executives from eleven major economic sectors of Singapore. It aims to create an advanced nationwide information infrastructure which will interconnect computers in virtually every home, office, school, and factory. A major milestone in realizing this vision is Singapore ONE, a national initiative to deliver new levels of interactive, multimedia applications and services to homes, businesses and schools throughout Singapore using high bandwidth broadband technology. Domestic and multinational companies participate in the project to deliver information technology training courses.

Two other programs have recently been started by academic institutions—the *Virtual College* of the Singapore Polytechnic and the *Online Learning Environment* of Temasek Polytechnic.

Virtual College of Singapore Polytechnic University

Singapore Polytechnic's Virtual College began in 1997 and currently offers a postgraduate program, as well as short courses. When a student registers for a course, s/he receives a password allowing entrance to that course's homepage. Each student is assigned an instructor who also teaches the same module at the school's conventional campus, as well as a tutor for additional help. The students and tutors communicate via telephone, bulletin boards and e-mail. On-campus tutorials are available for students in the postgraduate program, while students in any other courses communicate with tutors and classmates online. Most courses are in the fields of multimedia and business, with many university departments offering courses. Students take written examinations after completing all the virtual lectures and course work, and testing is conducted on-site at the Singapore Polytechnic

Except for one postgraduate program, the virtual college does not currently award diplomas. Most programs are of a few months' duration or less. We do not regard this institution as a degree-granting institution at this time, but it has the potential to become a real college in the future.

Online Learning Environment of Temasek Polytechnic

The Online Learning Environment began in 1998. Originally proposed by the school's principal in early 1997, it was connected with Singapore ONE in late 1997 after a year-long experiment. This effort offers computer-based learning courses for both students of Temasek Polytechnic and the public who have registered as members.

There are two types of courses offered via the Internet and Singapore ONE: Internal Courses, primarily in business and engineering, which are open to Temasak Polytechnic students only. External Courses, open to the general public, include: Introduction to the Law in Singapore, Intercultural Communication, and AutoCAD.

Courses are of two types: Self-Learning and Instructor-Led. In the former, students log on to course work at any time. In the latter, laboratory work or classroom activities are also required. Students are able to meet with the course instructor in person, although usual communication is done online.

Virtual learning has also just begun in Singapore, and other conventional universities plan to connect with Singapore ONE to develop on-campus information technology. Independent virtual universities in Singapore do not seem to exist at this time.

Like the MSC in Malaysia, Singapore ONE is a key factor in the realization of the IT2000 Vision to develop information technology, and in the subsequent development of a virtual university in Singapore. Both MSC and IT2000 are government-initiated programs intended to extend computer network nationwide.

Korea

CAIS, Virtual University

To date, no virtual university has been established in Korea. However, plans to do so have been underway for several years. The Korea Advanced Institute of Science & Technology (KAIST), which has been spearheading the development of scientific technologies in Korea since 1971, leads this effort. KAIST intends to build an Intelligent Campus in order to make itself a world-class research oriented university. To this end it also established the Center for Advanced Information System (CAIS) in 1993.

CAIS managed to develop the Intelligent Campus Project from 1994 to 1996 and built all infrastructure necessary for the Virtual University. With this infrastructure, CAIS launched the three-year Virtual University Project in 1997.

Supporting these efforts is Korea's strong desire to catch up with the integrated campus information systems of developed countries. Although Seoul National University, Inha University, and other universities have on-campus information systems, the technological level is far inferior to the state-of-the-art networks in the US and other developed countries. This project, therefore, was proposed to introduce, evaluate and apply the world's most advanced technologies to the process of information systems embodiment. The Koreans are modeling their systems after MIT's Athena Project and Carnegie Mellon University's Andrew Project.

The goal of the Virtual University Project is to establish infrastructures based on systems such as the Electronic Teaching Assistant System, the Integrated Workflow System and the Distance Learning System, which will foster both virtual educational and administrative activities. Electronic conferencing, virtual laboratories, lectures-on-demand, and virtual classrooms and universities are some of the activities being developed.

In Korea, the virtual university project is underway and seems to be modeled after systems in US universities. There seems to be no preconceived ideas of how educational programs and course-delivery systems will be organized. However, since infrastructures and systems are prerequisites for virtual universities, this project may eventually lead to the establishment of true virtual universities in Korea.

Korea Multimedia Education Center (KMEC)

The Korean Multimedia Education Center was established in 1998, originally as a division of the Korean Educational Development Institution based on the recommendation of the Ministry of Education. In 1977 this institute was attached to the Educational Broadcasting System. The aim of this institute was to support the extensive use of information technology in school education, as well as in and life-long- and open-learning situations. To realize this goal, it is now creating the Cyber Learning System through EDUNET, an IT educational network. The system's major objectives in 1998 are to develop: 1) electronic texts for middle schools, 2) educational software to guide multimedia education, and 3) a dictionary database and a cyber-teacher support system.

This institute currently focuses on middle schools to promote information technology literacy in both students and teachers. In the near future, another primary objective might be to introduce information technology in higher education, and to create a virtual university.

Joint venture model – Hong Kong

OnLine education

Founded in 1993 to create new distance-learning opportunities for working professionals, OnLine Education is the first educational organization in Hong Kong to offer online courses of study leading to degrees. It has partnered with the following universities in the United Kingdom: Oxford Brookes University, the University of Paisley, the University of Lincolnshire & Humberside, and Charles Stuart University. These universities deliver courses and grant awards to OnLine Education students. The targeted students are working professionals who are familiar with computers and want additional training and/or academic degrees for their jobs.

Given the working nature of the student body, most educational programs offer MBA degrees, although Bachelors in Business Administration Management and a Bachelors of Science in Health Studies are also offered. Many of the courses presume three or more years of relevant work experience.

Students communicate with tutors, support stuff, and other students via e-mail, bulletin boards, teleconferencing, and the file library. Although classroom attendance is not required, informal face-to face meeting with tutors is possible. When finishing course work, students sit for examinations in Hong Kong.

This virtual university separates educational and administrative functions. Degrees are awarded by the UK universities based on their educational programs, while the Hong Kong office handles tutorials and administrative duties. Even though the university is located in Hong Kong, as are most of its students, the educational programs come from overseas universities. This is the major organizing characteristic of this type of institution.

China

China USA Business University (CUBU)

This is China's first independent university which has established partnership agreements with the US universities. Through these partnerships, CUBU has offered courses leading to an MBA degree since 1997, with educational programs approved by the Beijing Commission and filed with the State Education Commission.

The Chinese partner organization is Beijing Construction University, a non-governmental, full-time institution with degree programs accredited by the State Education Commission. The curriculum was developed in collaboration with

Newport University in the Unites States. Newport University has many study centers in 29 countries throughout the world, and is accredited by the Council for Private Postsecondary and Vocational Education.

CUBU's curriculum focuses on training senior- and middle-level managers and personnel in international business. All courses are part-time and can be completed within two years. Course work is self-paced through a combination of distance learning using independent directed study and faculty tutoring. While classes meet on weekday evenings and weekends for those students who reside in the Beijing area, all courses are delivered by computer to students living elsewhere. Because courses are taught in English, applicants need to pass an English proficiency test, and must also have a college degree or equivalent with a minimum of two years' work experience.

CUBU is negotiating with other universities in the US to provide additional courses in real estate management, rural/ urban planning and development, international business and computer science.

Hunan Multimedia Information Education College

The Hunan Multimedia Information Education College (HMIEC) began in 1997 with an agreement between Hunan University and the Hunan Province Post and Telecommunications Administration to explore ways to set standards for distance learning procedures, administration methods, technologies and market-demand pricing. This college was affiliated with Hunan University in the setting up of the Campus Network Project. HMIEC is the first online college in China which grants diplomas after three years of study. The majors include computer applications, construction engineering, and business administration, and the courses will be delivered via the China Internet.

This college has agreements with more than 10 universities in China and plans to enroll about 3000 Internet students. This college also has an agreement with Western Governors University in the US, which extends its ability to reach new markets with their distance learning courses and degree programs. They seek to develop international and joint delivery of courses, as well as joint ventures, consortia and other cooperative efforts.

Thus, while Hong Kong and China have started virtual universities as joint ventures, the "university" is mainly responsible for administration, while overseas universities provide educational programs and curricula. This method of establishing virtual universities is common in the US In China and Hong Kong, however, educational programs are not necessarily developed domestically.

Educational programs focus on business administration and information technology because both fields are designed to meet the new demands of working adults. Most students seek MBA degrees or business training because it is related to their occupations, and it costs less to study in a foreign-based program via computer while remaining in their own countries.

Information technology is very useful for delivering courses from overseas universities to students remaining in their home countries. Technology may make virtual classroom environments possible, but two-way communication may be done only virtually. Since courses are provided in English, a certain level of English proficiency is prerequisite for enrolment.

International cooperation model

Virtual University of the Asia Pacific (VUAP)

The Virtual University of the Asia Pacific will begin under the sponsorship of the APEC Education Foundation, a private funding organization affiliated with the Asia Pacific Economic Cooperation. VUAP has been developed to meet corporate training and academic educational needs of students worldwide. Courses and programs are to be delivered via the Internet.

VUAP is currently seeking educational providers for the creation of its first set of professional development modules. Representative curricular concentrations for professional development are: 1) business communication, 2) technology in the workplace, 3) multicultural workplaces, and 4) multinational and multicultural marketing. All modules must be completed in no more than six months from the date of registration. VUAP will focus on professional development modules, and certificates are currently being negotiated with individual educational providers. Online academic advising and tutoring service will be available. Further advanced tutoring and translation services will also available for an additional fee (Oshima, 1997).

The courses are open to all who have completed secondary education and demonstrated English language proficiency, as all courses are currently offered in English. However, this will only be the case until learning modules in other languages are developed.

This virtual university is now at the project planning stage and is currently looking for educational providers who offer specific professional-development courses. The VUAP will be an international consortium in which providers from all over the world may participate. Although it is expected that educational providers will be primarily universities, private companies and other organizations may become project providers. Accordingly, this type of virtual university differs from the conventional university concept in which faculty are charged with teaching students and evaluating their achievements. Many types of organizations, including conventional universities, private companies, non-university research institutions, etc., provide courses under the auspices of this university. A key factor in any of these virtual universities is the finding of ways to integrate the wide variety of courses, especially if any degree programs are offered.

Satellite delivery model – Japan

Japan is one of the most advanced countries in Asia, and it has a developed technology infrastructure. However, information technology is not widely employed in distance higher education. There is little demand for distance education itself in Japan and there has been little effort to establish a virtual university, although highly-developed computer networks are available on campus. In this network, satellite campuses are connected to the main campus via satellite.

Space Collaboration System (SCS)

The Space Collaboration System is an inter-university satellite network designed especially for the exchange of video among universities, colleges, and national institutes. It is financed by the Ministry of Education and has operated since 1996. Stations at 89 sites in 73 institutions, including private universities, were connected by 1997. The National Institute of Multimedia Education—one of these national inter-university institutions—operates the HUB station and controls the VSAT stations in other institutions. NIME is responsible for providing satellite channels and supporting various events or activities using SCS. The SCS system makes two-way synchronous communication possible and may be used for a wide variety of purposes: joint classes, seminars, symposia, academic and study meetings, etc.

Because of Ministry of Education restrictions, SCS cannot be used for exchanging credits at the present time, and there are no plans to use it to deliver higher education credit courses. However, in 1997 the University Council recommended changing the regulations promoting distance education using multimedia. As a consequence, in the future, the SCS system could be used for the exchange of credits among universities, or to deliver credit courses.

PINE-NET, EDC

In 1991 the Education Development College (EDC) began operating a two-way education system using the PINE-NET satellite, making it the first college in Japan to offer university courses via satellite. EDC has established ten two- or three-year technical colleges which provide a junior college level education, as well as the Hokkaido Information University, a four-year institution.

The satellite is used in university education to deliver synchronous two-way lectures. In order to satisfy degree requirements, students must earn at least 30 required credits (out of a required 124) in face-to-face instruction at study centers. This is needed to be in compliance with regulations governing institutions of higher education. The curriculum focuses on information technology and class time is shared with students at technical colleges.

EDC also established a Master's degree program in 1997, which also offers its courses by the satellite.

Academic Network for Distance Education by Satellite (ANDES)

The ANDES system has been operated by Tokyo Institute of Technology since 1995. It is used for special lectures to the public, for an exchange program between Tokyo Institute of Technology and Hitotsubashi University, and for campus lectures between the Tokyo Institute of Technology's Nagatsuda Campus and O-okayama Campus via fiber-optic transmission.

The first program is called Refresher Education by Communication Satellite. Educational programs are delivered to employees of specific companies in a series of open lectures, by lecturers who are not necessarily on the Tokyo Institute of Technology faculty. These lecturers respond to faxed and e-mailed questions from the audience. In this sense synchronous two-way communication cannot be done via this system.

The second program is Tele-Lecture Exchange, whereby classroom lectures are exchanged between two universities via digital satellite. Students see their counterparts at the other university via a video camera remote control system. Lecturers ask questions of students at the other university, who then respond using a unique system in which students press the button on a machine and the system transmits the answers to the instructor.

The third system is also a type of Tele-Lecture linking two campuses by optical fiber. All the lectures are offered as a synchronous tele-conference system via satellite.

Besides these examples, satellites are also used at Tokai University, Shinsyu University, and others, to connect their multiple campuses. The transmissions are considered equivalent to normal classroom lectures and awarded full credit. Private preparatory schools such as Kawaijuku Educational Institution and TOSHIN High School have also begun broadcasting classes using digital satellite television. However, these satellite lectures are limited to students affiliated with the sponsoring institutions.

However, these three examples differ from the rest of the cases. Not only do they provide educational programs, but exchange programs with other institutions of higher education via satellite. In a sense, the satellite connects independent institutions in order as to exchange the educational programs of each institution.

Because of Japan's established IT infrastructure, the Internet and e-mail are easily accessible and are frequently used as teaching and research tools in higher education. Keio University, Osaka City University, Tezukayama Gakuin University, and others, also deliver some university courses via the Internet. However, these are not accepted for credit but considered supplements to classroom lectures. Again, this is due to Ministry of Education regulations.

Currently there is no virtual university in Japan, in part because of Ministry regulations, although this might change in the near future. Since the University Council recommended the promotion of distance higher education using multimedia, such regulations could be revised.

There are two other issues to consider regarding distance higher education in Japan. There is not a great demand for lifelong learning or recurrent education at "universities." Although many Japanese universities provide night courses or programs for working adults, the number of students is very limited. Approximately 60 percent of the students of the University of the Air, for example, are housewives.

Also, the number of 18–year-olds in the general population has been steadily decreasing since its peak of 1992, so that by 2009 the number of openings in four-year universities will equal the number of applicants. In order to survive, universities must increase their applicant pool. This is the opposite of the situation in the rest of Asia, causing even less incentive to establish a virtual university in Japan.

Discussion

The features of open universities in Asia

The primary form of distance education in Asia is the open university. According to Moore and Kearsley (1996, p.41), the majority of open universities adhere to most of these general principles.

1 Anyone can enroll, regardless of previous education.
2 Students can begin a course at any time.
3 Course work is performed wherever the student chooses.
4 Course materials are developed by specialists.
5 Tutoring is provided by other specialists.
6 The enterprise is national in scope.
7 The enterprise enrolls large numbers and enjoys economies of scale.

Points 6 and 7 especially distinguish open universities in Asia from those in the rest of the world. Because of the economic situation, many developing countries in Asia cannot establish privately-funded open universities (see point 6).

Point 7 refers to "mega-universities." Open universities enjoy economies of scale, which provide affordable higher education, and open universities have helped increase the number of students attending college. When STOU was established in Thailand, the percentage of those enrolled in higher education programs increased from around 12% in 1980 to approximately 23% in 1983. In 1987 those enrolled in STOU were approximately one-fourth of all the college students in the country. (Yoshida, 1991).

Courses were generally delivered by mail or broadcast, combined with face-to-face instruction or tutoring, which is also due to the economies of scale. Although the majority of students are working adults who missed the opportunity for higher education, school dropouts have also entered open universities. Open universities have given a second chance at earning a university degree. In terms of curricula, open universities have provided a wide variety of both liberal arts and vocational programs, although programs have focused on vocational and technical orientation rather than the liberal arts.

Features of virtual universities in Asia

Virtual universities have begun emerging in some countries in Asia, with many efforts still at project levels. While we cannot evaluate them currently because of the early stages of these efforts, we can attempt to predict the future of those new efforts. We can observe features of virtual universities which are different from open universities, and look at how they fit into our current conception of universities. The following examines five aspects of virtual universities in Asia.

Governance

Some projects, such as the Multimedia Super Corridor in Malaysia and IT 2000 in Singapore, involve virtual universities that have resulted from government initiatives. Governments designate certain regions as "super corridors" for media and communication networks and infrastructure. The governments then integrate information technology into all aspects of the economy, including education and training, and have the power to compel cultural institutions, corporations, industries and individuals to participate. Education is then part of an overall industrial plan, and is even subordinate to the economic and industrial needs of these countries (Cunningham et al, 1998). In these situations, education for the development of society could be considered more important than education for individual well-being.

Generally, this type of governance is a characteristic of developing countries with strong governments and does not apply to countries such as Japan, where some university systems tend to be in the private sector.

Educational programs

Educational programs in virtual universities are likely to focus on vocational and technical training, rather than on the liberal arts. While this is also true of open universities, the tendency is even more strongly and narrowly focused in virtual universities. Business administration and computer science are two main programs of virtual universities, and we expect most students to be middle-management white collar workers.

Open universities award mainly bachelor's degrees, as well as undergraduate certificates prior to the completion of bachelor's degrees. Degrees from virtual universities in Asia, however, may be more diversified. Some institutions plan to award postgraduate degrees such as the MBA. The MBA curricula in virtual universities are often imported or modeled after overseas MBA programs. Other institutions offer short, non-degree training programs.

MBA programs have recently been very popular in Asia, with many students enrolling in MBA programs abroad. Developing countries in Asia, therefore, are expected to be good potential markets for virtual universities offering MBA degrees. In order for virtual universities to survive and thrive in Asia it is imperative that the educational quality and prestige of these degrees be maintained.

Student profile

Virtual universities are targeting working adults, as they are able to study on a part-time basis. Although open universities also serve working adults, virtual universities specifically focus on middle-management workers in the private sector. This group needs to continue developing their skills, as they need to keep abreast of new developments in business and information technology (Witherspoon, 1998).

Virtual university students need computer literacy before beginning their studies, as well as access to computers with network capabilities in order to access on-line course material. Relatively few people today are familiar with or have access to their own computers. If this situation continues in the future, it may limit the number of students. While theoretically and technically, virtual universities can serve huge numbers of students, in most Asian countries, the numbers of computers are still insufficient.

The meaning and purpose of education will differ for students based on the types of programs they enroll in. Degree candidates will want the degree to increase their chances for better jobs. Those students enrolled in short programs are seeking practical training and skills. In order to meet these demands, higher education institutions cannot limit themselves to a single type of student.

Cost-effectiveness

The relative cost of establishing the IT infrastructure necessary for virtual universities is in dispute. Some argue that virtual universities are less expensive than conventional universities. If there is established infrastructure, establishing virtual universities is not necessarily expensive. When a sufficient number of students enroll and educational materials are used repeatedly during a certain period, the cost of establishing a virtual university is offset. Governments, private companies, and other organizations often invest in the establishment of virtual universities in Asia. Support from both foreign and domestic business is important in building the necessary infrastructure.

Another expense for virtual university students is purchasing computer equipment. Relatively few working adults have access to their companies' computers for study—either at home or in the workplace. Without some sort of computer rental or loan system this new type of educational opportunity might be available only to privileged groups.

How well can virtual universities maintain their program's academic standards, as well as monitor student progress? The usual student dropout ratio is much higher for distance education institutions than in conventional universities. Even though virtual universities have advisory systems which students frequently use, it not easy to finish course work through independent study alone. However, if the advisory system of virtual universities is operated only through online, the retention ratio may not increase, because virtual environments still lack a strong physical socialization process.

High-quality educational materials must, of course, be offered, but education depends not only on course materials, but also on student achievement. It is more difficult to monitor student progress in virtual, rather than conventional universities because the more diverse student body in the former will have more diverse goals.

The method of education

Distance education is done asynchronously in time and space, while classroom education is synchronous. By using existing technology, distance education has gone a long way toward making the asynchronous teaching/learning experience close to a synchronous situation, but maintaining two-way or face-to-face communication remains an important aspect of educational technology. Behind this is the notion that face-to-face communication provides the best educational environment. When courses are delivered via mail or broadcast, schooling or tutorial sessions at local study centers is provided and academic advice by mail or telephone is offered. Even though these efforts make two-way or face-to-face communication possible, they cannot usually be done in real time.

However, information technology makes real-time communication possible. Education using new IT may make virtual learning more flexible, more convenient and more interactive. Students can ask questions of tutors via e-mail anytime and anywhere, and hold discussions with each other at any time through bulletin board systems. Some hold that the major benefit of computer telecommunication is its high degree of interaction and flexibility (Moore & Kearsley, 1996, p. 57.)

However, students do not necessarily communicate with tutors nor participate in discussion with other students. Even when local study centers are available, they are not used by all students. In fact, it is possible for students to finish their entire course of study in a completely virtual environment. Flexible learning is, in a sense, individualized learning, which may cause fragmented learning, which is always criticized in huge universities.

Traditionally, education is meant not only to impart cognitive knowledge to students but also to foster socialization skill through interaction with others. What will happen in virtual interaction?

We must consider one more aspect of virtual environment: student and campus cultures, which are important aspects of traditional educational institutions and significantly influence the socialization process of students. They provide the institution's inheritance. How will the culture of institutions occur without people interacting?

Campus without Walls—A Tentative Conclusion

In the book *The Knowledge Economy* (Neef, ed., 1998), authors argue that the emergence of the knowledge economy has had a dual effect on higher education:

First, information technology has deeply impacted teaching, learning and administrative practices. The establishment of virtual universities and the rapid proliferation of distance education practices are clear evidence of this (Saba, 1998).

Second, business, industrial, and even big agricultural concerns see themselves as knowledge generators and disseminators, thus ending the semi-monopoly of higher education over creation and dissemination of new knowledge.

This trend is now impacting on the Asian region and leading to the creation of virtual universities. The adoption of information technology will change the environment of higher education. However, we can not tell how IT will change higher education in Asia. Information technology, like all technologies, is itself neutral. We measure the level of technology on one scale—as developing or less developing. Although technology has the same content anywhere in the world, the way it works varies for each society. The function of technology depends how each society accepts it. Even though information technology will be widespread in Asia, it will not have necessarily the same function as in the US or other Western societies.

Four major factors influence the future of higher education in Asia. How information technology will change higher education or how successful virtual universities in Asia will be depends on these factors.

1 *The public demand for higher education.* This is a basic factor for developing universities. In Asia there continues to be a great public demand for higher education, which is expected to continue in the future. Although the economic situation of developing countries has prevented enrollment ratios in higher education from increasing, there are quite a number of high school dropouts and working adults seeking advanced schooling, and societal needs for a well-trained labor force are high. Given this situation, virtual universities have the potential to grow, as they provide higher education which is not time- or location-dependent, but which meets the general demand for continuing higher education.
2 *Funding.* Given the economies of developing countries, establishing virtual universities is expensive, even though it may cost less than conventional universities. The important ways of obtaining funding in developing Asian countries can vary widely, ranging from government, private companies and industries, international organizations, to foreign companies. If the latter regards Asia as a good market, they may invest in establishing and supporting

virtual universities. As we have seen, the power of government is relatively strong in Asian countries, yet it is difficult for governments alone to maintain all areas of information technology, including education. One of the keys to success is reducing costs by involving the private sector, both foreign and domestic.

3 *Educational providers.* In the US and other Western countries, business and industries become some of the content providers for virtual universities, i.e. they become knowledge generators and disseminators. This effectively erases the borders between universities and other organizations. In Asia, foreign universities become the educational providers for Asian virtual universities, while the involvement of business and industrial organization is not as clear.
4 *The national culture.* Traditionally higher education has its own mission, and that mission differs by country. Higher education and society are closely related, yet that relationship might be severed by information technology. Virtual universities in the US or other Western countries are based on networks beyond the borders of countries. It may make economic sense for some smaller countries to cooperatively develop (as a consortium) a single virtual university. Do we have the possibility to build regional networks in Asia? It depends on the degree to which Asian virtual universities are involved in each social culture. In Asia, some countries suffer from significant ethnic, religious, or political domestic disputes.

How do we combine these four factors? If they work well in developing virtual universities, we can realize the campus without walls. Virtual universities are borderless organizations because they may develop beyond campus and national borders. As argued in *The Knowledge Economy*, business and industry may become educational providers, a trend which may end the semi-monopoly of higher education over the creation and dissemination of new knowledge.

This could change our traditional concept of universities. Traditionally, universities are protected under the name of the academic freedom of the faculty. The faculty was the university's central policy-making central body and maintained control in terms of creating knowledge. In order to create knowledge, the faculty conducted research. New knowledge has been the based on the research in modern universities. However, in virtual universities, new knowledge may also come from business and industry.

Traditionally, universities have had two other functions: teaching knowledge and skills, and socializing students. In the virtual environment, the teaching function can be fulfilled, and there is a shift from teacher-centered to learner-centered learning. But how will the socialization function occur in virtual universities? Theoretically, students can earn degrees without communicating face-to-face with teachers or other students. This may change our concept of universities.

Virtual universities are products of technological development. However, they may revolutionize the functions of higher education and change our concept of a university. Asia is a good case to monitor whether the "campus without walls" works well all over the world in the next century.

References

Cunningham, S. et al (1998) *New Media and Borderless Education: A Review of the Convergence between Global Media Networks and Higher Education*, Department of Employment, Education, Training and Youth Affairs, Australia.

Daniel, S. John (1996) *Mega-Universities and Knowledge Media-Technology Strategies for Higher Education*, Kogan Page.

ICDL database (1998) Institutional Database. http://www.icdl.ac.uk.

IGNOU (1998a) Intend-Openet Programme.

IGNOU (1998b) Openet Programme.

IGNOU (1998c) A Profile.

Moore, M. G. and Kearsley, G. (1996) *Distance Education –A Systems View*, Wadsworth Publishing Company.

National Institute of Multimedia Education (1992) *Distance Education in Asia and the Pacific* Vol. **I**, **II**, UNESCO Higher Education Series No. 7.

National Institute of Multimedia Education (1994) Distance Education in Asia and the Pacific (Revised Edition), NIME Research Report No.62, National Institute of Multimedia Education.

Neef, D. (ed) (1998) *The Knowledge Economy*, Butterworth-Heineman.

Oshima, A. (1997) The Advent of the Virtual Learning Era, The Japanese Economy in 1998 Trends and Forecasts, Kodansha, pp.244–247.

Witherspoon, J. (1998) Higher Education in the Age of the Knowledge Economy, Distance Education Report, Vol. 2, No. 10) A Perspective on the Virtual University, Distance Education Report, Vol. 2, No. 6, June 1998.

Yoshida, A. (1991) Gakureki-syugi no Kakucyoki ni okeru Hitobito no Ishiki (The Consciousness of Students in the Period of Increasing Educational Credentialism – A Comparative Study between Japan and Thai NIME Research Report No. 36, National Institute of Multimedia Education, pp. 121–144.

Yoshida, A. et al. (1995) Elite Dankai ni okeru Chugoku Enkaku Koto Kyoiku (Distance Education in China at the Stage of Elite Higher Education) NIME Research Report No. 77, National Institute of Multimedia Education.

14

The Internet and Virtual Universities: Towards Learning and Knowledge Systems

F. TED TSCHANG

In the recent debates about virtual education, it has been quite common to see highly polarized views about the virtues and dangers of the medium. One view suggests that the current trend in virtual education will lead to inferior education because of the institutional forces (Noble, 1998), while other perspectives suggest that the current educational systems are broken, and predict that virtual education will spell the demise of traditional campuses and education. An intermediate view is one that suggests that virtual universities are better able to cater to students who cannot afford a conventional education, and that the effectiveness of technology depends on how it is incorporated into educational practice. These views in part embody different assumptions about *how technology will be used* or integrated into educational systems, and whether we look at the worst or best outcomes, and assumptions about the motivations (e.g. what happens when VUs are treated as a business). "It has been said that the university's value 'lies in the complex relationship it creates between knowledge, communities, and credentials', and that 'Changes contemplated in either the institutional structure or technological infrastructure of the university should recognize this relationship (Brown and Duiguid, 1996).

The wide range of views and the nascent state of the practice raises many questions about what educational systems will emerge, what institutions could best provide for needs, and what the optimal outcomes for a society might be. Since it is difficult if not impossible to tell which of these scenarios will hold, we have sought to provide a range of perspectives in this book. By providing a variety of analyses of trends, concepts, and possible scenarios, we have sought to provide a more balanced view of what is happening, and what could happen, including the technological trends and thinking in developed countries, and the experiences in both developed and developing regions. The various chapters indicated that what is best varies considerably, as different countries and regions face different issues, and have different backgrounds or capacities to solve problems or create systems.

To answer the question of *what is happening*, the book examined how VUs could be viewed in terms of their basic elements, namely, technologies, systems, actors, and institutions. The systems include infrastructure, delivery systems for contents, digital libraries, research systems, as well as the contents, thinking and skills to be taught. Many chapters illustrated (either implicitly or explicitly) the shifts in learning modes to student-centered, teacher-guided, and community-oriented modes, and to the use of multiple information sources.

In this conceptualization, supported by the descriptions of various institutions, it is clear that VUs can be broadly construed. This is necessary because the emerging systems have different origins, e.g. some from campus-based universities, others from open universities. They also employ a variety of technological and teaching practices and service a variety of educational needs, e.g. independent study students, full-time students and so on. The nature of the successful institutions, including their motivations, could also dictate the quality of learning eventually achieved.

The question of *what could happen* is more hypothetical, since in general, it is too early to determine the *success* of the many ongoing experiments with virtual education. We can gauge the *appropriateness* of certain technologies and their uses through their characteristics, certain stylized experiences, and the experiences we have examined. Selected scenarios of education and its subsystems can also be presented, such as the scenario in Chapter 3 for learning environments, and the possibilities for collaborative education within the Social Web concept (Chapter 8).

The rest of this chapter synthesizes evidence from the book on various issues concerning technology and virtual education. At this point, it is useful to recall the discussion of propositions regarding technology and its impacts from Chapter 1. Following loosely from this, we can examine technology from the following perspectives:

- Technology that is added on as a multiplier and used to enable new learning functions—this technology-push view determines outcomes that may or may not be "good".
- Environmental reactions to technology (possible outcomes)—while the possible *positive* outcomes are already well known, there is also the possibility of institutional and other forms of resistance to technology.
- Technology that is integrated into learning systems—to avoid negative consequences, this perspective examines the conditions for the successful integration of technology.

While there is a somewhat fine difference between the first perspectives, untangling the evidence this way helps us to better understand how divergent scenarios of the success or failure of virtual education could occur.

In later sections, we also look at two issues that cut across the book: the nature of knowledge on the Internet, and the implications for virtual education in developing countries.

Technology as an Enabler and Multiplier for Learning: Basic Costs and Benefits

First, we examine what happens if technology is added onto existing educational systems, and what the expected multiplier effects might be, independent of other considerations. This is typically evaluated in terms of basic costs and benefits. The chapters illustrated how technologies could deliver a quality education, assuming they are appropriately used. As to how effective technologies actually are depends on other considerations, some of which will be reviewed in further detail in the next sections.

Whether or not institutions decide to adopt technology, or whether or not these technologies and learning systems are successful in the economy, depends on the costs and benefits to institutions and consumers. Many advocates believe that IT and virtual education will lower costs, increase productivity, and facilitate access to more knowledge, but as shown in this book, this is not a matter of simple calculation. Other costs such as opportunity costs also contribute to making decision-making a complex process. Furthermore, educational systems must take into account the need for changes in social attitudes and commitments in the learners, teachers, and bureaucracies that administer education. All these hidden costs go beyond the traditional input costs. There are also larger costs (e.g. costs of getting access to the Internet) to societies that do not yet have the basic infrastructure. (These are also addressed later in this chapter.)

Thus, one of the key tradeoffs concerns the accessibility of education (i.e. affordability and convenience) compared with the quality of education achieved. Perhaps the greatest benefits that a virtual education can provide are those of increased interactive learning (Schank, 1997), in particular interactivity with information, and greatly enhanced access to information.

The quality of education can also vary depending on the technology that is used and how it is used. For instance, VUs that rely on low cost but low quality learning systems, e.g. codifying and communicating online knowledge in the form of simple video or text, would probably not be any better than video correspondence courses, or the kind of teaching currently given in large lecture halls on-campuses (noted in Chapter 2). Indeed, this appears to be a common fate of many of the current online distance education systems in the U.S. and elsewhere. Furthermore, the higher costs of tailoring systems will put pressure on the economies of scale that might be realized from technology (Chapter 7).

Technology design and implementation issues

In the design and implementation of virtual education systems, many technology-specific issues need to be faced. The issues are too many to list, so we will only provide a brief summary of the issues:

Technology can have many impacts on the learning process, as shown in the various chapters of this book (see especially Chapters 3 and 4). Since many things can go wrong in the application of technology, uncertainty is prevalent. To address this, a trial and error approach is useful, as are methods for seeking and adapting to feedback at the institutional and learner levels (e.g. Chapters 12 and 13).

Another issue pertaining to technology is that of standardization of the technology. Some schools such as the Monterrey Institute of Technology have chosen to standardize on one platform such as Lotus Notes. This clearly provides economies of scale and integration, but at the same time could make institutions less able to incorporate new technologies and other changes. These needs will also occur at the level of the Web, e.g. the need to ensure the technical interoperability of systems, such as means of storing and making knowledge globally accessible across different platforms (Chapter 5).

In this era of rapid technological advance, it is even more critical to consider broader "design" issues such as socio-technical interoperability, and social and cognitive interfaces with systems, that is, making systems easier to use and access, both for individuals and their unique needs, and across groups of different individuals (Chapter 7).

Integrating Technology with Learning Systems

A slightly different perspective on technology from the one above considers how to integrate technology with other aspects of learning. The new educational landscape involves changes in learners, the learning process, and the content to be learnt. In other words, while much of the advocacy on virtual education has focused on the issues of *where* learning can take place and *when*, it is critical to ask *what* is to be learnt and *how*. With technology enhancements to learning, we have to ask *how well* technologies are integrated into the broader learning context, that is, if they are appropriately combined with appropriate pedagogy and effective teaching (as was noted in many of the chapters). This involves drastic changes in how learners and teachers work, with the ability to achieve greater rewards coming from greater changes in these patterns of work.

While many systems are now capable of delivering interactive learning (see Chapter 6 for examples), major constraints still exist, such as bandwidth limitations on the Internet (Chapter 5), teachers attitudes and institutional resistance. Furthermore, technology is evolving quickly, as these chapters have pointed out, and given the high costs of codifying knowledge (as noted in Chapter 6), it is clear that virtual education is not a one-time fix-all solution. With regards to knowledge and content, in this era of electronic information, the ease of codifying and replicating knowledge may lead to the tendency to simply disseminate knowledge via automatic means, and to downplay the *learning* of knowledge. Thus, it is important to keep in mind at all times that education is *not* equivalent to information dissemination or retrieval.

What to learn

As far as the needs for learning or *what* is to be learnt, various chapters motivated this by recognizing how the knowledge-based economy is transforming the economy's needs for worker skills. As a consequence, many campus curricula are being transformed to meet those needs, including an understanding of technology (specifically, the Internet) and its impacts (Chapter 4). Many other forms of knowledge need to be included, such as global issues and cultures, and the ability to work cross-functionally, across disciplines, and in teams. Other chapters illustrated how, at least in developing countries, virtual universities were preparing learners to meet these new challenges, perhaps more effectively than their campus-based counterparts (see for e.g. Chapters 11 and 12).

Implications for learners and learning

The role of social processes in virtual learning was an important point highlighted in many chapters, particularly illustrated in the case of collaborative learning (Chapter 8) and digital libraries (Chapter 7). Without a well thought out and implemented social learning process, virtual education will not be that much different from a correspondence education, albeit one with electronic bells and whistles.

It was also shown that in a virtual educational program, the role of students has to change, at the very least, with students taking more responsibility for learning and social behavior (Chapter 12), and in one scenario, possibly gearing up for more independent, self-directed and self-constructed learning (Chapter 3).

Implications for teachers

The implications for teachers are even more far-reaching, and are perhaps not as well addressed by institutions as they should be (see Chapters 3 and 12). Perhaps there is also a role for national policies in setting standards and guidance for teachers. Due to a lack of space, we did not focus too much on the new roles that teachers have to play. Aside from teachers' new mentoring roles, and the knowledge and skills that they must possess, some may also have to confront new working arrangements and conditions being imposed by institutions and other employers. For instance, many teachers are concerned that the implementation of systems that attempt to codify knowledge too much will take teachers out of the learning loop, adversely affecting their livelihoods as well as the learning process (Noble, 1998). In other cases, mass changes in teaching may take generations, since many teachers will only be able to change their thinking and skills incrementally.

Towards learning environments

Our approach also focused on how learning could be conceptualized as occurring not simply within institutional boundaries, but in *learning environments*. The term *learning environment* was broadly construed to include the many ways of combining learners, other actors (e.g. teachers) and learning resources. Three views of learning environments were illustrated: one illustrated a scenario where educational functions become blurred as institutional boundaries are erased by technology (Chapter 2); one where students are operating in learning environments independently of institutions (Chapter 3); and one where current technologies and their limits for developing learning environments were examined (Chapter 6). In the broadest sense, technology has the capability to cause convergence, that is, to break down boundaries between institutions and fields of knowledge (Chapters 1 and 2).

Changes in institutions and institutional boundaries

The global knowledge-based economy is putting pressure on educational systems through its demand for highly skilled "knowledge workers", as well as through increased competitive pressures at national and institutional levels. Thus, regardless of the quality of education they may eventually offer, institutions will have to change, in some cases, by themselves, in others, by force of the market. This will in turn lead to the restructuring of the educational industry (Chapter 2). There are many "models" of how virtual universities may be constructed, and the most appropriate model may vary from region to region, depending on the unique starting conditions (e.g. resource endowments), historical role of the government, social needs, available resources and so on (e.g. Chapters 9 and 13). A number of institutions are virtualizing their operations quite effectively, as shown by the examples in part three of this book. To become a true virtual university, an institution has to completely change itself, by committing a tremendous amount of institutional resources, and putting all its energy into a substantial revamping of the teaching/learning process (as shown in Chapter 11).

Technology can present both a threat and opportunity through its ability to impact on institutions and reshape their boundaries, as discussed in Chapters 2 and 3. Indeed, the notion of certain traditional institutions can be completely revised, such as the notion of a library as a place to gather and access information. Chapter 7 provides an example of an appropriate evolutionary strategy for digital libraries, which combines technological advances together with an understanding of institutional functions, i.e. the functions found in traditional libraries. While technologies can increase our abilities to manage information many-fold, many traditional library functions that people are used to may have to be preserved, such as people's needs for social spaces and social contexts for information. Thus, a variety of fields of knowledge is needed to design a digital library, and the social architecture of traditional libraries will have to be blended into the new digital spaces.

Changes at the institutional level also require changing the way in which resources are viewed. Many schools are finding it necessary to develop more complete and "renewable" systems for creating knowledge, ranging from teacher development programs (Chapter 12) to doing research and development on pedagogy by way of graduate student research (Chapter 13).

Other kinds of institutions can also play an important role in shaping the playing ground for VUs, such as associations and governmental bodies and their influence. For instance, in the U.S., the national association Educause has sought to influence debate on educational technology in favor of technology. In other cases, national bodies or governments have adversely affected progress, as shown in the case of distance education in Brazil, where national bodies resisted or restricted reforms in distance education (Chapter 12), and in Japan, where the government's restrictive position on accreditation has limited the opportunities for their development (Chapter 14).

Knowledge on the Internet: How Can We Access and Make Sense of it?

Some of the issues that need to be resolved involve how the character of knowledge is changing because of the Internet. Some of the salient issues seen in this book included:

The value of knowledge

With the advent of the Web, information has become even more important as a commodity than ever. In the race to obtain access to useful or economically valuable knowledge, institutions and students alike will confront many new issues. One is the tradeoff between (creating and learning) locally valued knowledge (e.g. indigenous knowledge), and globally valued knowledge. Indigenous knowledge may have less value in the global economy, but is still important for the growth of local content and software industries, and even for its contributions to social well-being.

Information overload and "Infosense"

Another issue seen in the chapters concerns information overload and filtering in the virtual learning environment. The Internet has led to a proliferation of information and data, and a consequent need to make sense of it (what has been termed *infosense*, or turning information into knowledge (Devlin, 1999)). At the same time, it is important to screen out that which is useless, such as the "junk" or incorrect knowledge found all over the Internet. The growth of information resources on the Internet is quite directionless, and much that is on the Web is not even useful for educational purposes. Some of the chapters (e.g. Chapters 3, 4 and 5) pointed out this problem. Information filters will have to be provided to discriminate between useful knowledge and junk. While software tools are being developed to create these distinctions, making the broader Web a "learning environment", we have to recognize that this is not an automated process, and humans still have an important role to play. People, organizations or even communities of practice can help users and students to make sense of information by structuring it, standardizing it, or catering it to individual needs. Thus, solutions must also take into consideration various issues at not only the level of the individual learner but also at the systems level.

As the transition to Web-based information is made, systems such as digital libraries will have to take into account historical structures of institutions and legacies, such as the accumulated "knowledge about knowledge" (Chapter 7). As with indigenous knowledge, centuries of historical human knowledge are irreplaceable and thus, need to be preserved and made accessible. Knowledge is not simply information codified in text or individual minds, but it is information that has been socially constructed and contextualized at specific times in unique circumstances, and thus, can also be located in societal, institutional and organizational culture and routines, and in many other social "places". Legacy systems are also a problem, since knowledge that is codified in one system's format may not necessarily be able to migrate (or be moved) to newer systems using a different electronic media or format.

Finally, one other issue that inhibits the broader access to knowledge is the set of intellectual property laws and other means for allocating or restricting the access to information, if not also providing access to it. Much of the more economically valuable knowledge is being locked-up in private hands or networks. There is a greater need to examine this broader trend, and of the socio-economic means of regulating information and making it accessible, in greater detail. This is only the tip of the iceberg, and much remains to be done.

Balkanization

As we noted, information technology has the potential to make all local knowledge potentially "global", but through private desires to protect intellectual property or private advantages, it can lead to what some have called the "balkanization" of communities, or the locking away of knowledge within privately-accessible networks. The rapid increase in the number of Intranets in the U.S. is one testament to this. Systems such as digital libraries could certainly stay in private hands, becoming inaccessible to the people who may need them the most. This augurs for international means (e.g. conventions or mechanisms) for making access to basic information, knowledge and learning a "right" for every individual.

Whose knowledge?

As suggested above, knowledge has socio-cultural and political contexts. This leads us to ask "whose knowledge" is being created or promoted by institutions. In traditional institutions, teachers were charged with being the creators of knowledge, and there were professional standards for curricula as well as rigorous academic review processes. In a constructivist environment like the Internet, students have to be responsible for large parts of their own learning, but the supporting environments must also be there. While some bodies of academic knowledge, particularly scientific ones, will remain the same, other types of knowledge may be socially constructed and laden with context and perspectives. In these environments, the practices for socially validating and disseminating knowledge could be more democratic and community-based.

At the same time, technologies are outpacing human abilities to master them. This situation requires the development of improved means for humans to easily learn or evolve capabilities to deal with information during technological change. For instance, the availability of newer forms of databases and techniques for information manipulation threatens to outstrip the general users' abilities to deal with them.

Towards Broadly Accessible Educational Systems

The Internet, and in particular Web-based education, has the potential to either greatly enhance access to knowledge (leading to technological leapfrogging) (Chapter 11), and to help break out of outmoded and poor teaching modes (Chapter 12). On the other hand, it can also exacerbate social inequality, if the poor do not have basic access or abilities to access the rich resources on the Web.

Sources of funding for virtual education systems can also vary considerably, with some regions (typically the more industrialized ones) being able to depend more on private sector funding and resources, such as campus-based university initiatives and corporate funding.

The Internet and knowledge gaps in development

The discussion of Internet and other technologies in Part 2 of the book focused mainly on activities in the developed countries. There is still a great need to study how the Internet is influencing or will influence developing societies and economies. Different levels of development can be seen as gaps in the ability to acquire, use, and create knowledge (World Bank, 1998). Disparities in these abilities can further exacerbate economic disparities. Thus, despite the overwhelming needs of many developing countries, there is a view that suggests that a country's participation in the information revolution is a *necessity*, rather than a luxury. This is based on the presumption that the information revolution will have far-reaching impacts on economic growth, and countries that do not "catch the wave", will be consigned to increasingly bleaker futures (relative to the rest of the world's).

However, as indicated by the various chapters, there is a wide range of financial situations and preparedness for the Internet across countries and regions. Poorer countries or poorer segments of the population usually have little or no access to the Internet, computer equipment, and trained teachers. Poorer students will generally also have inadequate prior educational preparation, including IT skills. This initial base of knowledge is necessary to be able to learn or accumulate further knowledge.

Because of this state of poor infrastructure and access, even the introduction of virtual education could further exacerbate inequality within a country, especially for those who cannot afford a virtual education, or who do not have the ability to use the technology. Another possibility is that virtual universities could put financial pressure on campus-based universities by causing them to lose some of their self-supporting students—which many campus-based universities traditionally take in so as to be able to provide financial aid to deserving students who desire a traditional education. In light of these dangers, there needs to be a better systemic review of how virtual education could reshape the educational opportunities for peoples of all socio-economic backgrounds.

Facilitating access to resources

One of the basic needs is to solve the problem of student access to educational opportunities. These require the addressing of several connected issues, including financial resources, infrastructure, and knowledge (e.g. skills). This is particularly serious for marginalized students and populations, particularly those without the means to afford an education. Apart from financial resources, imaginative use will have to be made of technologies, practices and teaching resources in order to achieve this, such as the use of inexpensive computers and Internet connections, public domain software, and networks of volunteers. A few representative solutions are as follows:

To address the lack of access to financial resources, basic technologies and skills as a package, institution-to-institution programs may be more effective. One example of a more complex solution was the World Bank's African Virtual University, which used multilateral aid to transfer infrastructure and knowledge by linking domestic institutions with foreign ones. This also allowed programs to become self-sustaining over time (as shown in Chapter 10).

In general, the globalization of markets and the Internet is allowing VUs, particularly those located in developed countries, to make learning opportunities available to non-traditional students in developing countries.

For countries or institutions with minimum levels of Internet infrastructure, "soft" technical aid and knowledge is also available through the Internet, including free resources like public domain software, data, documents, digital libraries, course curricula, and other items already available on the Web. Systems that provide knowledge on research and practical experiences may eventually become available. Already, many universities and institutes provide access to their working papers on the Web. There is a potential need for institutions to ensure that a broad set of knowledge is widely or freely available, for instance, by going beyond the concept of a public library, and onto that of a global knowledge- and learning-oriented library that caters to the broader needs of development.

Technology transfer also involves many cultural and social issues. While the institutional cases in this volume illustrated how universities in countries like Mexico adapted systems to their needs, these institutions all had strong research and teaching bases to begin with. It is clear that simply having access to courses and software on the Web will not solve the individual learners' problems in developing countries, and may even harm them. Systems developed in developed countries or regions don't always fit well with traditional cultural, economic and other conditions in developing countries. Furthermore, while it is often assumed that IT and other technologies have no cultural context, many technologies and their uses actually have dehumanizing or destabilizing effects on traditional cultures.

Finally, another major unsolved problem that can limit the accessibility of information across countries is that of language. While language translation technologies are being heavily researched, we are still far from a translator that can translate languages in many languages or across multiple contexts. Most current translator research tends to focus on only a handful of languages.

Access is also a matter of design, and systems can be designed for broader access (e.g. Chapter 7). Learning and information systems are generally created for providing access to specific groups and their needs (e.g. research or teaching at a particular level), and at the individual query level. In the broader interest of providing widespread access and saving on resources, systems will have to be designed to provide *broader*, more general, accessibility (e.g. to the public), that could vary across individual *capabilities* and *levels of preparation*. One example is that of user interfaces, which need to be accessible to a range of users, including ones from different cultural, linguistic and other backgrounds.

Reforming Educational Systems

In many countries, many higher educational systems will have to be reformed. In that process, virtual education stands not as a full substitute, but rather as a complementary part of educational reform.

The rapid development of markets and proliferation of different types of systems for virtual education, particularly in developed countries, can make policy-making confusing. Furthermore, the nature of the Internet is such that it involves myriad actions and decisions taken at micro or individual levels. It requires allowing students and other participants in the learning process to self-organize themselves to achieve individual and social objectives.

The uncertainty and self-organized nature of this system suggests that a prudent course of action is to design educational systems to be not only highly innovative and experimental, but also adaptive.

How do we reconcile the nature of the "culture" of the Internet and its medium, with the culture of the traditional governmental establishment? This requires new forms of governance that provide "engaged reflection", that involve policy which is neither too restrictive, but that encourages the goals of innovation, diffusion of best practice, and so on. Furthermore, as noted earlier, systems will have to account for individual and cultural differences and sensitivities. Means for standardizing or approving practices need to be incorporated. This raises many significant challenges for public policy. A systems approach will need to be taken to policy-making.

In the context of systems, one challenge involves how to integrate the different levels of learning, keeping in mind that technology now allows individuals to transcend these levels, or the lack of facility with technology can also hold back individuals at a particular level. At each step of the educational scale, we should ensure that the appropriate thinking and learning skills, as well as content and pedagogies, are being employed. This suggests that a flexible "ladder" of educational opportunities has to be considered for taking students from one level to the next, or allowing them to leapfrog levels of learning.

Future Agendas

As highlighted in this synthesis, the research in this book suggests that much work needs to be done. Promising areas or needs for educational research on virtual education could include:

- Finding new ways to enhance the most disenfranchised or impoverished peoples' access to knowledge, and studying what works. Issues of access are the hardest to solve, requiring concerted efforts along a variety of dimensions, including financial, technological and human ones. A variety of models, both public and private, must be examined and brought to bear, taking into account the varying conditions across countries, particularly developing ones.
- Understanding better the nature of knowledge and learning in these new and evolving contexts. In the Web environment, contextual knowledge, as well as the social processes that influence it, changes fleetingly. As a result, new ways of understanding and teaching knowledge are needed.
- Improving the understanding of the benefits of Web-based education, in relation to the new economics created by technology.

- Fully extending the capabilities of the Web to realize its potential for social interaction and cognitive training.
- Developing the means for students to self-construct their own learning environments, as well as for teachers to keep up with the changes that will come with them.
- Developing a practice of knowledge management in teaching, including the means for disseminating and ensuring best practices.
- Greater understanding of the cross-cultural aspects of learning and knowledge.

References

Brown, J. S. and Duguid P. (1996) The University in a Digital Age. *Change* July/August 1996: 12-19.

Devlin K. (1999) *Infosense: Turning Information into Knowledge*. New York, NY: W. H. Freeman.

Noble D. F. (1998) Technology in Education: the Fight for the Future. *Educom Review*, **33**(3), May/Jun 1998.

Schank, R. (1997) *Virtual Learning: A Revolutionary Approach to Building a Highly Skilled Workforce*. New York, NY: McGraw Hill.

World Bank (1998) World Development Report: Knowledge for Development, Washington, DC: the World Bank.

AUTHOR INDEX

SUBJECT INDEX